Fast analytical techniques for electrical and electronic circuits

Today, the only method of circuit analysis known to most engineers and students is nodal or loop analysis. Although this works well for obtaining numerical solutions, it is almost useless for obtaining analytical solutions in all but the simplest cases.

In this unique book, Vorpérian describes remarkable alternative techniques to solve, almost by inspection, complicated linear circuits in symbolic form and obtain meaningful analytical answers for any transfer function or impedance.

Although not intended to replace traditional computer-based methods, these techniques provide engineers with a powerful set of tools for tackling circuit design problems. They also have great value in enhancing students' understanding of circuit operation. The numerous problems and worked examples in this book make it an ideal textbook for senior/graduate courses or a reference book.

This book will show you how to:

- use less algebra and do most of it directly on the circuit diagram,
- obtain meaningful analytical solutions to complex circuits with reactive elements and dependent sources by reducing them to a set of simple and purely resistive circuits which can be analyzed by inspection,
- analyze feedback amplifiers easily using the simplest and most natural formulation,
- analyze PWM converters easily using the model of the PWM switch.

Originally developed and taught at institutions and companies around the world by Professor David Middlebrook at Caltech, the extended and new techniques described in this book are an indispensable set of tools for linear electronic circuit analysis and design.

Vatché Vorpérian received his PhD in Electrical Engineering in 1984 from the California Institute of Technology and joined the faculty of Electrical Engineering at Virginia Tech in the same year. In 1991 he joined the Jet Propulsion Laboratory where he is currently a senior member of the technical staff. He has published over 35 conference and journal papers in the field of power electronics and has taught many professional advancement courses to industry.

Fast analytical techniques for electrical and electronic circuits

Vatché Vorpérian

Jet Propulsion Laboratory
California Institute of Technology

CAMBRIDGE
UNIVERSITY PRESS

PUBLISHED BY THE PRESS SYNDICATE OF THE UNIVERSITY OF CAMBRIDGE
The Pitt Building, Trumpington Street, Cambridge, United Kingdom

CAMBRIDGE UNIVERSITY PRESS
The Edinburgh Building, Cambridge CB2 2RU, UK,
40 West 20th Street, New York, NY 10011-4211, USA
477 Wiliamstown Road, Port Melbourne VIC 3207, Australia
Ruiz de Alarcón 13, 28014 Madrid, Spain
Dock House, The Waterfront, Cape Town 8001, South Africa

http://www.cambridge.org

© Cambridge University Press 2002

First published 2002

Printed in the United Kingdom at the University Press, Cambridge

Typeface Times 10.5/14pt   *System* Poltype® [VN]

A catalogue record for this book is available from the British Library

ISBN 0 521 62442 8 hardback

To my parents Edward and Azadouhi Vorpérian

Contents

Preface   xi

1   Introduction   1

1.1   Fast analytical methods   1
1.2   Input impedance of a bridge circuit   2
1.3   Input impedance of a bridge circuit with a dependent source   4
1.4   Input impedance of a reactive bridge circuit with a dependent source   8
1.5   Review   11
      Problems   11
      References   14

2   Transfer functions   15

2.1   Definition of a transfer function   15
2.2   The six types of transfer functions of an electrical circuit   17
2.3   Determination of the poles of a network   19
2.4   Determination of the zeros of a transfer function   24
2.5   The complete response, stability and transfer functions   34
2.6   Magnitude and phase response   41
2.7   First-order transfer functions   43
2.8   Second-order transfer functions   48
2.9   Review   52
      Problems   53

3   The extra element theorem   61

3.1   Introduction   61
3.2   Null double injection   62

3.3   The EET for impedance elements   74
3.4   The EET for dependent sources   88
3.5   Review   98
      Problems   99
      References   106

4   **The *N*-extra element theorem**   107

4.1   Introduction   107
4.2   The 2-EET for impedance elements   108
4.3   The 2-EET for dependent sources   130
4.4   The NEET   137
4.5   A proof of the NEET   147
4.6   Review   153
      Problems   154
      References   162

5   **Electronic negative feedback**   163

5.1   Introduction   163
5.2   The EET for dependent sources and formulation of electronic feedback 164
      5.2.1 Gain analysis   164
      5.2.2 Driving-point analysis   170
      5.2.3 Loop gain   175
5.3   Does this *circuit* have feedback or not? This is *not* the question 179
5.4   Gain analysis of feedback amplifiers   180
5.5   Driving-point analysis of feedback amplifiers   195
      5.5.1 Input *impedance* for current mixing   196
      5.5.2 Output *impedance* for voltage sensing   200
      5.5.3 Input *admittance* for voltage mixing   204
      5.5.4 Output *admittance* for current sensing   209
5.6   Loop gain: a more detailed look   213
5.7   Stability   218
5.8   Phase and gain margins   226
5.9   Review   233
      Problems   234
      References   251

6   High-frequency and microwave circuits   252

6.1   Introduction   252
6.2   Cascode MOS amplifier   252
6.3   Fifth-order Chebyshev low-pass filter   261
6.4   MESFET amplifier   265
6.5   Review   310
      Problems   311
      References   316

7   Passive filters   317

7.1   Introduction   317
7.2   *RC* filters with gain   317
7.3   Lattice filters   327
7.4   Resonant filters   335
      7.4.1 Parallel resonant filters   336
      7.4.2 Tapped parallel resonant filter   339
      7.4.3 The three-winding transformer   344
7.5   Infinite scaling networks   349
      7.5.1 Infinite grid   349
      7.5.2 Infinite scaling networks   351
      7.5.3 A generalized linear element and a unified R, L and C model   356
7.6   Review   358
      Problems   358
      References   364

8   PWM switching dc-to-dc converters   365

8.1   Introduction   365
8.2   Basic characteristics of dc-to-dc converters   366
8.3   The buck converter   370
8.4   The boost converter   386
8.5   The buck-boost converter   392
8.6   The Cuk converter   397
8.7   The PWM switch and its invariant terminal characteristics   400

8.8   Average large-signal and small-signal equivalent circuit models of the PWM switch   402
8.9   The PWM switch in other converter topologies   411
8.10 The effect of parasitic elements on the model of the PWM switch   426
8.11 Feedback control of dc-to-dc converters   432
      8.11.1 Single-loop voltage feedback control   433
      8.11.2 Current feedback control   440
      8.11.3 Voltage feedback control with peak current control   453
8.12 Review   460
      Problems   461
      References   470

      Index   472

Preface

The title of this book could easily have been called *Variations on a Theme by Middlebrook*, or *Applications of The Extra Element Theorem and its Extensions*. Neither title, however, would have captured the unique message of this book that one can solve very complicated linear circuits in symbolic form almost by inspection and obtain more than one meaningful analytical answer for any transfer function or impedance. The well-known and universally practiced method of nodal or loop analysis not only becomes intractable when applied to a complicated linear circuit in symbolic form, but also yields unintelligible answers consisting of a massive collection of symbols. In a meaningful analytical answer, the symbols must be grouped together in *low-entropy* form – a term coined by R. D. Middlebrook – clearly indicating series and parallel combination of circuit elements, and sums and products of time constants. The illustrative examples in Chapter 1 serve as a quick and informal introduction to the basic concepts behind the radically different approach to network analysis presented in this book.

Today, the only method of circuit analysis known to most engineers, students and professors is the method of nodal or loop analysis. Although this method is an excellent general tool for obtaining *numerical solutions*, it is almost useless for obtaining *analytical solutions* in all but the simplest cases. Anyone who has attempted inverting a matrix with symbolic entries – sometimes as low as second-order – knows how tedious the algebra can get and how ridiculous the resulting high-entropy expressions can look. The purpose of this book is not to eliminate the linear algebra approach to network analysis, but instead to provide additional new and efficient tools for obtaining analytical solutions with great ease and without letting the algebra run into a brick wall.

Among the most important techniques discussed in this book are the extra element theorem (EET) and its extension the *N*-extra element theorem (NEET). These two theorems are discussed in Chapters 3 and 4 after a brief and essential review of transfer functions given in Chapter 2. The EET and its proof were given by R. D. Middlebrook. The NEET was given without proof by Sarabjit Sabharwal, an undergraduate at Caltech in 1979. In Chapter 4, a completely original treatment of the NEET is given, where it is stated in its most general form using a new compact notation and, for the first time, proven directly using matrix analysis.

The subject of electronic feedback is treated in Chapter 5 using the EET for dependent sources, and another theorem by R. D. Middlebrook called simply "the feedback theorem". Both methods lead to a much more *natural* formulation of electronic feedback than the well-known block diagram approach found in most textbooks. Block diagrams are useful tools in linear system theory to help visualize abstract concepts, but they tend to be very awkward tools in network analysis. For instance, in an electronic feedback circuit neither the impedance loading nor the bi-directional transmission of the feedback network are easily captured by the single-loop feedback block diagram unless the feedback network and the amplifier circuit are both manipulated and *forced* to fit the block diagram. The fact is block diagrams bear little resemblance to circuits and their use in network analysis mainly results in loss of time and insight.

The examples presented in Chapters 6 and 7 are a *tour de force* in analysis of complicated circuits which demonstrate the efficacy of the fast analytical techniques developed in the previous chapters. Among the examples discussed in these chapters are higher-order passive filters and a MESFET amplifier. Some infinite networks, including fractal networks, are discussed in Chapter 7 where an interesting, and possibly new, result is presented. It is shown that a resistor, an inductor and a capacitor are all special cases of a single, two-terminal, linear element whose voltage and current are related by a fractional derivative or its inverse, the Riemann–Liouville fractional integral.

Pulse-width-modulated (PWM) switching dc-to-dc power converters are introduced in Chapter 8 to illustrate further the applications of the fast analytical techniques presented in this book. The analysis of PWM converters has been one of the hot topics of nearly every conference in power electronics since the early 1970s, and many specialized analytical techniques have been developed since. The simplest and fastest of these techniques is based on the equivalent circuit model of the PWM switch, which is introduced after a discussion of basic PWM converters. The PWM switch is a three-terminal nonlinear device which is solely responsible for the dc-to-dc conversion function inside a PWM converter. Hence, the PWM switch and its equivalent circuit model are to a PWM converter what the transistor and its equivalent circuit are to an amplifier. To analyze the dynamics of a PWM converter, one simply replaces the PWM switch with its equivalent circuit model and proceeds in exactly the same way as in an amplifier circuit analysis.

This book is based on my experience in electronic circuit analysis as a student, design engineer, teacher and researcher. The limitations of the "standard" circuit analysis I studied as an undergraduate soon became apparent on my first job as a power supply design engineer at Digital Equipment Corporation, Maynard, MA. I spent inordinate amounts of time deriving various small-signal transfer functions of switching converters in order to understand and improve their stability and dynamic behavior. Most of the senior engineers around me had acquired excellent design skills mostly by experience and did not rely too much on analysis. When I returned to graduate school at Caltech, I took Middlebrook's course which engendered a complete turn around: I learned how to handle complicated linear networks and obtain transfer functions, in low-entropy form, using very simple and elegant techniques. I gradually adopted these techniques in my seven years of teaching at Virginia Polytechnic Institute and State University confirming the adage, "the best way to learn something is to teach it."

Logically, Middlebrook's book, which is still in preparation, should have preceded mine. I began writing this book in the summer of 1996 with the intention of completing it by the winter of 1997. Clearly, I did not realize that writing a book at nights and on weekends would be considerably more difficult and time consuming than I had ever imagined. Fortunately, I had the constant support and encouragement of family, friends and colleagues. I would especially like to thank Gene Wester and Dave Rogers, both at the Jet Propulsion Laboratory, for their careful review and corrections of some of the chapters of this book. I would also like to thank my former supervisor Robert Detwiler; my current supervisor Mark Underwood; my colleagues Chris Stell, Tony Tang, Roman Gutierrez, Avo Demirjian, Dan Karmon, Mario Matal, Joseph Toczylowsky, Karl Yee, James Gittens, Mike Newell, David Hykes, Chuck Derksen and Tien Nguyen for making JPL an enjoyable place to work. Although this book is dedicated to my parents for their countless sacrifices, I would not have been able to write it without the enduring support, love and care of my favorite mezzo-soprano, best friend and wife Shoghig.

Vatché Vorpérian
June 2000

1 Introduction
The joys of network analysis

1.1 Fast analytical methods

The universally adopted method of teaching network theory is the formal and systematic method of nodal or loop analysis. Although the matrix algebra of formal network analysis is ideal for obtaining *numerical* answers by a computer, it fails hopelessly for obtaining *analytical* answers which provide physical insight into the operation of the circuit. It is not hard to see that, when numerical values of circuit components are not given, inverting a 3×3, or higher-order, matrix with symbolic entries can be very time consuming. This is only part of the problem of matrix analysis because even if one were to survive the algebra of inverting a matrix symbolically, the answer could be an unintelligible and lengthy symbolic expression. It is important to realize that an analytical answer is not merely a symbolic expression, but an expression in which various circuit elements are grouped together in one or more of the following ways:

(a) series and parallel combinations of resistances

   Example: $R_1 + R_2 \parallel (R_3 + R_4)$

(b) ratios of resistances, time constants and gains

   Example: $1 + \dfrac{R}{R_3 \parallel R_4}$, $1 + \dfrac{g_m R_L}{A_o}$, $A_m \left(1 + \dfrac{\tau_1}{\tau_2}\right)$

(c) polynomials in the frequency variable, s, with a unity leading term and coefficients in terms of sums and products of time constants

   Example: $1 + s(\tau_1 + \tau_2) + s^2 \tau_1 \tau_3$

Such analytical expressions have been called low-entropy expressions by R. D. Middlebrook[1] because they reveal useful and *recognizable* information (low noise or entropy) about the performance of the circuit. Another extremely important advantage of low-entropy expressions is that they can be easily approximated into simpler expressions which are useful for design purposes. For instance, a series-parallel combination of resistances, as in (a), can be simplified by ignoring the smaller of two resistances in a series combination and the larger of two resistances in a parallel combination. When ratios are used as in (b), they can be simplified depending on their relative magnitude to unity. Depending on the relative magnitude of time constants, frequency response characteristics as in (c) can be simplified and either factored into two real roots, with simple analytical expressions, or remain as a complex quadratic factor.

In light of the above, the aim of fast analytical techniques can be stated as follows: fast derivation of low-entropy analytical expressions for electrical circuits. The following examples illustrate the power of this new approach to circuit analysis.

1.2   Input impedance of a bridge circuit

We will determine the input resistance, R_{in}, of the bridge circuit[2] in Fig.1.1 in a few simple steps using the extra element theorem (EET). The EET[3] and its extension, the N-extra element theorem[4] (NEET), are the main basic tools of fast network analysis discussed in this book. Both of these theorems will be introduced, derived and stated in their general form in later chapters, but since the EET for an impedance function is so trivial, we will use it now to obtain an early glimpse of what lies ahead.

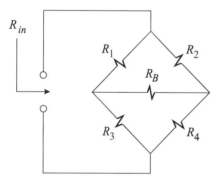

Figure 1.1

We see in Fig. 1.1 that if any one of the resistors of the bridge is zero or infinite, we can write R_{in} immediately by inspection. For instance, if we designate R_B as the extra element and let $R_B \to \infty$, as shown in Fig. 1.2a, we can immediately write:

$$R_{in}\,|_{R_B \to \infty} = (R_1 + R_3)\,\|\,(R_2 + R_4) \tag{1.1}$$

The EET now requires us to perform two additional calculations as shown in Figs. 1.2b and c. We denote the port across which the extra element is connected by (B).

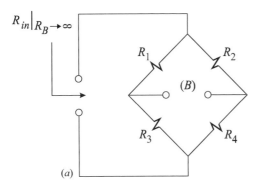

Figure 1.2

In Fig. 1.2*b*, we determine the resistance looking into the network from port (B) with the *input port short* and obtain by inspection:

$$\mathcal{R}^{(B)} = R_1 \| R_3 + R_2 \| R_4 \tag{1.2}$$

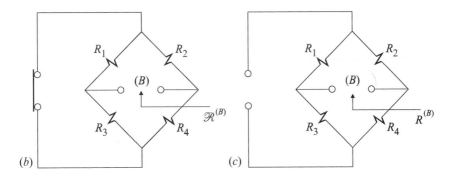

Figure 1.2 (*cont.*)

In Fig. 1.2*c*, we determine the resistance looking into the network from port (B) with the *input port open* and obtain by inspection:

$$R^{(B)} = (R_1 + R_2) \| (R_3 + R_4) \tag{1.3}$$

We now assemble these three separate and independent calculations to obtain the input resistance R_{in} in Fig. 1.1 using the following formula given by the EET:

$$R_{in} = R_{in}|_{R_B \to \infty} \frac{1 + \dfrac{\mathcal{R}^{(B)}}{R_B}}{1 + \dfrac{R^{(B)}}{R_B}} \tag{1.4}$$

Upon substituting Eqs. (1.1), (1.2) and (1.3) in (1.4):

$$R_{in} = (R_1 + R_3) \,\|\, (R_2 + R_4) \frac{1 + \dfrac{R_1 \,\|\, R_3 + R_2 \,\|\, R_4}{R_B}}{1 + \dfrac{(R_1 + R_2) \,\|\, (R_3 + R_4)}{R_B}} \tag{1.5}$$

Equation (1.5) is a low-entropy result because in it R_{in} is expressed in terms of series and parallel combinations of resistances and ratios of such resistances added to unity. Such an expression, for a given set of typical element values, can be easily approximated using rules of series and parallel combinations wherever applicable. In this expression, we can also see the contribution of the bridge resistance, R_B, to the input resistance, R_{in}, directly.

We can also appreciate two important advantages of the method of EET used in deriving R_{in} above. First, since the method of EET requires far less algebra than nodal analysis, it is considerably faster and simpler. Second, since the EET requires three separate and *independent* calculations, any *error in the analysis does not spread* and remains confined to a portion of the final answer. In a sense, this kind of analysis yields modular answers – if there is anything wrong with a particular module, it can be replaced without affecting the entire answer. This not only makes the analysis faster, but also the debugging of the analysis faster as well.

1.3   Input impedance of a bridge circuit with a dependent source

In this section we consider the effect of a dependent current source,[2,5] $g_m v_1$, in Fig. 1.3, on the input resistance R_{in}. This circuit is borrowed from a well-known

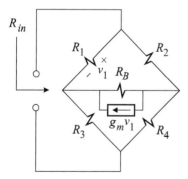

Figure 1.3

textbook by L. O. Chua and Pen-Min Lin[5] in which the authors determine the contribution of the transconductance, g_m, to the input resistance, R_{in}, using the parameter-extraction method. Because of the considerable amount of matrix algebra required by the parameter-extraction method, which would become prohibitively complex if all elements were in symbolic form, Chua and Lin have assigned numerical values ($R_1 = 1\,\Omega$, $R_2 = 0.2\,\Omega$, $R_3 = 0.5\,\Omega$, $R_4 = 10\,\Omega$ and $R_B = 0.1\,\Omega$) to all the resistors and determined:

$$R_{in} = \frac{96.3 + 5.1 g_m}{137.7 + 10.5 g_m}\,\Omega \tag{1.6}$$

We will now show how to determine R_{in} in three simple steps by applying the EET to the dependent current source $g_m v_1$. To demonstrate the superior power of this method of analysis, we will keep all circuit elements in symbolic form.

In Fig. 1.3, we designate the dependent current source as the extra element and set it to zero by letting $g_m = 0$. This reduces the circuit to the bridge circuit in Section 1.2, as shown in Fig. 1.4a. Hence, we have from Eq. (1.5):

$$R_{in}\big|_{g_m \to 0} = (R_1 + R_3) \,\|\, (R_2 + R_4) \frac{1 + \dfrac{R_1 \,\|\, R_3 + R_2 \,\|\, R_4}{R_B}}{1 + \dfrac{(R_1 + R_2) \,\|\, (R_3 + R_4)}{R_B}} \tag{1.7}$$

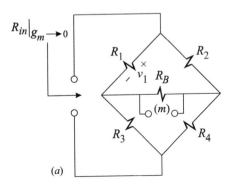

(a)

Figure 1.4

The EET now requires us to perform two additional calculations as shown in Figs. 1.4b and c in which the dependent current source is replaced with an independent one, i_m, pointing in the opposite direction. In Fig. 1.4b we determine the transresistance, v_1/i_m, which is the *inverse* of the transconductance gain g_m of the dependent source, with the input port short. Inspecting Fig. 1.4b, we see that $R_1 \,\|\, R_3$ and $R_2 \,\|\, R_4$ form a voltage divider connected across an equivalent Thevinin voltage source, $i_m R_B$, in series with a Thevinin resistance, R_B, so that we have:

$$\frac{v_1}{i_m R_B} = \frac{R_1 \parallel R_3}{R_B + R_2 \parallel R_4 + R_1 \parallel R_3} \tag{1.8}$$

It follows that the inverse gain, with the input port short, is given by:

$$\mathcal{G}^{(m)} = \frac{v_1}{i_m}\bigg|_{(in)\to short} = \frac{R_1 \parallel R_3}{R_B + R_2 \parallel R_4 + R_1 \parallel R_3} R_B \tag{1.9}$$

Similarly, we can determine in Fig. 1.4c that the inverse gain, with the input port open, is given by:

$$\bar{G}^{(m)} = \frac{v_1}{i_m}\bigg|_{(in)\to open} = \frac{R_B \parallel (R_3 + R_4)}{R_1 + R_2 + R_B \parallel (R_3 + R_4)} R_1 \tag{1.10}$$

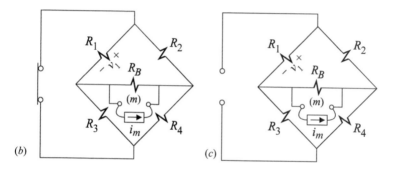

(b)   (c)

Figure 1.4 (*cont.*)

We can now assemble the final answer using the three separate calculations in Eqs. (1.7), (1.9) and (1.10) according to the following formula given by the EET:

$$R_{in} = R_{in}\big|_{g_m\to 0} \frac{1 + g_m \mathcal{G}^{(m)}}{1 + g_m \bar{G}^{(m)}} \tag{1.11}$$

Upon substituting, we get:

$$R_{in} = (R_1 + R_3) \parallel (R_2 + R_4) \frac{1 + \dfrac{R_1 \parallel R_3 + R_2 \parallel R_4}{R_B}}{1 + \dfrac{(R_1 + R_2) \parallel (R_3 + R_4)}{R_B}} \tag{1.12}$$

$$\times \frac{1 + \dfrac{g_m R_B}{1 + (R_B + R_2 \parallel R_4)/R_1 \parallel R_3}}{1 + \dfrac{g_m R_1}{1 + (R_1 + R_2)/R_B \parallel (R_3 + R_4)}}$$

Hence, by doing far less algebra than that required by the parameter-extraction method, we have obtained a low-entropy symbolic expression which is far superior to the one given in Eq. (1.6)

The EET, quite naturally, also allows for the value of a dependent source to become infinite so that a particular transfer becomes simplified in the same manner as that of an ideal operational amplifier circuit. In the case of R_{in} in Fig. 1.3, the EET allows us to write:

$$R_{in} = R_{in}\big|_{g_m \to \infty} \frac{1 + \dfrac{1}{g_m \overline{\mathscr{G}}^{(m)}}}{1 + \dfrac{1}{g_m \overline{G}^{(m)}}} \tag{1.13}$$

in which $\overline{G}^{(m)}$ and $\overline{\mathscr{G}}^{(m)}$ are the same as before and $R_{in}\big|_{g_m \to \infty}$ is determined in Fig. 1.5. The gain from v_1 to $g_m v_1$ reminds us of an opamp connected in some kind of

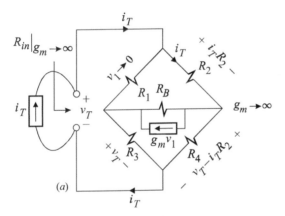

Figure 1.5

feedback fashion whose details we do not need to know at all. Now, if we let g_m become infinite, then $v_1 \to 0$ very much in the same manner as the differential input voltage of an opamp tends to zero when the gain becomes infinite and the output voltage stays finite. We can see in Fig. 1.5 that, with $g_m \to \infty$ and $v_1 \to 0$, the current through R_1 becomes zero and i_T flows entirely through R_2 creating a voltage drop $i_T R_2$ across it. At the same time, v_T appears across R_3 causing a current v_T/R_3 to flow through it. We can also see that the voltage drop across R_4, when $v_1 = 0$, is equal to $v_T - i_T R_2$ so that the current through it is simply $(v_T - i_T R_2)/R_4$. Summing the currents at the lower node of the bridge, we obtain:

$$i_T = \frac{v_T}{R_3} + \frac{v_T - i_T R_2}{R_4} \tag{1.14}$$

It follows from Eq. (1.14) that:

$$\frac{v_T}{i_T} = R_{in}\big|_{g_m \to \infty} = \frac{R_3 \parallel R_4}{1 + \dfrac{R_2}{R_4}} \qquad (1.15)$$

Substituting Eq. (1.15) in (1.13) we obtain another expression for R_{in} given by:

$$R_{in} = \frac{R_3 \parallel R_4}{1 + \dfrac{R_2}{R_4}} \; \frac{1 + \dfrac{1 + R_2 \parallel R_4/R_B}{g_m(R_B + R_2 \parallel R_4) \parallel R_1 \parallel R_3}}{1 + \dfrac{1 + R_2/R_1}{g_m(R_1 + R_2) \parallel R_B \parallel (R_3 + R_4)}} \qquad (1.16)$$

Although Eq. (1.16) looks simpler than Eq. (1.12), both are very useful analytical expressions. For very small values of g_m, Eq. (1.12) is a better expression because the bilinear factor containing g_m is close to unity and R_{in} is mostly dictated by the bridge circuit. If on the other hand g_m is very large, Eq. (1.16) is a better expression because R_{in} is mostly given by Eq. (1.15), and the bilinear function of g_m in Eq. (1.16) is close to unity.

1.4   Input impedance of a reactive bridge circuit with a dependent source

Consider now the reactive bridge circuit in Fig. 1.6 for which the input impedance[2] is to be determined. By designating the capacitor as the extra element, we will show how easily $Z_{in}(s)$ can be determined by simply analyzing a few purely resistive

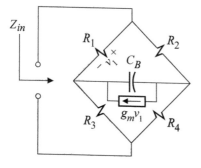

Figure 1.6

circuits. In other words, we will see how the EET allows one to determine a reactive transfer function, such as $Z_{in}(s)$, without ever having to deal with a reactive component such as $1/sC_B$. In fact, as we will see later, the most natural application of the EET and NEET is in the reduction of a circuit with N reactive elements to a set of purely resistive circuits.

If we designate $Z_B = 1/sC_B$ as the extra element and let $Z_B \to \infty$, we obtain the circuit in Fig. 1.7a, which is a special case of the circuit in Fig. 1.3 whose input impedance is given by Eq. (1.12). The derivation of the input impedance of the circuits in Figs. 1.3 and 1.7a are identical, with the exception that $R_B \to \infty$ in Fig. 1.7a. Hence, by letting $R_B \to \infty$ in Eq. (1.12) we obtain for Fig. 1.7a:

$$Z_{in}(s)|_{Z_B \to \infty} = (R_1 + R_3) \| (R_2 + R_4) \frac{1 + g_m R_1 \| R_3}{1 + \dfrac{g_m R_1}{1 + (R_1 + R_2)/(R_3 + R_4)}} \tag{1.17}$$

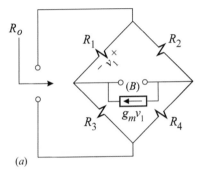

(a)

Figure 1.7

To obtain $Z_{in}(s)$, all we need to do is determine $\mathscr{R}^{(B)}$ and $R^{(B)}$, shown in Figs. 1.7b and c, respectively, and apply the EET:

$$Z_{in}(s) = Z_{in}(s)|_{Z_B \to \infty} \frac{1 + \dfrac{\mathscr{R}^{(B)}}{Z_B}}{1 + \dfrac{R^{(B)}}{Z_B}} \tag{1.18}$$

$$= R_o \frac{1 + sC_B \mathscr{R}^{(B)}}{1 + sC_B R^{(B)}}$$

in which $R_o = Z_{in}(s)|_{Z_B \to \infty}$ and is given by Eq. (1.17).

In Fig. 1.7b, the current i_T is given by the sum of $g_m v_1$ and the current through the branch $R_1 \| R_3 + R_2 \| R_4$, so that we have:

$$i_T = g_m v_1 + \frac{v_T}{R_1 \| R_3 + R_2 \| R_4} \tag{1.19}$$

In Fig. 1.7b we can also see that:

$$v_1 = v_T \frac{R_1 \| R_3}{R_1 \| R_3 + R_2 \| R_4} \tag{1.20}$$

Substituting Eq. (1.20) in (1.19), we obtain:

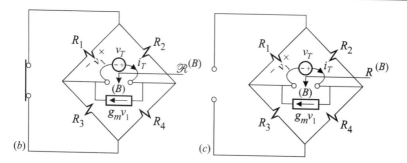

Figure 1.7 (*cont.*)

$$\mathscr{R}^{(B)} = \frac{v_T}{i_T} = \frac{R_1 \parallel R_3 + R_2 \parallel R_4}{1 + g_m R_1 \parallel R_3} \tag{1.21}$$

In Fig. 1.7c, the current i_T consists of the sum of $g_m v_1$ and the current through the branches $(R_1 + R_2)$ and $(R_3 + R_4)$ so that we have:

$$i_T = g_m v_1 + \frac{v_T}{R_1 + R_2} + \frac{v_T}{R_3 + R_4} \tag{1.22}$$

In Fig. 1.7c we can also see that:

$$v_1 = v_T \frac{R_1}{R_1 + R_2} \tag{1.23}$$

Substituting Eq. (1.23) in (1.22) we obtain:

$$i_T = \frac{v_T(g_m R_1 + 1)}{R_1 + R_2} + \frac{v_T}{R_3 + R_4} \tag{1.24}$$

whence it follows that:

$$R^{(B)} = \frac{v_T}{i_T} = \frac{R_1 + R_2}{1 + g_m R_1} \parallel (R_3 + R_4) \tag{1.25}$$

With $\mathscr{R}^{(B)}$ and $R^{(B)}$ determined, we can write $Z_{in}(s)$ in Eq. (1.18) in pole-zero form:

$$Z_{in}(s) = R_o \frac{1 + s/\omega_z}{1 + s/\omega_p} \tag{1.26}$$

in which:

$$\omega_z = \frac{1}{C_B \mathscr{R}^{(B)}} = \frac{1 + g_m R_1 \parallel R_3}{C_B(R_1 \parallel R_3 + R_2 \parallel R_4)} \tag{1.27}$$

$$\omega_p = \frac{1}{C_B R^{(B)}} = \frac{1}{C_B \dfrac{R_1 + R_2}{1 + g_m R_1} \,\|\, (R_3 + R_4)} \tag{1.28}$$

And such are the joys of network analysis!

1.5   Review

Although the matrix algebra of nodal or loop analysis is useful in obtaining numerical solutions of linear electrical circuits, it is not useful in obtaining *meaningful* analytical results in symbolic form. An analytical answer is not a mere collection of symbols but an answer in which the symbols are arranged in useful, or low-entropy, forms such as series-parallel combinations and ratios of various elements and time constants. This book presents efficient analytical tools for fast derivation of low-entropy results for electrical circuits. One such analytical tool is the extra element theorem (EET) which we have introduced in this chapter by way of examples in which the input impedance of various bridge circuits is determined.

Problems

1.1 High entropy versus low entropy. In order to appreciate the difference between high- and low-entropy expressions, consider the following for the input impedance of the circuit in the black box:

$$R_{in} = \frac{R_4 R_1 R_2 + R_4 R_1 R_3 + R_4 R_2 R_3}{R_4 R_2 + R_3 R_4 + R_1 R_2 + R_1 R_3 + R_2 R_3} \tag{1.29}$$

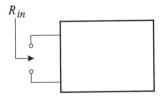

Figure 1.8

Are you able to make anything out of this expression? How does this expression simplify if $R_2 \ll R_3$? Consider now:

$$R_{in} = R_4 \,\|\, (R_1 + R_2 \,\|\, R_3) \tag{1.30}$$

Show that the two expressions above are equivalent. Which of the two is more meaningful? Using Eq. (1.30) show that when $R_2 \ll R_3$, we have the following simplification:

$$R_{in} \approx R_4 \| (R_1 + R_2) \tag{1.31}$$

1.2 Impedance using the EET. Following the example in Section 1.2, show in a few steps that the input impedance of the circuit below is given by:

$$Z_{in} = R_o \frac{1 + s/\omega_1}{1 + s/\omega_2} \tag{1.32}$$

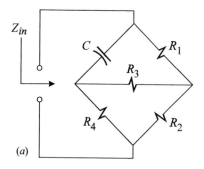

(a)

Figure 1.9

where:

$$R_o = R_1 + R_2 \| (R_3 + R_4)$$

$$\omega_1 = \frac{1}{CR_4 \| (R_3 + R_1 \| R_2)} \tag{1.33a, b, c}$$

$$\omega_2 = \frac{1}{C[R_1 + R_3 \| (R_4 + R_2)]}$$

Hint: Refer to Figs. 1.9b, c and d below and apply the EET in Eq. (1.4).

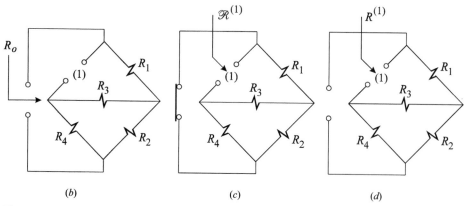

(b)   (c)   (d)

Figure 1.9 (*cont.*)

1.3 Output resistance of a current source using the EET. Show that the output resistance of the BJT current source in Fig. 1.10a, using the equivalent circuit model in Fig. 1.10b, is given by:

$$R_{out} = \frac{r_\mu + R_s}{1 + \dfrac{R_s}{R_E}} \cdot \frac{1 + \dfrac{1}{g_m r_\pi}\left(1 + \dfrac{r_\pi + R_s \| r_\mu}{R_E \| r_o}\right)}{1 + \dfrac{1}{g_m r_o}\left[1 + \dfrac{r_\mu + r_o}{r_\pi \| (R_s + R_E)}\right]} \qquad (1.34)$$

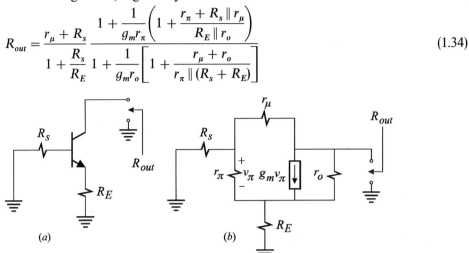

Figure 1.10

Hint: Refer to the example in Section 1.3 and to Figs. 1.10c–e below.

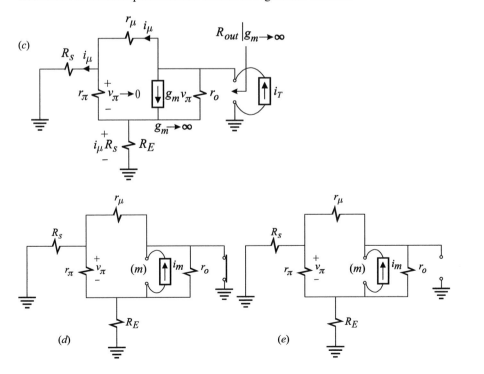

Figure 1.10 (*cont.*)

REFERENCES

1. R. D. Middlebrook, "Low-entropy expressions: the key to design-oriented analysis", *IEEE Frontiers in Education, Twenty-First Annual Conference*, Purdue University, Sept. 21–24, 1991, pp. 399–403.
2. V. Vorpérian, "Improved circuit analysis techniques require minimum algebra", *Electronic Design News*, August 3, 1995, pp. 125–134.
3. R. D. Middlebrook, "Null double injection and the extra element theorem", *IEEE Transactions on Education*, Vol. 32, No. 3, August 1989, pp. 167–180.
4. R. D. Middlebrook, V. Vorpérian, J. Lindal, "The N Extra Element Theorem", *IEEE Transactions on Circuits and Systems – I: Fundamental Theory and Applications*, Vol. 45, No. 9, Sept. 1998, pp. 919–935.
5. L. O. Chua and Pen-Min Lin, *Computer Aided Analysis of Electronic Circuits: Algorithms and Computational Techniques*, Prentice Hall, New York, 1975, pp. 568–569.

2 Transfer functions
Getting physical

2.1 Definition of a transfer function

Let $R(s)$ be the Laplace transform of the response of a linear time invariant (LTI) to an independent excitation whose Laplace transform is $E(s)$. The ratio of $R(s)$ to $E(s)$, with all initial conditions set to zero, is defined as a transfer function of the system:

$$H(s) \equiv \frac{Response(s)}{Excitation(s)} = \frac{R(s)}{E(s)} \tag{2.1}$$

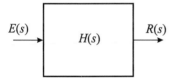

Figure 2.1

This equation can be represented by the block diagram shown in Fig. 2.1. When initial conditions are not zero, the Laplace transformation will properly transform the initial conditions into excitation functions so that the overall response, $R(s)$, is given by the linear superposition of the response due to the initial conditions $X_i(s)$ and the excitation function $E(s)$:

$$R(s) = H(s)E(s) + \sum_{i=1}^{n} X_i(s)h_i(s) \tag{2.2}$$

This equation can be represented by the block diagram in Fig. 2.2. Since initial conditions in the transform domain behave as any other excitation function, no further special consideration will be given to them.

Example 2.1 A simple circuit and its transform are shown in Figs. 2.3*a* and *b* respectively. The input voltage is applied at $t = 0$ and the inductor has an initial current $i_L(0)$ through it which, by the Laplace transformation, transforms into a voltage source $i_L(0)L$ as shown in Fig. 2.3*b*. The Laplace transform of the output is given by:

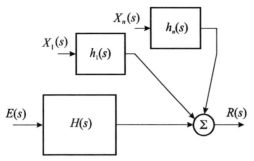

Figure 2.2

$$v_o(s) = H_1(s)v_{in}(s) - i_L(0)LH_2(s) \qquad (2.3)$$

where:

$$\left.\begin{array}{l} H_1(s) = \dfrac{1}{1 + \dfrac{R_1}{R_2 \parallel R_3}} \dfrac{1 + \dfrac{s}{\omega_1}}{1 + \dfrac{s}{\omega_2}} \\[3em] H_2(s) = \dfrac{1}{1 + \dfrac{R_2}{R_1 \parallel R_3}} \dfrac{1}{1 + \dfrac{s}{\omega_2}} \end{array}\right\} \qquad (2.4a, b)$$

in which:

$$\left.\begin{array}{l} \omega_1 = \dfrac{R_2}{L} \\[2em] \omega_2 = \dfrac{R_2 + R_1 \parallel R_3}{L} \end{array}\right\} \qquad (2.5a, b)$$

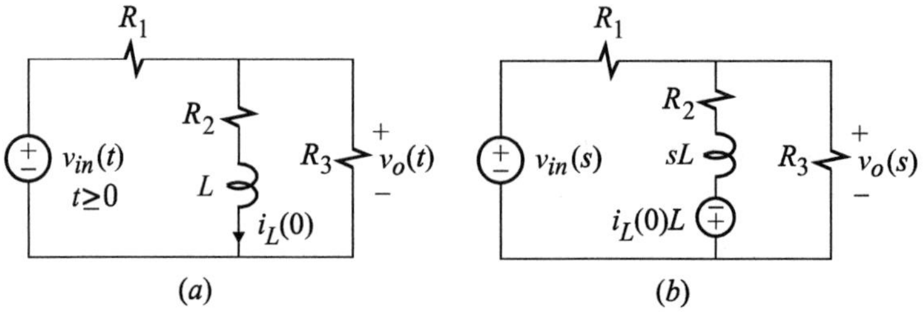

Figure 2.3

(a)   (b)

In Eq. (2.3), $H_1(s)$ is the transfer function from $v_{in}(s)$ to the output and $H_2(s)$ is the transfer function from $i_L(0)L$ to the output. Hence, if $i_L(0) = 0$ we simply have:

$$\frac{v_o(s)}{v_{in}(s)} = H(s) \tag{2.6}$$

in which $H(s)$ is the transfer function relating $v_{in}(s)$ to $v_o(s)$.   □

Block diagrams, such as those in Figs. 2.1 and 2.2, are useful visual representations of a general linear system governed by a set of linear differential equations with no specific physical system in consideration. Network analysis deals with electrical *circuit diagrams* which are approximate visual representations of physical circuits designed, built and tested in a laboratory. It is precisely in this respect that network theory and linear system theory differ from each other. Hence, analyzing an electrical circuit by transforming it into block diagrams or nodes and branches on rootless trees is a waste of analytical effort and precious time (see Problems 2.1 and 2.2).

Since an LTI system with no delays is governed by constant coefficient linear differential equations, $H(s)$ is given by the ratio of polynomials:

$$H(s) = \frac{N(s)}{D(s)} \tag{2.7}$$

The roots of $N(s)$ are called the *zeros* of $H(s)$, while the roots of $D(s)$ are called the *poles* of $H(s)$. $D(s)$ is also known as the characteristic equation of the system. In this chapter we will give a physical interpretation of $N(s)$ and $D(s)$ and show how they can be determined by manipulating the transform circuit rather than long algebraic equations.

2.2   The six types of transfer functions of an electrical circuit

Since in an electrical circuit a response or an excitation can be either a voltage or a current, we have the following six types of transfer functions shown in Figs. 2.4a–f:

(a) Voltage gain: $A_v(s) = v_2(s)/v_1(s)$
(b) Current gain: $A_i(s) = i_2(s)/i_1(s)$
(c) Transadmittance: $Y_t(s) = i_2(s)/v_1(s)$
(d) Transimpedance: $Z_t(s) = v_2(s)/i_1(s)$
(e) Driving-point admittance: $Y_{dp}(s) = i_1(s)/v_1(s)$
(f) Driving-point impedance: $Z_{dp}(s) = v_1(s)/i_1(s)$

We can easily visualize the first four of these to be transfer functions because they relate two quantities at two *different* places in a circuit. We are less accustomed, however, to think of the impedance or admittance looking into a port (driving-point) of a network as a transfer function simply because $v_1(s)$ and $i_1(s)$

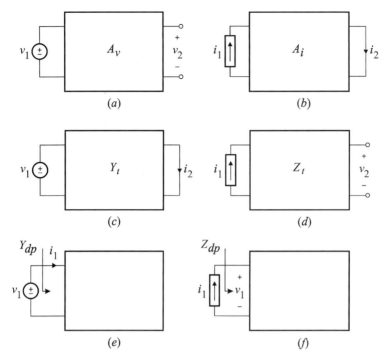

Figure 2.4

occur at the *same* place. There are other reasons as well. In simple cases, we can determine a driving-point impedance without being concerned with either $v_1(s)$ or with $i_1(s)$ altogether. In more complicated cases, we are free to connect *either* a test voltage source *or* a current source at a port and determine the ratio $v_1(s)/i_1(s)$ without distinguishing between response and excitation as in Eq. (2.1). In what follows, we emphasize the distinction between the response and excitation of driving-point impedance and driving-point admittance functions, and explain the importance of this distinction.

The driving-point impedance function shown in Fig. 2.4*f* is defined as:

$$Z_{dp}(s) = \frac{v_1(s)}{i_1(s)} \tag{2.8}$$

It follows from the definition of a transfer function in Eq. (2.1) that $v_1(s)$ is the response to the excitation $i_1(s)$. This implies that *the excitation function of a driving-point impedance is a current source*, as shown in Fig. 2.4*f*, and not a voltage source. Similarly, for a driving-point admittance function we have:

$$Y_{dp}(s) = \frac{i_1(s)}{v_1(s)} \tag{2.9}$$

so that $i_1(s)$ is the response to the excitation $v_1(s)$ as required by Eq. (2.1), which implies that *the excitation function of a driving-point admittance is a voltage source,* as shown in Fig. 2.4*e*, and not a current source. The reason why we tend to be careless in distinguishing between the nature of the excitation and response of these two functions is that they are reciprocals of each other and the determination of one is tantamount to the determination of the other.

The discussion above can now be summarized: if $v_1(s)/i_1(s)$ and $i_1(s)/v_1(s)$ are to be treated as *ratios*, the distinction between $v_1(s)$ and $i_1(s)$ as either a response or an excitation is not important, but if these are to be treated as *transfer functions*, then the distinction is important.

As we shall see next, the importance of putting the six different types of transfer functions of an electrical circuit on the same footing as the general transfer function in Eq. (2.1) lies in the determination of the poles and zeros of a transfer function.

2.3   Determination of the poles of a network

The poles of $H(s)$ in Eq. (2.1) are given by the roots of the characteristic equation $D(s)$, and the first thing we need to know about $D(s)$ is that it is completely determined by the physical structure and not by the excitation or the response. In other words, regardless of how a system is excited or which one of its responses is observed, $D(s)$ can be determined by setting $E(s) = 0$ and studying just the structure of the system by itself. The question is, how do we study an Nth-order network just by itself without any independent voltage or current sources in it? The answer is, there are two ways: we can either determine the coefficient matrix A of the state vector and expand the determinant $|sI - A| = D(s)$, or we can apply the N-extra element theorem (NEET). Whereas the expansion of $|sI - A|$ in symbolic form is pure, time-consuming algebraic torture, the application of the NEET is expeditious and quite rewarding. As we shall see later, NEET completely avoids the algebra of expanding determinants, and instead provides simple rules of manipulating the electrical network with all of its reactive elements removed! Since the principle behind the determination of $D(s)$ of an Nth-order network is exactly the same as that of a first-order network, we will use first-order networks in this section which do not require any knowledge of the NEET.

Before proceeding with these examples, we must thoroughly understand how an excitation, $E(s)$, is introduced to an electrical network or removed from it ($E(s) = 0$). Consider the electrical system shown in Fig. 2.5 which is excited by a current source, I_e, and a voltage source, V_e. Five different responses are monitored: two branch currents and three port voltages. Since $D(s)$ is a property of the structure and has nothing to do with excitation or response, the following ten transfer functions must have the same $D(s)$:

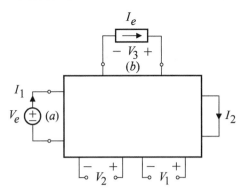

Figure 2.5

$$A_{V_n}(s) = \frac{V_n(s)}{V_e(s)}\bigg|_{I_e=0} = \frac{N_{A_n}(s)}{D(s)}; n = 1,2,3$$

$$G_{t_n}(s) = \frac{I_n(s)}{V_e(s)}\bigg|_{I_e=0} = \frac{N_{G_n}(s)}{D(s)}; n = 1,2$$

$$A_{I_n}(s) = \frac{I_n(s)}{I_e(s)}\bigg|_{V_e=0} = \frac{N_{I_n}(s)}{D(s)}; n = 1,2$$

$$Z_{t_n}(s) = \frac{V_n(s)}{I_e(s)}\bigg|_{V_e=0} = \frac{N_{Z_n}(s)}{D(s)}; n = 1,2,3$$

(2.10a–d)

To remove all the excitations from this system in Fig. 2.5, we must set $V_e = 0$ and $I_e = 0$. To set an independent voltage source to zero, we replace it with a short circuit; and to set an independent current source to zero, we replace it with an open circuit. Following this procedure, we obtain the structure shown in Fig. 2.6, from which the characteristic equation $D(s)$ can be determined directly using the NEET.

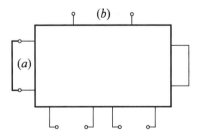

Figure 2.6

We must be equally clear about introducing an excitation to a network. To introduce an independent voltage excitation to a network *without* changing its structure, we must introduce it inside a branch (series insertion) and not across a branch. The reason of course is clear; if an independent voltage excitation is connected across a branch, then upon setting it to zero that branch gets shorted out and the structure changes. If on the other hand an independent voltage source is introduced inside a branch, then the network will revert to its original form when the voltage source is shorted. The dual argument applies when an independent current source is introduced: in order for the network, or the structure, to remain unchanged, an independent current source must be introduced across, and not inside, a branch.

For a first-order network the pole is given by the negative of the reciprocal of the time constant formed by the reactive element and the resistance seen by it. Thus the characteristic equation of a first-order network is given by:

$$D(s) = 1 + s/\omega_1 \tag{2.11}$$

in which ω_1 is related to the time constant by:

$$\tau = \frac{1}{\omega_1} \tag{2.12}$$

The root of $D(s)$ in Eq. (2.11) is $-\omega_1$, which is a negative, or a left-half plane (LHP), pole. When there is no reason for confusion, we shall refer to ω_1 rather than $-\omega_1$ as the pole.

Example 2.2 Determine $D(s)$ of the input impedance of the bridge circuit in Fig. 2.7a using the concept of a transfer function. To do so, we connect a test current source $I(s)$ to the input port and determine the response $V(s)$ as shown in Fig. 2.7b:

$$Z_{in}(s) = \frac{V(s)}{I(s)} \tag{2.13}$$

The characteristic equation of this transfer function is determined from the structure obtained by setting the excitation to zero, i.e. $I(s) = 0$. This is shown in Fig. 2.7c in which the time constant is given by the product of C and the effective resistance connected to it, which is seen to be:

$$\tau = C[R_1 + R_3 \| (R_4 + R_2)] \tag{2.14}$$

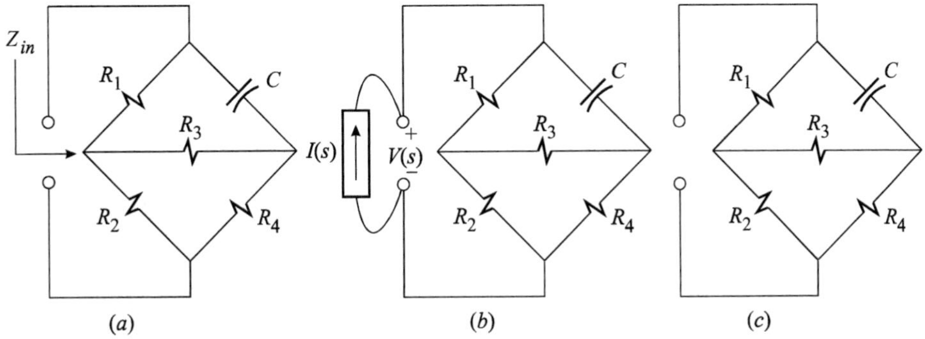

Figure 2.7

It follows that the magnitude, or the frequency, of the pole is given by:

$$\omega_a = \frac{1}{\tau} = \frac{1}{C[R_1 + R_3 \parallel (R_4 + R_2)]} \tag{2.15}$$

Since this is the negative root of the characteristic equation we have:

$$D_a(s) = s + \omega_a \tag{2.16a}$$

Note that Eq. (2.16a) has the units of rad/s. As will be discussed later, it is preferable to have the leading constant in a frequency polynomial set at unity so that Eq. (2.16a) can be rewritten as:

$$D(s) = 1 + s/\omega_a \tag{2.16b}$$

Observe that in Eq. (2.16b) we have effectively applied the EET as described in Problem 1.2.   □

The purpose of the next example is to clarify a subtle point of confusion that can arise when the excitations of admittance and impedance functions are not carefully distinguished.

Example 2.3   Consider now the same circuit of Example 2.2 for which the pole of the input admittance, $Y(s)$, is to be determined as indicated in Fig. 2.8a. We may now argue, *incorrectly*, as follows: since both circuits are the same, $Y(s)$ must have the same $D(s)$ as $Z(s)$ because all transfer functions defined in this circuit must have the same $D(s)$. Of course we know this is not true because $Y(s) = 1/Z(s)$. Correct application of the concept of a transfer function to $Y(s)$ requires that we connect a test voltage source $V(s)$ to the input as shown in Fig. 2.8b and determine the response $I(s)$:

$$Y(s) = \frac{I(s)}{V(s)} \tag{2.17}$$

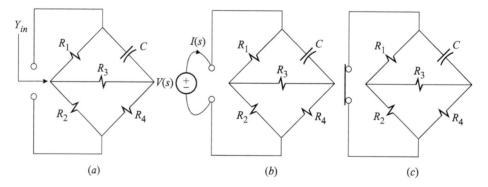

(a)   (b)   (c)

Figure 2.8

To set the excitation of this transfer function to zero, we short the input port and obtain the circuit (the structure or the system without excitation) in Fig. 2.8c, from which the characteristic equation is determined. Comparison of Figs. 2.8c and 2.7c shows that the structures in which each of these transfer functions is defined is different (although closely related).

The time constant of the circuit in Fig. 2.8c is given by the product of C and the effective resistance connected to it which in this case is seen to be:

$$\tau = CR_4 \parallel (R_3 + R_1 \parallel R_2) \tag{2.18}$$

It follows that the pole and the characteristic equation are given by:

$$\omega_b = \frac{1}{\tau} = \frac{1}{CR_4 \parallel (R_3 + R_1 \parallel R_2)} \tag{2.19}$$

$$D(s) = 1 + s/\omega_b \tag{2.20}$$

Once again, as in Example 2.1, observe that we have effectively applied the EET as described in Problem 1.2.   □

Example 2.4   Consider now the transimpedance function of the bridge circuit in Fig. 2.9. According to Example 2.1, any transfer function in this circuit relating any response to the excitation $I_1(s)$ has a pole given by ω_a in Eq. (2.15). Hence, the transimpedance can be written as:

$$Z_t(s) \equiv \frac{V_3(s)}{I_1(s)} = \frac{N(s)}{1 + s/\omega_a} \tag{2.21}$$

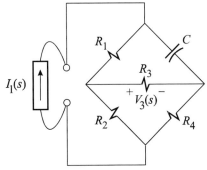

Figure 2.9

At low frequencies and in the limit $s \to 0$, we can easily verify, using current division between the R_2 and the $(R_3 + R_4)$ branches, that:

$$Z_t(0) \equiv R_0 = \frac{R_3}{1 + (R_3 + R_4)/R_2} \tag{2.22}$$

According to Eqs. (2.21) and (2.22), $Z_t(0) = N(0) = R_0$. Since it is preferable to have the leading constant term in the frequency polynomial equal to unity, we rewrite Eq. (2.21) in the following form:

$$Z_t(s) = R_0 \frac{N_a(s)}{1 + s/\omega_a} \qquad (2.23)$$

in which $N_a(0) = 1$. A method for the determination of $N_a(s)$ will be discussed in Section 2.4 (see also Problem 2.3). ☐

Example 2.5   Consider now the bridge circuit in Fig. 2.10. According to Example 2.3, any transfer function relating $V_1(s)$ to any voltage or current in this circuit has a pole ω_b given by Eq. (2.19). In particular, we can write the voltage gain:

$$A_V(s) \equiv \frac{V_3(s)}{V_1(s)} = A_{V0} \frac{N_b(s)}{1 + s/\omega_b} \qquad (2.24)$$

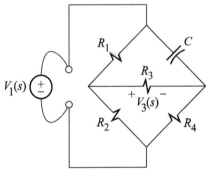

Figure 2.10

in which $N_b(0) = 1$ and A_{V0} is the dc voltage gain given by:

$$A_{V0} = \frac{1}{1 + R_1/R_2} \frac{1}{1 + (R_4 + R_1 \| R_2)/R_3} \qquad (2.25)$$

Equation (2.25) can be easily verified by examining the circuit in Fig. 2.10 at dc. A method for the determination of $N_b(s)$ will be discussed in Section 2.4 (see also Problem 2.4). ☐

2.4   Determination of the zeros of a transfer function

Lemma: The zeros of a transfer function correspond to conditions in the transform circuit which yield a null in the response of that transfer function.

The simple proposition stated above provides a very quick way of determining zeros of a transfer function directly from the circuit diagram. Assume that the numerator of a transfer function has *exact* analytical factors $n_i(s)$, each of order m_i, for *arbitrary* values of circuit elements:

$$H(s) = \frac{R(s)}{E(s)} = \frac{N(s)}{D(s)} = \frac{\prod_i n_i(s)}{D(s)} \tag{2.26}$$

The zeros of $H(s)$ in Eq. (2.26) are given by the roots of the factors $n_i(s)$:

$$n_i(s_k) = 0; k = 1, 2, \ldots, m_i \tag{2.27}$$

In Eq. (2.27), neither the order of the polynomial factors $n_i(s)$ nor the nature of their roots (real or complex) are relevant to the present discussion – all we care to discover are the conditions of the transform network which yield the individual factors $n_i(s)$. It is quite clear from Eq. (2.27) that for $s = s_k$ we have:

$$\frac{R(s_k)}{E(s_k)} = 0 \tag{2.28}$$

Since $E(s_k) \neq 0$, it follows that:

$$R(s_k) = 0 \tag{2.29}$$

Equation (2.29) represents *a null (or a zero) in the response in the transform domain in the presence of the excitation evaluated at $s = s_k$.* The interesting thing is that a null in the response can be most easily studied on the circuit diagram itself. As it turns out, each factor $n_i(s)$ corresponds to a condition in the transform circuit which prevents the excitation from reaching the response. Furthermore, for each null condition, the network equations simplify immensely so that the zeros, or the factors $n_i(s)$, can be fished out with a line or two of algebra in most cases.

If the preceding discussion sounds "Greek," the following examples will show the simplicity and clarity of it all.

Example 2.6 Suppose we are interested in determining the zeros, or the numerator, of the transfer function of the circuit in Fig. 2.11a. We begin by assuming that the transform response $v_o(s)$ is zero, or a null, for some $s = s_k$, where s_k is a zero of

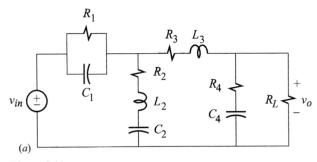

(a)

Figure 2.11

the numerator, as in Eqs. (2.27) and (2.29), i.e. $v_o(s_k) = 0$. This is shown in Fig. 2.11b, in which we see that a null in the response across the resistor R_L must be accompanied by a null in the current through R_L. However, the only way this can happen is if *all* the transform current $i_3(s_k)$ flows through the impedance branch $Z_4(s_k) = R_4 + 1/s_k C_4$, which in turn can happen if $Z_4(s_k)$ acts like a "short" in the transform circuit, or a *transform short*, i.e. $Z_4(s_k) = 0$. It follows that the zeros of $Z_4(s)$ are zeros of the desired function because they correspond to a condition of the transform network, in this case a transform short across the response, which prevents the excitation from reaching the response. We now have:

$$Z_4(s) = R_4 + \frac{1}{sC_4} = \frac{1 + sC_4 R_4}{sC_4} \tag{2.30}$$

from which it immediately follows that:

$$Z_4(s_1) = 0 \Rightarrow 1 + s_1 C_4 R_4 = 0 \Rightarrow s_1 = -\frac{1}{R_4 C_4} \tag{2.31}$$

Hence, the transfer function has an LHP zero at $-1/R_4 C_4$ and one of the factors of the numerator is:

$$n_1(s) = 1 + sC_4 R_4 \tag{2.32}$$

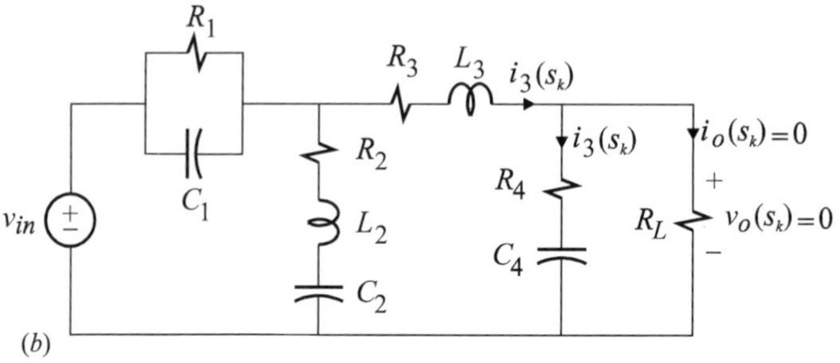

(b)

Figure 2.11 (*cont.*)

We continue to look for other conditions in the transform network which may cause a null in the response. Turning our attention to $i_3(s)$, we wonder if there is a condition which may result in $i_3(s_k) = 0$. According to Fig. 2.11c, this can happen if $i_1(s_k)$ flows entirely through $Z_2(s_k)$, which in turn can happen if $Z_2(s_k)$ acts like a transform short, i.e. $Z_2(s_k) = 0$. We have for $Z_2(s)$:

$$Z_2(s) = R_2 + sL_2 + \frac{1}{sC_2} \tag{2.33}$$

$$= \frac{1 + sC_2 R_2 + s^2 L_2 C_2}{sC_2}$$

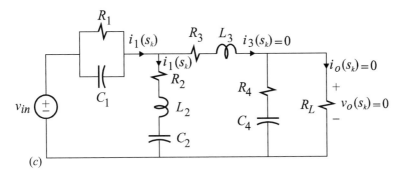

(c)

Figure 2.11 (*cont.*)

Since the zeros of $Z_2(s_k) = 0$ correspond to another null condition in the response of the transfer function, its numerator corresponds to the second factor in the numerator of the transfer function (see Eq. (2.26)):

$$n_2(s) = 1 + sC_2R_2 + s^2L_2C_2 \tag{2.34}$$

Finally, we look for a condition which may cause a null in $i_1(s)$, which in turn will cause a null in the output voltage. According to Fig. 2.11d, this can happen if the impedance, $Z_1(s)$, encountered by $i_1(s)$ becomes infinite, or a *transform open*, for some $s = s_k$, i.e. $Z_1(s_k) \to \infty$. We have for $Z_1(s)$:

$$Z_1(s) = \frac{R_1}{1 + sC_1R_1} \tag{2.35}$$

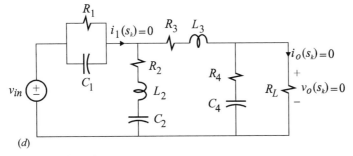

(d)

Figure 2.11 (*cont.*)

It is clear that when $Z_1(s)$ is evaluated at its poles or the roots of its denominator, then it becomes infinite:

$$Z_1(s_k) \to \infty \Rightarrow 1 + sR_1C_1 = 0 \tag{2.36}$$

It follows that the third factor in the numerator of the transfer function is:

$$n_3(s) = 1 + sR_1C_1 \tag{2.37}$$

Since there are no more conditions which can yield a null in the response, the numerator of the transfer function consists of the product of the three factors found above:

$$\frac{v_o(s)}{v_{in}(s)} = A_o \frac{n_1(s)n_2(s)n_3(s)}{D(s)} \qquad (2.38)$$

The results of this example can be easily generalized to an arbitrary passive ladder network shown in Fig. 2.12. The zeros of the input–output transfer function $v_o(s)/v_{in}(s)$ of such a network are given by the *zeros* of all the *shunt impedances*, $(Z_i(s); i = 1, 3, 5, \ldots)$ and the *poles* of the *series impedances*, $(Z_i(s); i = 2, 4, 6, \ldots)$. $\square$

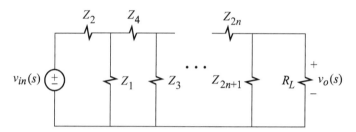

Figure 2.12

Example 2.7   For the simple equivalent circuit model of the common-emitter amplifier circuit shown in Fig. 2.13a, we would like to determine the zeros of the voltage gain transfer function. We can see from Fig. 2.13b that a null in the transform output, i.e. $v_o(s) = 0$, requires a null in the collector current, $i_c(s) = 0$, which in turn requires a null in the emitter current, $i_e(s) = 0$, because the collector and emitter currents are related by a constant $\alpha = \beta/(1 + \beta)$, i.e. $i_c(s) = \alpha i_e(s)$. We can see in Fig. 2.13b that the emitter current encounters the emitter impedance,

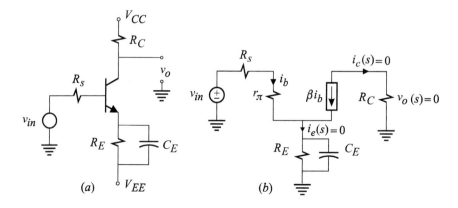

Figure 2.13

$Z_E(s)$, which acts as a *transform open* at its poles. It follows that the poles of $Z_E(s)$ correspond to a null condition of $i_e(s)$ and hence of $v_o(s)$. We have:

$$Z_E(s) = \frac{R_E}{1 + sR_EC_E} \tag{2.39}$$

Since $Z_E(s) \to \infty$ as $s \to -1/R_EC_E$, one of the factors of the numerator of the voltage gain is $1 + sR_EC_E$. Further investigation of the circuit reveals no other conditions that would result in a null in $v_o(s)$, so that the transfer function is given by:

$$\frac{v_o(s)}{v_{in}(s)} = A_o \frac{1 + sC_ER_E}{D(s)} \tag{2.40}$$

in which A_o is the low-frequency gain and $D(s)$ is the denominator. These are given (see Problems 2.4 and 2.5):

$$\left. \begin{array}{l} A_o = -\dfrac{R_C}{R_E/\alpha + (r_\pi + R_s)/\beta} \\[4mm] D(s) = 1 + sC_ER_E \left\| \dfrac{r_\pi + R_s}{1 + \beta} \right. \end{array} \right\} \tag{2.41a, b}$$

By comparing the low-entropy expressions of the zero and the pole in Eqs. 2.40 and 2.41b, we can see that the zero always comes before the pole. □

Example 2.8 What happens to the numerator, or the zero, in the previous example if a feedback resistance is added as shown in Fig. 2.14a? We can see in Fig. 2.14b that a null in $v_o(s)$, and hence in $i_o(s)$, for certain $s = s_k$, implies that $\beta i_b(s_k)$ flows entirely through R_f. Also, we can see that the voltage drop across R_f with $v_o(s_k) = 0$ equals the voltage drop across r_π and the emitter impedance $Z_E(s_k)$, so that we have:

$$\beta i_b(s_k)R_f = i_b(s_k)r_\pi + (1 + \beta)i_b(s_k)\frac{R_E}{1 + s_kR_EC_E} \tag{2.42}$$

Eliminating $i_b(s_k)$ from both sides in Eq. (2.42) and collecting terms we get:

$$\frac{\beta R_f - r_\pi}{(1 + \beta)R_E} = \frac{1}{1 + s_kR_EC_E} \tag{2.43}$$

This equation corresponds to a first-order polynomial in s which has a single root ($k = 1$), which we can easily solve for:

$$s_1 = -\frac{1}{C_E}\left(\frac{1}{R_E} - \frac{1 + \beta}{\beta R_f - r_\pi}\right) \tag{2.44}$$

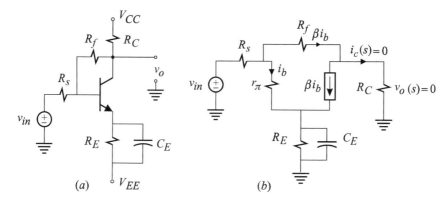

Figure 2.14

With $\alpha = \beta/(\beta + 1)$ and $r_e = r_\pi/(1 + \beta)$ this zero can be written as:

$$
\left.
\begin{aligned}
s_1 &= -\frac{1}{C_E}\left(\frac{1}{R_E} - \frac{1}{\alpha R_f - r_e}\right) \\[2mm]
&= -\frac{1}{C_E R_E \| (r_e - \alpha R_f)}
\end{aligned}
\right\}
\qquad (2.45a, b)
$$

We can see from this expression that if we let $R_f \to \infty$ we obtain the LHP zero of Example 2.7. In the presence of R_f, we can see from Eq. (2.45) that this zero begins to move towards and into the right-half plane (RHP) with decreasing value of R_f. (For very small values of $R_f < r_e/\alpha$ the zero moves back into the LHP, however this case is of little practical value.) According to Eq. (2.45b), the numerator of the transfer function is:

$$
\frac{v_o(s)}{v_{in}(s)} = A_o \frac{1 + sC_E R_E \| (r_e - \alpha R_f)}{D(s)}
\qquad (2.46)
$$

The low-frequency gain is given by:

$$
A_o = -\frac{R_C}{\dfrac{R_E + r_e}{\alpha} + \dfrac{R_s}{\beta}} \cdot \frac{1 - \dfrac{1}{R_f}\dfrac{r_e + R_E}{\alpha}}{1 + \dfrac{R_C}{R_f}\dfrac{R_s \| [(R_E + r_e)(1 + \beta)]}{R_s \| R_C \| (r_e + R_E)}}
\qquad (2.47)
$$

The denominator (see Problem 2.6) is given by:

$$
D(s) = 1 + sC_E R_E \left\| \left[\frac{r_\pi + R_s}{1 + \beta} \cdot \frac{1 + \dfrac{R_C + (1 + g_m R_C)R_s \| r_\pi}{R_f}}{1 + \dfrac{R_s + R_C}{R_f}} \right] \right.
\qquad (2.48)
$$

The form of A_o above is due to the EET, which will be discussed in Chapter 3. □

Example 2.9   To determine the numerator of the transfer function of the circuit in Fig. 2.15a, we begin by assuming a null in the response, $v_o(s_k) = 0$, for certain $s = s_k$, and follow this null through the rest of the circuit as shown in Fig. 2.15b.

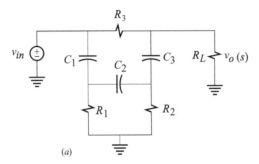

(a)

Figure 2.15

A step-by-step explanation of Fig. 2.15b is given:

1. Since $v_o(s_k) = 0$, the voltage drop across C_3, given by $i(1/s_kC_3)$, is equal and opposite to the voltage drop across R_2. Therefore the current through R_2 is given by $i(1/s_kC_3R_2)$.
2. The sum of the currents at node A flows through C_2 and is given by $i(1 + 1/s_kC_3R_2)$. Therefore, the voltage drop across C_2 is given by $i(1 + 1/s_kC_3R_2)(1/s_kC_2)$.
3. The voltage drop across R_1 is equal to the sum of the voltage drops across C_2 and R_2 and is given by $i(1/s_kC_3) + i(1 + 1/s_kC_3R_2)(1/s_kC_2)$. Therefore, the current in R_1 is given by:

$$[i(1/sC_3) + i(1 + 1/sC_3R_2)(1/sC_2)]/R_1$$

4. The sum of the currents through R_1 and C_2 flows through C_1 so that the voltage drop across C_1 is given by:

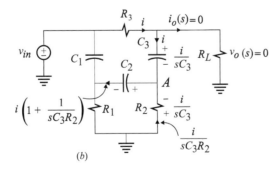

(b)

Figure 2.15 (cont.)

$$i\left\{1 + \frac{1}{s_k C_3 R_2} + \left[\frac{1}{s_k C_3} + \left(1 + \frac{1}{s_k C_3 R_2}\right)\frac{1}{s_k C_2}\right]\frac{1}{R_1}\right\}\frac{1}{s_k C_1} \qquad (2.49)$$

5. The sum of the voltage drops across C_1, R_3, C_3 and C_2 is zero so that we have:

$$iR_3 + i\frac{1}{s_k C_3} + i\left(1 + \frac{1}{s_k C_3 R_2}\right)\frac{1}{s_k C_2}$$

$$+ i\left\{1 + \frac{1}{s_k C_3 R_2} + \left[\frac{1}{s_k C_3} + \left(1 + \frac{1}{s_k C_3 R_2}\right)\frac{1}{s_k C_2}\right]\frac{1}{R_1}\right\}\frac{1}{s_k C_1} = 0 \qquad (2.50)$$

In the above, after canceling the current i and multiplying each term by $s_k^3 C_3 C_2 C_1 R_1 R_2$ we obtain the following third-degree polynomial whose roots ($k = 1, 2, 3$) correspond to the null condition in the response $v_o(s_k) = 0$:

$$N(s_k) = 1 + a_1 s_k + a_2 s_k^2 + a_3 s_k^3 = 0 \qquad (2.51)$$

in which:

$$\left.\begin{aligned}
a_1 &= C_1 R_1 + C_2(R_1 + R_2) + C_3 R_2 \\
a_2 &= R_1 R_2(C_1 C_2 + C_1 C_3 + C_2 C_3) \\
a_3 &= C_1 C_2 C_3 R_1 R_2 R_3
\end{aligned}\right\} \qquad (2.52a, b, c)$$

This completes the determination of the numerator. The transfer function can now be written as:

$$\frac{v_o(s)}{v_{in}(s)} = \frac{N(s)}{D(s)} = \frac{1 + a_1 s + a_2 s^2 + a_3 s^3}{D(s)} \qquad (2.53)$$

The determination of $D(s)$ will be discussed in Chapter 4 using the NEET. □

Example 2.10 It will be shown now that the circuit in Example 2.9 behaves as a notch filter for a particular choice of circuit elements. In order to have a notch response, the numerator in Eq. (2.53) must factor as follows:

$$\left.\begin{aligned}
N(s) &= (1 + s^2/\omega_o^2)(1 + s/\omega_z) \\
&= 1 + s/\omega_z + s^2/\omega_o^2 + s^3/\omega_z\omega_o^2
\end{aligned}\right\} \qquad (2.54a, b)$$

Comparison of the first two terms in Eqs. (2.54b) and (2.53) yields:

$$\left.\begin{aligned}
\omega_z &= \frac{1}{a_1} = \frac{1}{C_1 R_1 + C_2(R_1 + R_2) + C_3 R_2} \\
\omega_o^2 &= \frac{1}{a_2} = \frac{1}{R_1 R_2(C_1 C_2 + C_1 C_3 + C_2 C_3)}
\end{aligned}\right\} \qquad (2.55a, b)$$

Comparison of the last terms in Eqs. (2.54*b*) and (2.53) yields the following condition for the numerator to factor:

$$a_3 = a_2 a_1 \tag{2.56}$$

Substituting for the coefficients in Eq. (2.56) we get:

$$R_3 C_1 \parallel C_2 \parallel C_3 = R_1(C_1 + C_2) + R_2(C_2 + C_3) \tag{2.57}$$

Hence, if the circuit elements are chosen in such a way as to satisfy the time constant relation in Eq. (2.57), the numerator $N(s)$ will factor exactly as given in Eq. (2.54*a*). It is very important to realize now that each factor of $N(s)$ in Eq. (2.54*a*) does *not* correspond to a different null condition, as stated at the beginning of Section 2.4. The reason is that $N(s)$ in Eq. (2.54*a*) factors only for a *special choice* of circuit elements and *not for an arbitrary* choice.

A good choice of circuit components for practical design is:

$$C_1 = C_2 = C_3 \tag{2.58}$$

With this choice of capacitors, it follows from Eq. (2.57) that:

$$R_3 = 6(R_1 + R_2) \tag{2.59}$$

The notch frequency is now given by:

$$\omega_o = \frac{1}{C\sqrt{3R_1 R_2}} \tag{2.60}$$

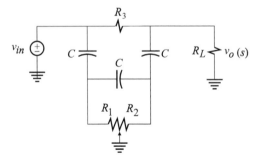

Figure 2.16

The excellent feature of this circuit is that as long as $R_3 = 6(R_1 + R_2)$, the notch frequency can be tuned by changing R_1 and R_2 using a potentiometer as shown in Fig. 2.16 in which the sum of R_1 and R_2 remains constant. This filter will be revisited later to determine $D(s)$ and study its complete response.   □

Although the method described above is very effective for the determination of the zeros, or the numerator, of a transfer function, it is not systematic. In

Chapters 3 and 4 we will show a systematic way of determining the numerator (and the denominator) of a transfer function using the EET and its extension the NEET. The concept of null response explained above will be essential to the derivation of these theorems.

2.5   The complete response, stability and transfer functions

This section presents a very brief review of the basic concepts in time-domain response, frequency-domain response and stability of a linear system. The concepts are illustrated using the simplest possible circuits and no formal derivations are given as these can be found in numerous textbooks.

The complete response of an LTI system, with zero initial conditions, to an arbitrary excitation, $e(t)$, applied at $t = 0$ is given by:

$$r(t) = \mathcal{L}^{-1}\{R(s)\} \tag{2.61}$$

$$= \mathcal{L}^{-1}\{H(s)E(s)\}$$

$$= \int_0^t h(\tau)e(t - \tau)d\tau$$

where $h(t)$ is the impulse response and is given by:

$$h(t) = \mathcal{L}^{-1}\{H(s)\} \tag{2.62}$$

The complete response in Eq. (2.61) consists in general of the sum of the natural response, $r_n(t)$, and the forced response, $r_f(t)$:

$$r(t) = r_n(t) + r_f(t) \tag{2.63}$$

A system is said to be stable if the natural response decays to zero:

$$\lim_{t \to \infty} r_n(t) \to 0 \Rightarrow \text{Stable} \tag{2.64}$$

If the natural response becomes unbounded then the system is unstable:

$$\lim_{t \to \infty} r_n(t) \to \infty \Rightarrow \text{Unstable} \tag{2.65}$$

A system is neither stable nor unstable if the natural response remains bounded for all time and does not decay to zero:

$$K_1 \leq \lim_{t \to \infty} r_n(t) \leq K_2 \tag{2.66}$$

It is important to realize that for an unstable system the forced solution, $r_f(t)$, in

Eq. (2.63) may very well be bounded and that it is the natural component which makes the complete response unbounded. As will be illustrated in Example 2.11, the only time the forced response becomes unbounded is when the system is excited at its natural frequencies.

In the preceding discussion, stability was characterized using time-domain criteria. Stability can also be described in the frequency domain by use of the characteristic equation $D(s)$. A system is stable if the poles of $H(s)$ are in the left-half plane. If s_i are the roots of the characteristic equation, then we can write:

$$\{\mathscr{R}e(s_i) < 0;\, D(s_i) = 0\} \Rightarrow \text{Stable} \tag{2.67}$$

A system is unstable if:

$$\{\mathscr{R}e(s_i) > 0;\, D(s_i) = 0\} \Rightarrow \text{Unstable} \tag{2.68}$$

A system is neither stable nor unstable if:

$$\mathscr{R}e(s_i) = 0;\, D(s_i) = 0 \tag{2.69}$$

The above is nothing more than the introductory material on the solution of constant coefficient differential equations stated in the jargon of system theory: pole, zero, response, excitation, stable and unstable. The following examples illustrate these concepts using the simplest circuits.

Example 2.11   In this example we will demonstrate the equivalence between frequency-domain and time-domain characterization of a stable and an unstable system. We will also comment on the nature of the forced response and its relationship to the excitation function.

The transfer function of the circuit in Fig. 2.17a is an impedance function given by:

$$\frac{v(s)}{i(s)} = \frac{R}{1 + s/\omega_1};\, \omega_1 = \frac{1}{RC} \tag{2.70}$$

The pole of this transfer function is in the left-half plane at $-\omega_1$ so that the system is stable and the natural response is expected to decay with increasing time. If the current source is a step function, i.e. $i(t) = I_o u(t)$ or $i(s) = I_o/s$, then:

$$v(s) = I_o R \frac{1}{s(1 + s/\omega_1)} \tag{2.71}$$

The complete response in time domain is given by:

$$v(t) = \mathscr{L}^{-1}\{v(s)\}$$

$$= I_o R(1 - e^{-\omega_1 t})u(t) \tag{2.72}$$

$$= I_o R u(t) - I_o R e^{-\omega_1 t}u(t)$$

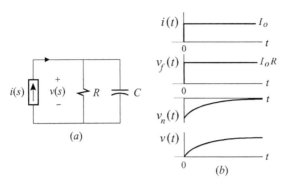

Figure 2.17

The natural solution in Eq. (2.72) is given by:

$$v_n(t) = -I_o Re^{-\omega_1 t}u(t) \tag{2.73}$$

which can be seen to decay to zero with $t \to \infty$ as expected, because the pole of the system is in the LHP. The forced response in Eq. (2.72) is:

$$v_f(t) = I_o Ru(t) \tag{2.74}$$

which is a step function of height I_o. For large times, a step function is analogous to a dc signal, which can be thought of as a signal with a spectral line at zero frequency. Now if we compare the forced response and the excitation, we see that they are both dc or at the same frequency (zero).

If we let $R \to -R$, then the complete solution in Eq. (2.62) becomes:

$$v(t) = I_o Re^{\omega_1 t}u(t) - I_o Ru(t) \tag{2.75}$$

The transfer function in Eq. (2.70) now has a RHP pole at ω_1:

$$\frac{v(s)}{i(s)} = \frac{-R}{1 - s/\omega_1}; \quad \omega_1 = \frac{1}{RC} \tag{2.76}$$

We can see from Eq. (2.75) that the natural solution, $I_o Re^{\omega_1 t}u(t)$, becomes unbounded, which is consistent with the RHP pole. Therefore, the frequency-domain and time-domain stability criteria of the system are equivalent. It is very important

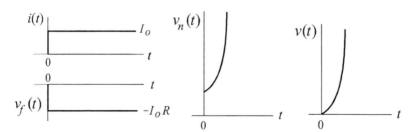

Figure 2.18

to see that, even though the system is unstable, the forced solution in Eq. (2.75), $-I_oRu(t)$, remains bounded and well-behaved as before. The forced, natural and the complete solutions in this case are shown in Fig. 2.18.   □

Example 2.12   If we let $R \to 0$ in Example 2.11, we obtain the circuit in Fig. 2.19a. This extremely simple circuit illustrates one which is neither stable nor unstable, whose natural solution neither becomes unbounded nor decays as time increases, and whose forced solution becomes unbounded as a linear function of time when excited at the frequency of the pole.

The transfer function now is simply the impedance of the capacitor:

$$\frac{v(s)}{i(s)} = \frac{1}{sC} \tag{2.77}$$

The pole is seen to be at the origin and the complete response in time domain for an arbitrary excitation is given by the well-known relation (assuming a zero initial condition):

$$v(t) = \frac{1}{C}\int_0^t i(t)dt \tag{2.78}$$

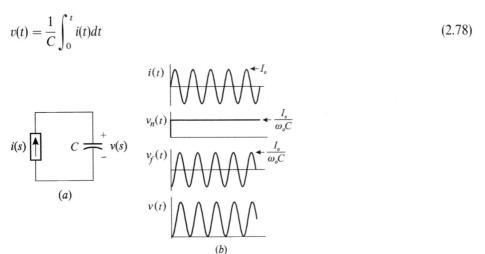

(a)

(b)

Figure 2.19

Let us first excite this circuit with $i(t) = I_o u(t)\sin \omega_o t$, so that the complete response according to Eq. (2.78) is given by:

$$v(t) = \frac{I_o}{\omega_o C}(1 - \cos \omega_o t)u(t) \tag{2.79}$$

$$= \frac{I_o}{\omega_o C}u(t) + \frac{I_o}{\omega_o C}\sin\left(\omega_o t - \frac{\pi}{2}\right)u(t)$$

The natural component in Eq. (2.79) is:

$$v_n(t) = \frac{I_o}{\omega_o} u(t) \tag{2.80}$$

while the forced solution is given by:

$$v_f(t) = \frac{I_o}{\omega_o C} \sin\left(\omega_o t - \frac{\pi}{2}\right) u(t) \tag{2.81}$$

The natural solution is seen to be a step function which neither decays nor grows in time as expected because of the pole at the origin. The forced solution is a sinusoidal function of the same frequency as that of the excitation except for a phase shift of $\pi/2$. The various waveforms are shown in Fig. 2.19b.

Now let us excite the circuit with a step function current source which for large times behaves essentially as a dc signal. This is shown in Fig. 2.20. Since the frequency of the pole is also zero, or dc, we should expect the solution to become unbounded in time because the system is being excited at its natural frequency. Indeed with $i(t) = I_o u(t)$, we find $v(t)$ from Eq. (2.78) to be:

$$v(t) = u(t)\frac{I_o}{C} t \tag{2.82}$$

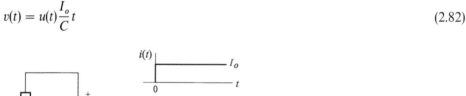

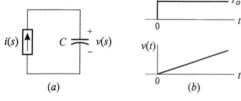

(a)   (b)

Figure 2.20

The complete response in Eq. (2.82) consists only of the forced response, which is seen to become unbounded as a linear function of time and not follow the excitation. (The natural response in Eq. (2.82) is zero because the forced solution by itself satisfies the initial condition $v(0) = 0$.) If the capacitor had an initial voltage across it, then the complete response would have been:

$$v(t) = u(t)\left[V_C(0) + t\frac{I_o}{C}\right]$$

in which $u(t)V_C(0)$ would have been the natural response.   □

Example 2.13 The same concepts as in Example 2.12 are demonstrated here using a system which has a pair of imaginary poles with zero real parts. For the circuit in Fig. 2.21a we have:

$$\frac{v_o(s)}{v_i(s)} = \frac{1}{1 + (s/\omega_o)^2}; \ \omega_o = \frac{1}{\sqrt{LC}} \tag{2.83}$$

The poles of this system are on the imaginary axis at $\pm j\omega_o$, at an angular frequency of ω_o. Since the real part of the poles is zero, we expect the natural solutions neither to decay nor become unbounded as a function of time.

Let the excitation be $v_i(t) = V_a u(t)$ so that the complete response is given by:

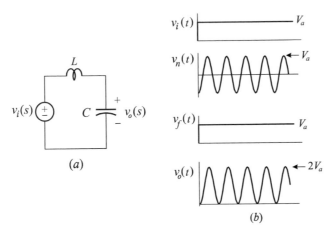

(a)

(b)

Figure 2.21

$$v_o(t) = \mathcal{L}^{-1} \left\{ \frac{V_a}{s} \frac{1}{1 + (s/\omega_o)^2} \right\} \tag{2.84}$$

$$= -V_a u(t)\cos \omega_o t + V_a u(t)$$

This and other waveforms are shown in Fig. 2.21b. The natural solution in Eq. (2.84) is given by:

$$v_n(t) = -V_a u(t)\cos \omega_o t \tag{2.85}$$

which is a bounded function as in Eq. (2.66):

$$-V_a \le v_n(t) \le V_a$$

This is a pure oscillation with fixed amplitude V_a. Hence, the natural solution neither decays nor becomes unbounded as a function of time, and the system is neither stable nor unstable. The forced solution is given by:

$$v_f(t) = V_a u(t) \tag{2.86}$$

which is a step function just like the excitation. Once again we see that the forced response and the excitation are at the same frequency (zero or dc).

The frequency of the pole of this system is ω_o, so that if we excite this system with a sinusoidal signal of frequency ω_o, we should expect the solution to become unbounded in time. Let the excitation now be $v_i(t) = V_a u(t) \sin \omega_o t$, as shown in Fig. 2.22, so that the complete response, $v_o(t)$ is now given by:

$$
\left. \begin{aligned}
v_o(t) &= \mathscr{L}^{-1}\left\{ \frac{V_a}{\omega_o} \frac{1}{[1 + (s/\omega_o)^2]^2} \right\} \\
&= \left(\frac{V_a}{2} \right) u(t) \sin \omega_o t - \left(\frac{V_a}{2} \right) u(t) \omega_o t \cos \omega_o t
\end{aligned} \right\}
\tag{2.87a, b}
$$

The natural response in Eq. (2.87b) is:

$$
v_n(t) = (V_a/2)u(t)\sin \omega_o t
\tag{2.88}
$$

which, just like before, is a bounded sinusoid as in Eq. (2.85). The forced response in Eq. (2.87b) is:

$$
v_f(t) = -(V_a/2)u(t)\omega_o t \cos \omega_o t
\tag{2.89}
$$

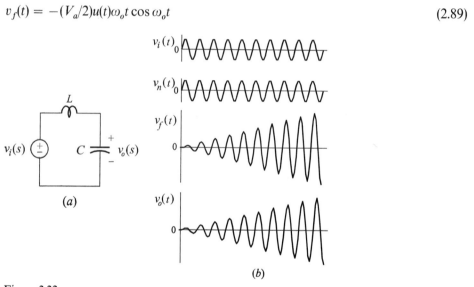

Figure 2.22

which becomes unbounded as a linear function of t. The natural, forced and complete responses are shown in Fig. 2.22b. Hence, the complete response in Eq. (2.87) becomes unbounded too! The important point here is that one must distinguish between the unbounded response of an *unstable* system from that of an *undamped* system (poles on the imaginary axis) excited at its resonance. In the former it is the natural response which becomes unbounded exponentially with time, whereas in the latter it is the forced response which becomes unbounded linearly in time (in general, algebraically in time, i.e. $t, t^2, \ldots$, which can happen if identical undamped resonant circuits are cascaded with buffers. The natural response of such a cascaded system comprises the natural response of each stage.

The natural response of the first circuit will be a bounded sinusoid which will serve as an excitation to the second circuit. The second circuit will in turn respond by a linearly growing sinusoid and the third circuit will respond with a quadratically growing sinusoid (see Problem 2.8)).   □

Thus far, we have shown how the complete response in time domain of an LTI system to an arbitrary excitation can be determined from the transfer function relating the response to the excitation. The stability of the system has been discussed in terms of the natural component of the complete response in time domain as well as the poles of the transfer function in frequency domain.

2.6   Magnitude and phase response

In the previous section we considered the complete response of an LTI system to an arbitrary excitation. In this section we will give a brief and complete discussion of the response of *stable* systems to *sinusoidal* excitations in *steady state*. As we found out in Section 2.5, in a stable system transients decay to zero so that in steady state the complete response is given by the forced response only.

Consider the stagle LTI system in Fig. 2.23*a* to which a sinusoidal excitation has been applied for a long time:

$$e(t) = E_a \cos(\omega t + \phi_o) \tag{2.90}$$

In Eq. (2.90), we have omitted the unit step function because we are only interested in the steady-state response and not the initial response when the excitation is applied. The steady-state response can be determined very easily *without* the use of Laplace transformation simply by letting $s = j\omega$ in $H(s)$:

$$r(t) = [E_a | H(j\omega)|]\cos[\omega t + \phi_o + \angle H(j\omega)] \tag{2.91}$$

in which:

$$\left.\begin{array}{l} |H(j\omega)| \equiv \text{Magnitude response} \\[2mm] \angle H(j\omega) \equiv \text{(Relative) phase response} \end{array}\right\} \tag{2.92a, b}$$

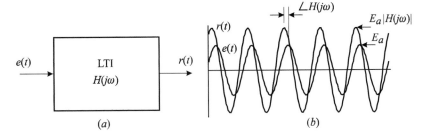

(a)   *(b)*

Figure 2.23

Equation (2.91) tells us that the steady-state response to a sinusoidal excitation is a sinusoid with the same frequency as the excitation and an amplitude and a phase (relative to $e(t)$) given by $E_a|H(j\omega)|$ and $\angle H(j\omega)$, respectively. The waveforms $e(t)$ and $r(t)$ are shown in Fig. 2.23b. By letting $s = j\omega$ in $H(s)$, we simply obtain a complex number $H(j\omega)$ whose magnitude, $|H(j\omega)|$, and angle, $\angle H(j\omega)$, are functions of frequency so that we can write $H(j\omega)$ as:

$$H(j\omega) = |H(j\omega)|e^{j\angle H(j\omega)} \tag{2.93}$$

The response $r(t)$ in Eq. (2.91) can now be written directly in terms of $H(j\omega)$:

$$\left.\begin{aligned} r(t) &= \mathcal{R}e\{(E_a e^{j\phi_0})H(j\omega)e^{j\omega t}\} \\ &= \mathcal{R}e\{R(j\omega)e^{j\omega t}\} \end{aligned}\right\} \tag{2.94a, b}$$

in which we have introduced the response phasor $R(j\omega)$:

$$R(j\omega) = (E_a e^{j\phi_0})H(j\omega) \tag{2.95}$$

The excitation $e(t)$ in Eq. (2.90) can also be written using phasor notation as:

$$\begin{aligned} e(t) &= \mathcal{R}e\{E_a e^{j\phi_0}e^{j\omega t}\} \\ &= \mathcal{R}e\{E(j\omega)e^{j\omega t}\} \end{aligned} \tag{2.96}$$

in which $E(j\omega)$ is the excitation phasor given by:

$$E(j\omega) = E_a e^{j\phi_0} \tag{2.97}$$

Note that, according to Eqs. (2.95) and (2.97), we have:

$$R(j\omega) = E(j\omega)H(j\omega) \tag{2.98}$$

which is consistent with $s = j\omega$ in $H(s) = R(s)/E(s)$.

In summary, to determine the steady-state response of a *stable* system to a sinusoidal excitation at a frequency ω, we determine the magnitude and phase of the transfer function by letting $s = j\omega$ in $H(s)$ and apply Eq. (2.91). In other words there is no need to find the inverse Laplace transform.

Example 2.14 For the reactive bridge with a dependent source discussed in Section 1.4, the input impedance was found to be:

$$Z_{in}(s) = R_o \frac{1 + s/\omega_z}{1 + s/\omega_p} \tag{2.99}$$

Let a sinusoidal current source be applied to the input port:

$$i(t) = I_o \sin(\omega_o t) \tag{2.100}$$

The input voltage will then be given by:

$$v(t) = V_o \sin[\omega_o t + \phi(\omega_o)] \tag{2.101}$$

in which the amplitude of the voltage is given by:

$$V_o = I_o |Z(j\omega_o)| \tag{2.102}$$

$$= I_o R_o \sqrt{\frac{1 + (\omega_o/\omega_z)^2}{1 + (\omega_o/\omega_p)^2}}$$

and the (relative) phase is given by:

$$\phi(\omega_o) = \tan^{-1}\frac{\omega_o}{\omega_z} - \tan^{-1}\frac{\omega_o}{\omega_p} \tag{2.103}$$

Although magnitude and phase functions look rather complicated, they have fairly simple plots, as will be discussed next. □

The magnitude response is often expressed in decibels:

$$|H(j\omega)|\, \mathrm{dB} = 20 \log(|H(j\omega)|) \tag{2.104}$$

Plots of the magnitude and phase response against the logarithmic frequency axis are known as Bode plots. Basic first-order and second-order transfer functions and their Bode plots will be given in Sections 2.7 and 2.8.

2.7   First-order transfer functions

In this section we review, briefly, first-order transfer functions and emphasize how to write them in a form which best describes their graphs. Figure 2.24a shows a magnitude response which decreases with frequency with a slope of $-20\,\mathrm{dB}\,\mathrm{dec}$. It is customary to write the transfer function of this graph as:

$$H(s) = \frac{K}{s} \tag{2.105}$$

The main drawback of this form is that K, in general, may not correspond to an important or interesting point on the graph. For example, in Eq. (2.105) the numerical value of K corresponds to the frequency at which the magnitude of $H(j\omega)$ is unity, which may or not be of any relevance. Another drawback of the

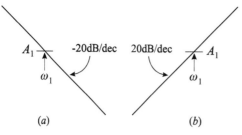

Figure 2.24

form in Eq. (2.105) is that the units of H are equal to the units of K multiplied by radians/s. Hence, a better way of writing Eq. (2.105) is to separate the constant K explicitly into a product of a frequency ω_1 and another constant A_1 which has the same units as H:

$$H(s) = A_1 \left(\frac{\omega_1}{s} \right) \tag{2.106}$$

In this equation, A_1 is chosen as a point of interest or relevance on the graph, as shown in Fig. 2.24a. One must realize that if there is no such point of interest on the graph, then there is not much point in writing an equation for it in the first place. (If we remove the point (ω_1, A_1) from the graph in Fig. 2.24a, then we are left with a featureless graph with a slope of -20 dB/dec.) Similarly, for the graph in Fig. 2.24b we have:

$$H(s) = A_1 \left(\frac{s}{\omega_1} \right) \tag{2.107}$$

The phase response of these transfer functions is independent of frequency and is $-90°$ for Eq. (2.106) and $90°$ for Eq. (2.107).

Example 2.15   For the negative impedance converter in Fig. 2.25a, it can be shown that the input impedance is given by:

$$Z_{in}(s) = -\frac{R_1}{sR_2C} \tag{2.108}$$

One way of writing this equation, using normalized frequency notation, is:

$$Z_{in}(s) = -R_1 \frac{\omega_1}{s} \tag{2.109}$$

in which $\omega_1 = 1/R_2C$. In this equation, emphasis is placed on the point (R_1, ω_1) as shown in Fig. 2.25b.

An alternative way of writing Eq. (2.108) is by choosing another normalizing frequency ω_a:

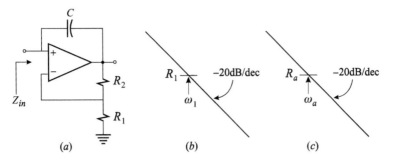

Figure 2.25

$$Z_{in}(s) = R_a \frac{\omega_a}{s} \tag{2.110a}$$

in which:

$$R_a = -\frac{1}{\omega_a(CR_2/R_1)} \tag{2.110b}$$

In Eq. (2.110a), emphasis is placed on the magnitude of Z_{in} at ω_a which is given by R_a. The effective transformation of C to $-CR_2/R_1$ can be clearly seen in Eq. (2.110b). The magnitude plot is shown in Fig. 2.25c. ☐

Example 2.16   For the integrator in Fig. 2.26a the transfer function is given by:

$$\frac{v_o(s)}{v_{in}(s)} = -\frac{1}{sRC} \tag{2.111}$$

When designing an integrator, one is usually concerned with the gain at a certain frequency ω_a so that the above can be normalized as:

$$\frac{v_o(s)}{v_{in}(s)} = -A_a \frac{\omega_a}{s} \tag{2.112}$$

in which:

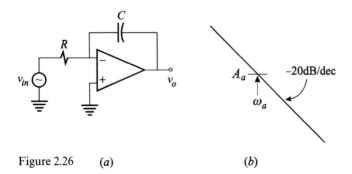

Figure 2.26   (a)   (b)

$$A_a = \frac{1}{\omega_a RC} \tag{2.113}$$

Hence, the time constant for a particular gain A_a at a particular frequency ω_a is given by $RC = 1/\omega_a A_a$. □

Consider next the first-order low-pass transfer function:

$$H(s) = \frac{A_1}{1 + s/\omega_1} \tag{2.114}$$

The asymptotic magnitude and phase plots are shown in Fig. 2.27a and are seen to belong to a group of four first-order transfer functions which possess mutual symmetry with respect to the magnitude and logarithmic frequency axes. Another form of writing the transfer function in Eq. (2.114) is:

$$H(s) = \frac{K}{s + \omega_1} \tag{2.115}$$

We shall avoid this form because in this expression the frequency variable, s, is not normalized and the units of the constant K are given by the units of H/s. In contrast, A_1 and H in Eq. (2.114) have the same units, while $1 + s/\omega_1$ is a *unitless* factor which describes the variation of the magnitude and phase as a function of frequency.

If we invert, or take the reciprocal of, the frequency factor in Eq. (2.114) we obtain:

$$H(s) = A_1(1 + s/\omega_1) \tag{2.116}$$

The asymptotic magnitude and phase plots of Eq. (2.116) are shown in Fig. 2.27b. These are seen to be the inversions of the magnitude and phase plots of Eq. (2.114) with respect to the magnitude and phase axes, respectively.

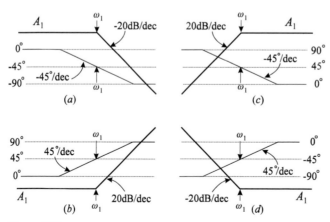

Figure 2.27

The magnitude response shown in Fig. 2.27c is seen to be an inversion of the magnitude plot in Fig. 2.27a with respect to the logarithmic frequency axis. The transfer function which corresponds to this graph is obtained by inverting the normalized frequency variable in the transfer function of the graph in Fig. 2.27a:

$$\left(\frac{s}{\omega_1}\right) \rightarrow \left(\frac{\omega_1}{s}\right) \tag{2.117}$$

Applying this transformation to Eq. (2.114) we obtain the transfer function of the graph in Fig. 2.27c:

$$H(s) = \frac{A_1}{1 + \omega_1/s} \tag{2.118}$$

Likewise, the magnitude response in Fig. 2.27d is seen to be an inversion of the magnitude response in Fig. 2.27b with respect to the logarithmic frequency axis, so that its transfer function is obtained by inverting the normalized frequency variable in Eq. (2.116):

$$H(s) = A_1(1 + \omega_1/s) \tag{2.119} \quad \square$$

Example 2.17   For the circuit in Fig. 2.28a we have:

$$\frac{v_o(s)}{v_{in}(s)} = -\frac{R_2}{R_1 + 1/sC} \tag{2.120}$$

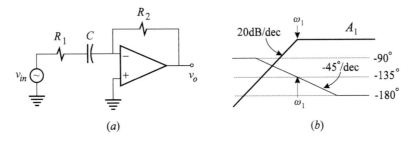

(a)   (b)

Figure 2.28

At high frequencies, the gain approaches $-R_2/R_1$ so that we can write the transfer function in Eq. (2.120) as:

$$\frac{v_o(s)}{v_{in}(s)} = -\frac{R_2/R_1}{1 + 1/sCR_1} \tag{2.121}$$

$$= \frac{A_1}{1 + \omega_1/s}$$

in which:

$$A_1 = -\frac{R_2}{R_1}$$ (2.122a, b)

$$\omega_1 = \frac{1}{R_1 C}$$

The asymptotic magnitude and phase plots are shown in Fig. 2.28b. □

Example 2.18   If the feedback and source branches in the previous example are interchanged, we obtain the circuit in Fig. 2.29a. The transfer function of this circuit is the reciprocal of the one in Eq. (2.120) and is given by:

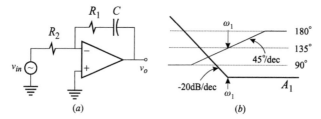

Figure 2.29

$$\frac{v_o(s)}{v_{in}(s)} = -\frac{R_1 + 1/sC}{R_2}$$ (2.123)

At frequencies above $\omega_1 = 1/R_1 C$, the transfer function approaches $A_1 = -R_1/R_2$ so that Eq. (2.123) is written as:

$$\frac{v_o(s)}{v_{in}(s)} = A_1(1 + \omega_1/s)$$ (2.124)

The asymptotic magnitude and phase plots are shown in Fig. 2.29b. □

2.8   Second-order transfer functions

We continue with a brief review of second-order transfer functions and emphasize the form in which they are best written. Figure 2.30a shows a magnitude response which decreases with frequency with a slope of $-40\,\text{dB/dec}$ and passes through the point (ω_o, A_o). As in first-order transfer functions, we shall avoid the form K/s^2 and normalize the frequency variable with respect to ω_o, so that A_o and ω_o appear explicitly in $H(s)$ as follows:

$$H(s) = A_o \left(\frac{\omega_o}{s}\right)^2$$ (2.125)

Similarly, the response in Fig. 2.30b is given by:

$$H(s) = A_o \left(\frac{s}{\omega_o} \right)^2 \tag{2.126}$$

Other second-order transfer functions with real roots can be formed by cascading any two of the first-order transfer functions considered in Section 2.7. For example, a cascade of the transfer functions in Figs. 27a and c results in a

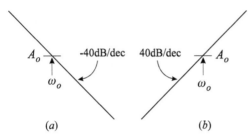

(a)   (b)

Figure 2.30

band-pass filter, whereas a cascade of the transfer functions in Figs. 2.27b and d results in a band-reject response. For the band-pass response, the transfer function is given by the product of Eqs. (2.114) and (2.118):

$$H(s) = \frac{A_m}{(1 + \omega_L/s)(1 + s/\omega_H)} \tag{2.127}$$

An asymptotic magnitude plot of this transfer function is shown in Fig. 2.31a. The transfer function of the band-reject response is given by the product of Eqs. (2.116) and (2.119):

$$H(s) = A_m(1 + \omega_L/s)(1 + s/\omega_H) \tag{2.128}$$

An asymptotic magnitude plot of this transfer function is shown in Fig. 2.31b. Note, once again, that these two transfer functions are reciprocals of each other and their graphs are symmetrical with respect to the magnitude axis.

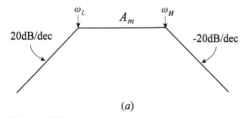

(a)

Figure 2.31

Next we consider second-order transfer functions with complex roots. These are expressed in terms of a Q-factor and a frequency ω_o. The magnitude response of

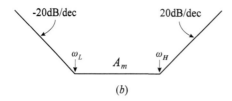

Figure 2.31 (*cont.*)

these transfer functions is characterized by a peaking at ω_o which is proportional to Q, while their phase response is characterized by a $180°$ transition centered at ω_o. The magnitude and phase response of the following transfer function is shown in Fig. 2.32a:

$$H(s) = \frac{A_o}{1 + \dfrac{1}{Q}\dfrac{s}{\omega_o} + \left(\dfrac{s}{\omega_o}\right)^2} \tag{2.129}$$

When Q is greater than 2 or 3, the peak occurs approximately at ω_o and is approximately given by A_oQ. Since the magnitude scale is in decibels, the peak is at a distance of $Q\,\mathrm{dB} \equiv 20\log Q$ above the low-frequency asymptote.

The four second-order transfer functions in Fig. 2.32, just like the first-order transfer functions in Fig. 2.27, possess mutual symmetry with respect to the magnitude and logarithmic frequency axes. Figure 2.32b is the inversion of Fig. 2.32a with respect to the magnitude axis, so that its response is given by:

$$H(s) = A_o\left[1 + \frac{1}{Q}\frac{s}{\omega_o} + \left(\frac{s}{\omega_o}\right)^2\right] \tag{2.130}$$

Figures 2.32c and d are the inversions of Figs. 32a and b, respectively, with respect to the logarithmic frequency axis, so that they are obtained from Eqs. (2.129) and (2.130) by the transformation $(s/\omega_o) \to (\omega_o/s)$, respectively. Hence, the response in Fig. 2.32c is given by:

$$H(s) = \frac{A_o}{1 + \dfrac{1}{Q}\dfrac{\omega_o}{s} + \left(\dfrac{\omega_o}{s}\right)^2} \tag{2.131}$$

The response in Fig. 2.32d is given by:

$$H(s) = A_o\left[1 + \frac{1}{Q}\frac{\omega_o}{s} + \left(\frac{\omega_o}{s}\right)^2\right] \tag{2.132}$$

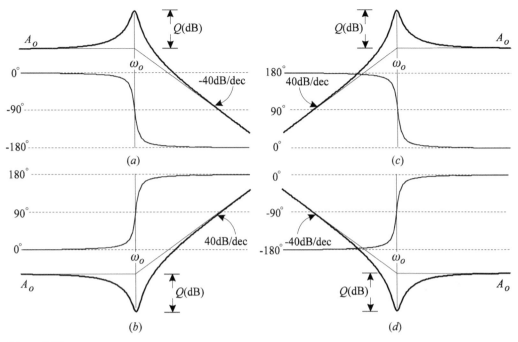

Figure 2.32

Finally, we consider the resonant band-pass and band-reject responses shown in Figs. 2.33a and b, respectively. The transfer function in Fig. 2.33a is given by:

$$H(s) = \frac{A_o}{1 + \left(\dfrac{\omega_o}{s} + \dfrac{s}{\omega_o}\right)Q} \tag{2.133}$$

The transfer function in Fig. 2.33b is given by:

$$H(s) = A_o\left[1 + Q\left(\frac{\omega_o}{s} + \frac{s}{\omega_o}\right)\right] \tag{2.134}$$

Since both transfer functions in Eqs. (2.133) and (2.134) are seen to be symmetrical with respect to the logarithmic frequency axis, they remain invariant under the transformation $(s/\omega_o) \rightarrow (\omega_o/s)$.

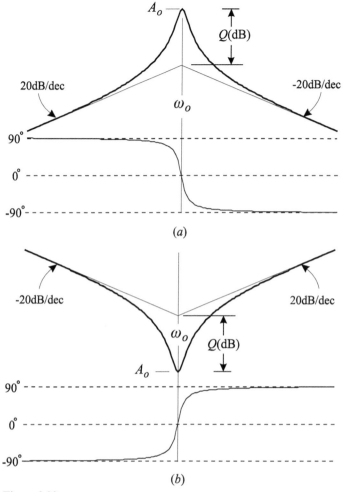

Figure 2.33

2.9   Review

A transfer function is the ratio of the Laplace transform of the response of a linear system to the Laplace transform of an excitation applied to that system as defined in Eq. (2.1). In this definition, it is assumed that all initial conditions are zero and only a single excitation is applied. In an electrical network, the excitation can be either a current source or a voltage source and the response can be either a current in a branch or a voltage across a port. Hence, there are six types of transfer functions in an electrical network: voltage gain, current gain, transimpedance, transadmittance, driving-point, or input, impedance and admittance.

The significance of the denominator, $D(s)$, of a transfer function is that it is entirely determined by the physical structure of the network in which no external excitation is applied and thus can be determined directly from the network. For a first-order network, all that one has to do is determine the time constant, τ, of the network with the excitation of the transfer function set to zero, i.e. if the excitation is a current source, it must be opened; if it is a voltage source, it must be shorted. The denominator is then given by $D(s) = 1 + s\tau$. For higher-order networks, $D(s)$ is determined using the N-extra element theorem, as will be explained in Chapter 4.

The physical significance of the numerator, $N(s)$, of a transfer function is that it corresponds to conditions in the transform network which result in a null in the response of the transfer function. Null conditions can be studied directly on the transform circuit diagram to determine $N(s)$ in analytically factored form. Each analytical factor of $N(s)$ corresponds to a particular null condition. An alternate method of determining $N(s)$ using the N-extra element theorem will be discussed in Chapter 4.

In writing a transfer function, the leading term in $N(s)$ and $D(s)$ must always be made to equal unity, so that the constant multiplying the transfer will have the same units as the transfer itself and will correspond either to a flat gain or a gain at a resonant peak. When a transfer function has neither a flat gain nor a resonant peak, then the multiplying constant must be made to correspond to a value on one of the asymptotes.

Problems

2.1 Circuit diagram versus block diagram. A common-emitter amplifier and its equivalent circuit model are shown in Figs. 2.34a and b. For purposes of illustrating the Miller effect, only the collector-to-base capacitance, C_μ, has been included while the base-to-emitter diffusion capacitance, C_π, has been ignored.

(a) Using the methods described in this chapter, determine the voltage gain:

$$\frac{v_o(s)}{v_{in}(s)} = -A_o \frac{N(s)}{D(s)} \tag{2.135}$$

in which A_o is the dc gain, and $N(s)$ and $D(s)$ are first-order polynomials given by:

$$\left. \begin{array}{l} N(s) = 1 - s/\omega_1 \\ D(s) = 1 + s/\omega_2 \end{array} \right\} \tag{2.136a, b}$$

The parameters A_o, ω_1 and ω_2 are given by:

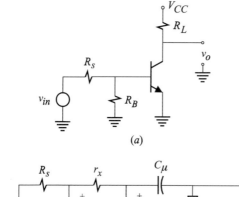

(a)

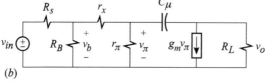

(b)

Figure 2.34

$$A_0 = \frac{1}{1 + \dfrac{R_s}{R_B}} \; \frac{g_m R_L}{1 + \dfrac{r_x + R_s \parallel R_B}{r_\pi}}$$

$$\omega_1 = \frac{g_m}{C_\mu}$$

$$\omega_2 = \frac{1}{C_\mu R_L + C'_\mu r_\pi \parallel (r_x + R_s \parallel R_B)}$$

(2.137a–c)

in which C'_μ is the Miller capacitance and is given by:

$$C'_\mu = C_\mu (1 + g_m R_L)$$

(2.138)

The bandwidth of the amplifier is given by ω_2, in which C_μ appears effectively multiplied by a factor $(1 + g_m R_L)$ which is proportional to the gain A_o. Note that $g_m R_L$ is exactly the voltage gain of the stage across which C_μ is connected, i.e. the gain $v_o/v_\pi = -g_m R_L$. Hence, increasing the gain is accompanied by a proportional decrease in the bandwidth. The maximum possible bandwidth for a given load R_L is $1/C_\mu R_L$, and is obtained when the source impedance R_s and base spread resistance r_x are zero. This is why the base spread resistance and source resistance are crucial parameters in the design of wide-band and high-gain amplifiers.

Hints: (*i*) Since A_o is the dc gain, it can be determined by removing C_μ from the equivalent circuit model in Fig. 2.34*b* and taking the Thevenin equivalent of the source side, as shown in Fig. 2.34*c*.

(*ii*) To determine the numerator $N(s)$, study the null response condition shown in Fig. 2.34*d*. In this figure, a null in the output voltage implies that the voltage across C_μ is v_π and the current through it is $g_m v_\pi$.

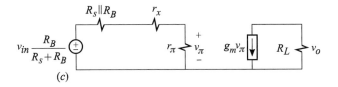

(c)

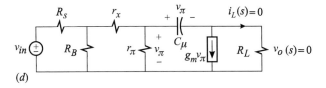

(d)

Figure 2.34 (cont.)

(iii) To determine $D(s)$, set the excitation to zero and determine the time constant $R^{(\mu)}C_\mu = \omega_2^{-1}$ in which $R^{(\mu)}$ is the driving-point resistance shown in Fig. 2.34e.

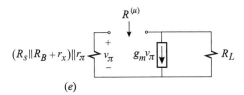

(e)

Figure 2.34 (cont.)

(b) Sketch a magnitude and phase plot of the voltage gain.

(c) Write the nodal equations at v_b, v_π and v_o and show that they correspond to the block diagram in Fig. 2.34f in which the various parameters are given in terms of conductances:

$$g_a = G_B + G_s + g_x$$

$$y_b = g_x + g_\pi + sC_\mu$$

$$y_o = G_L + sC_\mu$$

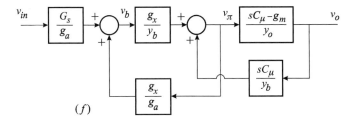

(f)

Figure 2.34 (cont.)

(d) Determine the voltage gain using Fig. 2.34f. In comparison to the original circuit diagram in Fig. 2.34b, does the block diagram provide more insight into the operation of the circuit? Is there any significance to the feedback block g_x/g_o? Do you have any idea about the output or input impedance of the circuit by looking at this block diagram? Compare what happens to the block diagram and the circuit diagram in the limit $r_x \to 0(g_x \to \infty)$?

(e) Determine the output impedance in Fig. 2.34g and show that it is given by:

$$Z_o(s) = R_L \frac{1 + s/\omega_3}{1 + s/\omega_2}$$

in which ω_2 is the same as above and ω_3 is given by:

$$\omega_3 = \frac{1}{C_\mu r_\pi \| (r_x + R_s \| R_B)}$$

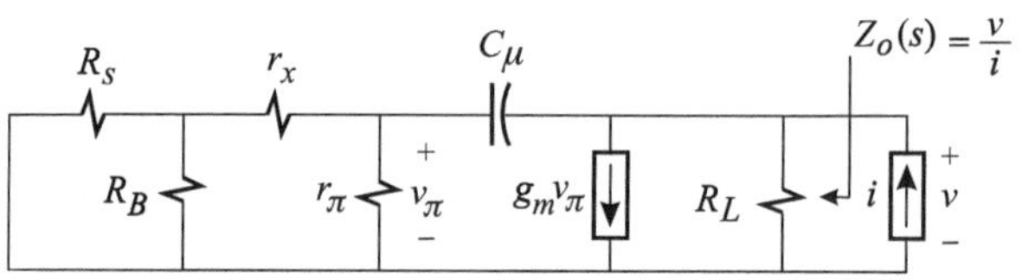

(g)

Figure 2.34 (cont.)

Hint: Since R_L is the low-frequency asymptote of $Z_o(s)$, it can be determined by removing C_μ from the circuit. Also, since the numerator of $Z_o(s)$ is the same as the denominator of the output admittance $Y_o = Z_o^{-1}(s)$, ω_3 can be determined by shorting the output.

The purpose of this problem was to show that it is easier to analyze a circuit using the techniques explained in this chapter rather than solve simultaneous equations and transform the circuit into a block diagram as is done in numerous textbooks.

2.2 Block diagram versus circuit diagram. A power amplifier with a gain A_o develops $\eta\%$ total harmonic distortion (THD) at its output stage at a certain frequency f_0. Determine the amount of loop gain T required to reduce the THD by a factor k. Assume flat gain characteristics independent of frequency.

Hint: The distortion can be modeled as an injected signal s_D at the output of the amplifier, as shown in Fig. 2.35 above. Using this block diagram, determine the output due to s_D in terms of the loop gain βA_o.

This problem illustrates the usefulness of a block diagram in visualizing a problem in which no specific circuit is given. It would have been considerably more time consuming if we had attempted to solve this problem using a specific circuit.

2.3 Determination of zeros in a reactive bridge circuit. For the bridge circuit in

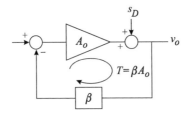

Figure 2.35

Examples 2.4 and 2.5, show that the numerators of the transfer functions in Eqs. (2.23) and (2.24) are the same and given by:

$$N_a(s) = N_b(s) = 1 - s/\omega_z$$

in which:

$$\omega_z = \frac{1}{R_1 C}\left(\frac{R_2}{R_4}\right)$$

Hint: A null in the response $v_3(s)$ is accompanied by a null in the current in R_3. It follows that (*i*) the voltage across C_1 is equal to the voltage across R_1 and the voltage across R_2 is equal to the voltage across R_4 and (*ii*) the current through C_1 is equal to the current through R_4 and the current through R_1 is equal to the current through R_2. The null conditions are illustrated in Fig. 2.36 and are seen to be independent of the nature of the excitation, i.e. current source or voltage source.

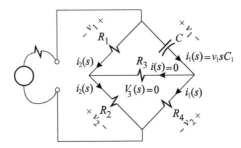

Figure 2.36

2.4 Low-frequency gain of the CE amplifier. Verify Eq. (2.41*a*) in Example 2.7.

2.5 Pole due to the emitter bypass capacitor. Verify Eq. (2.41*b*) in Example 2.7.

Hint: Let $v_{in} = 0$ and determine the effective resistance seen by C_E, which is given by R_E in parallel with the resistance seen looking directly into the emitter, $R^{(e)}$, as shown in Fig. 2.37.

2.6 Effect of shunt–shunt feedback using the EET. The resistor R_f in Example 2.8 feeds back a current proportional to the output voltage. This is known as shunt–shunt feedback or voltage sampling and current mixing. The extent to which the feedback current is subtracted from, or mixed with, the input current depends on

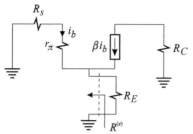

Figure 2.37

the source resistance R_s. If the source resistance R_s is zero, then the entire current through R_f flows through the source and the current entering the base of the transistor remains independent of R_f. This means that with $R_s = 0$, R_f loads the input and output of the amplifier without providing any output voltage feedback.

Verify Eq. (2.48) in Example 2.8 using the EET as described in Chapter 1 and show that the pole with R_f is at a lower frequency than without R_f provided $\beta > R_s/R_C$. Note that if $R_s = 0$, $D(s)$ becomes independent of R_f.

Hint: Following Section 1.2, the resistance looking into port (*e*) can be determined in three steps. First, determine the resistance looking into port (*e*) with $R_f \rightarrow \infty$ which, according to Problem 2.5, is given by $(r_\pi + R_s)/(1 + \beta)$. Second, determine the resistance $\mathcal{R}^{(f)}$ looking into port (*f*) with the emitter port short. Third, determine the resistance $R^{(f)}$ looking into port (*f*) with the emitter port open. These are shown in Figs. 2.38*a* and *b*.

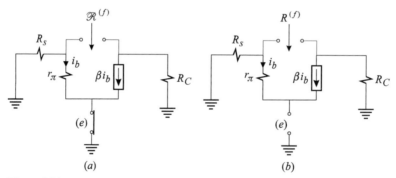

(a)　　　　　　　(b)

Figure 2.38

The resistance looking into the emitter is then given by:

$$R^{(e)} = \frac{r_\pi + R_s}{1 + \beta} \frac{1 + \mathcal{R}^{(f)}/R_f}{1 + R^{(f)}/R_f} \tag{2.139}$$

2.7 Nonideal operational amplifier with pole-zero compensation. For an ideal opamp, the voltage gain of the circuit in Fig. 2.39*a* is given by:

$$\frac{v_o(s)}{v_{in}(s)} = A_o \left(1 + \frac{\omega_1}{s}\right) \tag{2.140}$$

in which

$$A_o = -\frac{R}{R_s} \quad \Bigg\}$$

$$\omega_1 = 1/RC \quad \Bigg]$$

(2.141a, b)

(a) If the input impedance, output impedance and the gain of the operational amplifier are finite, as indicated in Fig. 2.39b, show that the gain is given by:

$$\frac{v_o(s)}{v_{in}(s)} = A_l \frac{1 + s/\omega_1}{1 + s/\omega_2}$$

(2.142)

in which A_l, ω_1 and ω_2 are determined independently and are given by:

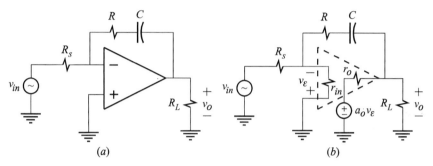

(a)   (b)

Figure 2.39

$$A_1 = -a_o \frac{1}{1 + R_s/r_{in}} \frac{1}{1 + r_o/R_L}$$

$$\omega_1 = \frac{1}{C(R - r_o/a_o)}$$

$$\omega_2 = \frac{1}{C\left[R + r_o \| R_L + R_s \| r_{in}\left(1 + \dfrac{a_o}{1 + r_o/R_L}\right)\right]}$$

(2.143a–c)

(b) Show that the gain can also be written as:

$$\frac{v_o(s)}{v_{in}(s)} = A_h \frac{1 + \omega_1/s}{1 + \omega_2/s}$$

(2.144)

in which:

$$A_h = \frac{\omega_2}{\omega_1} A_l$$

(2.145)

In this expression A_h is the high-frequency asymptote. If you were to substitute for ω_1, ω_2 and A_l in the expression above, you would get a rather long expression for A_h. You can determine A_h directly from the circuit by examining it at high frequencies in which the capacitor is replaced by a short. Derive the gain A_h using any method you know and save your derivation. In the next chapter we will show how to obtain the following expression for A_h in a few simple steps:

$$A_h = -\frac{R}{R_s} \frac{1 - \frac{1}{a_o}\frac{r_o}{R}}{1 + \frac{1}{a_o}\left(1 + \frac{r_o}{R_L}\right)\left(1 + \frac{R + r_o \| R_L}{R_s \| r_{in}}\right)} \tag{2.146}$$

How does your expression compare with this one? Are you ready for Chapter 3?

(c) Sketch an asymptotic magnitude and phase plot of the ideal and nonideal voltage gains.

2.8 Response of cascaded resonant circuits. Two identical and undamped resonant circuits are cascaded as shown in Fig. 2.40.

(a) If $v_i(t)$ is a step function as in Fig. 2.21 in Example 2.13, determine the forced and natural components of $v_{o1}(t)$ and $v_{o2}(t)$.

(b) Repeat part (a) if $v_i(t)$ is a sinusoid as in Fig. 2.22, Example 2.13.

The purpose of this problem is to show a natural response which grows linearly with time and a forced solution which grows quadratically in time. Such solutions occur in systems which have multiple pairs of identical imaginary poles (identical eigenvalues) and require special consideration in linear system (or differential equation) theory.

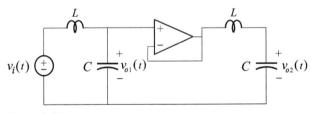

Figure 2.40

2.9 Bode plots. Sketch asymptotic magnitude and phase plots of the following transfer functions:

(a) $H(s) = A_o \dfrac{\omega_o}{s} \dfrac{1}{1 + s/\omega_a}$; $\omega_o < \omega_a$

(b) $H(s) = A \dfrac{\omega_o}{s} \dfrac{(1 + s/\omega_a)^2}{(1 + s/\omega_b)^2}$; $\omega_o < \omega_a < \omega_b$

3　The extra element theorem
A basic simplification tool

3.1　Introduction

The extra element theorem (EET) in its present and most useful form is due to R. D. Middlebrook.[1] The structure of this theorem is a consequence of linear networks and was given by Bode in his famous book, *Network Analysis and Feedback Amplifier Design*.[2] Starting with the loop equations of a general linear network and using determinant theory, Bode showed that any transfer function of that network is a bilinear transformation of any one of its elements, z. Specifically, Bode showed that when the loop equations are set up such that the impedance element z occurs only in the jth loop and none other, then the current in loop 2 is related to the excitation E_1 according to:

$$\frac{I_2}{E_1} = \frac{\Delta_{12}^0 + z\Delta_{12jj}}{\Delta^0 + z\Delta_{jj}} \tag{3.1}$$

This equation is written using Bode's notation in which Δ^0 is the determinant of the $[Z]$ matrix in $[I][Z] = [E]$ with $z = 0$, Δ_{12}^0 is the same as Δ^0 with row 1 and column 2 struck out, Δ_{jj} is the determinant of $[Z]$ with the jth row and column struck out (note that this determinant is independent of z because z occurs only in the jth row and column of $[Z]$), and Δ_{12jj} is the same as Δ_{jj} with row 1 and column 2 struck out. Clearly, the application of this theorem to the determination of transfer functions is no picnic. Bode's objective in deriving this equation, however, was not to determine transfer functions but rather to study the sensitivity of a transfer function with respect to an arbitrary element of a linear network. Useful adaptations of the bilinear transformation to the determination of transfer functions and sensitivities have been discussed in more recent literature[3–6] in which simpler forms of the various determinants are given. In this chapter, we will give the simplest possible interpretation and form of the bilinear transformation and call it the extra element theorem. The EET and the notation developed in this chapter extend most naturally to the more general case of the N-extra element theorem[7,8] discussed in Chapter 4.

3.2   Null double injection

The key ingredient in the EET is the null response calculation using double injection, or null double injection for short. Consider the linear time invariant circuit shown in Fig. 3.1 in which u_{i1} is the first independent excitation, which can be either a voltage source or a current source, and u_{o1} is the first arbitrary response, which can be either a voltage or current. Let I be the second independent excitation applied at port (1) and the voltage V across it be the second response to be studied. Each response is given by linear superposition of the contribution of each excitation to that response:

$$
\left.\begin{aligned}
u_{o1} &= a_{11} u_{i1} + a_{12} I \\
V &= a_{21} u_{i1} + a_{22} I
\end{aligned}\right\}
\qquad (3.2a, b)
$$

Each coefficient a_{ij} is a particular transfer function defined as:

$$
\left.\begin{aligned}
a_{11} &= \left.\frac{u_{o1}}{u_{i1}}\right|_{I=0} \\[2mm]
a_{12} &= \left.\frac{u_{o1}}{I}\right|_{u_{i1}=0} \\[2mm]
a_{21} &= \left.\frac{V}{u_{i1}}\right|_{I=0} \\[2mm]
a_{22} &= \left.\frac{V}{I}\right|_{u_{i1}=0}
\end{aligned}\right\}
\qquad (3.3a, b, c, d)
$$

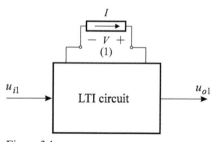

Figure 3.1

The transfer functions a_{11} and a_{22} have the following interpretations. Since setting $I = 0$ is the same as opening port (1), a_{11} can be written as:

$$a_{11} = \frac{u_{o1}}{u_{i1}}\bigg|_{port\,(1)\rightarrow OPEN} \tag{3.4}$$

This is shown in Fig. 3.2a. When $u_{i1} = 0$, the transfer function $a_{22} = V/I$ in Eq. (2.2d) is nothing more than the driving-point impedance looking into port (1), as shown in Fig. 3.2b:

$$Z^{(1)} \equiv a_{22} = \frac{V}{I}\bigg|_{u_{i1}=0} \tag{3.5}$$

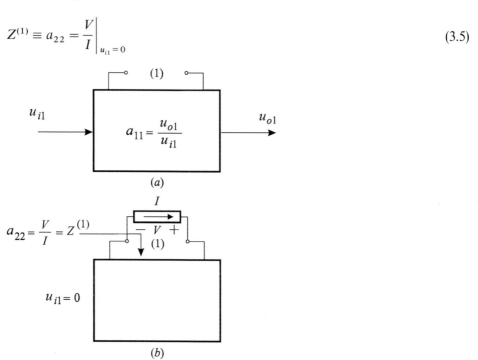

Figure 3.2

With the application of two excitations, or double injection, it is possible to null any response of the LTI network. In particular, if we choose to null u_{o1}, as shown in Fig. 3.2c, then we have from Eqs. (3.2a, b):

$$\left.\begin{array}{l} 0 = a_{11}u_{i1} + a_{12}I \\ V = a_{21}u_{i1} + a_{22}I \end{array}\right\} \tag{3.6a, b}$$

Equation (3.6a) gives the relationship between the two excitations which brings about the null in the response:

$$\frac{u_{i1}}{I}\bigg|_{u_{o1}=0} = -\frac{a_{12}}{a_{11}} \tag{3.7}$$

Substituting this result in Eq. (3.6b), we get:

$$V\big|_{u_{o1}=0} = -a_{21}\frac{a_{12}}{a_{11}}I\big|_{u_{o1}=0} + a_{22}I\big|_{u_{o1}=0} \tag{3.8}$$

whence the definition of the *null driving-point impedance* looking into port (1), $\mathscr{Z}^{(1)}$, follows:

$$\left.\begin{array}{l} \mathscr{Z}^{(1)} = \dfrac{V}{I}\bigg|_{u_{o1}=0} \\[2ex] = \dfrac{a_{11}a_{22} - a_{12}a_{21}}{a_{11}} \end{array}\right\} \tag{3.9a, b}$$

The double injection, the null response and $\mathscr{Z}^{(1)}$ are all shown in Fig. 3.2c.

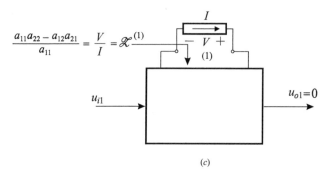

(c)

Figure 3.2 (*cont.*)

We now have two types of impedances looking into port (1): an ordinary driving-point impedance, $Z^{(1)}$, and a null driving-point impedance $\mathscr{Z}^{(1)}$. Remember that in the ordinary driving-point impedance, $Z^{(1)}$, I is the only excitation in the circuit, whereas in the null driving-point impedance, $\mathscr{Z}^{(1)}$, both excitations, I and u_{i1}, are applied simultaneously in such a way as to null the response u_{o1}, i.e. $u_{o1} = 0$.

Figure 3.2c is very important to the concept of null impedance because it shows an *operational method* for determining $\mathscr{Z}^{(1)}$ without having to compute it algebraically using Eq. (3.9b). Without Fig. 3.2c, there can be no useful form of the EET. The following examples illustrate the determination of $\mathscr{Z}^{(1)}$.

Example 3.1   Determine $Z^{(1)}$ and $\mathscr{Z}^{(1)}$ for the bridge circuit in Fig. 3.3 in which the response is taken across the resistance R_L.

The ordinary driving-point impedance looking into port (1) with the excitation $v_{in} = 0$ is shown in Fig. 3.4a in which $Z^{(1)}$ is real and given by:

$$Z^{(1)} = R^{(1)} = R_2 + R_1 \parallel (R_4 + R_L \parallel R_3) \tag{3.10}$$

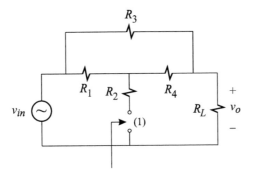

Figure 3.3

The null driving-point impedance is shown in Fig. 3.4b in which the voltage across R_L is nulled by the application of the two injections v_{in} and v_T. Since $\mathscr{Z}^{(1)} = R_2 + \mathscr{Z}''^{(1)}$, we shall work with the simpler circuit in Fig. 3.4c and determine $\mathscr{Z}''^{(1)}$. A null in the voltage across R_L implies that the current through R_L is

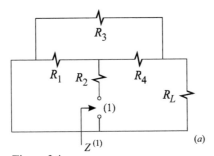

Figure 3.4

zero, which in turn implies that the entire current through R_4 flows through R_3. Also, a null in the voltage across R_L implies that v'_T appears directly across R_4, so that the current through R_4 is v'_T/R_4. It follows by Kirchhoff's voltage law (KVL) that the voltage across R_3 is $R_3(v'_T/R_4)$ and the voltage across R_1 is the sum of the voltages across R_4 and R_3, as shown in Fig. 3.4c. Application of Kirchhoff's current law (KCL) at node A yields for i'_T:

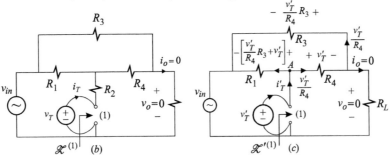

Figure 3.4 (*cont.*)

$$i'_T = \frac{v'_T}{R_4} + \frac{v'_T}{R_1}\left(1 + \frac{R_3}{R_4}\right) \tag{3.11}$$

from which we have:

$$\frac{v'_T}{i'_T} = \mathscr{Z}'^{(1)} = R_4 \left\|\left(\frac{R_1}{1 + R_3/R_4}\right)\right. \tag{3.12}$$

Hence, the null impedance $\mathscr{Z}^{(1)} = R_2 + \mathscr{Z}'^{(1)}$ is given by:

$$\mathscr{Z}^{(1)} = R_2 + R_4 \left\|\frac{R_1}{1 + R_3/R_4}\right. \tag{3.13}$$

In the next example, $\mathscr{Z}'^{(1)}$ will be determined with a different response nulled. $\square$

Example 3.2   For the same circuit in Example 3.1, determine $Z^{(1)}$ and $\mathscr{Z}^{(1)}$ when the response is taken across R_3 instead of R_L as shown in Fig. 3.5.

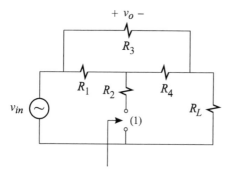

Figure 3.5

The ordinary driving-point impedance looking into port (1), with the excitation, $v_{in} = 0$, remains the same as before and is given by:

$$Z^{(1)} = R^{(1)} = R_2 + R_1 \| (R_4 + R_L \| R_3) \tag{3.14}$$

The null driving-point impedance, shown in Fig. 3.6a, is not the same as before because a different response is being nulled. As in the previous example, since $\mathscr{Z}^{(1)} = R_2 + \mathscr{Z}'^{(1)}$, we shall work with the simpler circuit in Fig. 3.6b and determine $\mathscr{Z}'^{(1)}$. In this case, when the voltage across R_3 is nulled, the voltages across R_1 and R_4 become identical. Also, a null in the voltage across R_3 implies that the current through it is nulled so that the entire current through R_4 flows through R_L and is simply given by $v'_T/(R_4 + R_L)$. All the remaining voltages and currents with v_o nulled are shown in Fig. 3.6b whence we have by KCL at node A:

$$i'_T = \frac{v'_T}{R_4 + R_L} + \frac{v'_T}{R_1}\left(\frac{R_4}{R_4 + R_L}\right) \tag{3.15}$$

This yields $\mathscr{L}'^{(1)}$ or $\mathscr{R}'^{(1)}$:

$$\frac{v'_T}{i'_T} = \mathscr{R}'^{(1)} = R_1 \frac{1 + \dfrac{R_L}{R_4}}{1 + \dfrac{R_1}{R_4}} \tag{3.16}$$

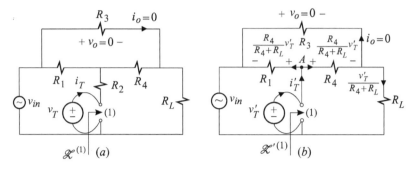

Figure 3.6

Hence $\mathscr{R}^{(1)}$ is given by:

$$\mathscr{R}^{(1)} = R_2 + R_1 \frac{1 + \dfrac{R_L}{R_4}}{1 + \dfrac{R_1}{R_4}} \tag{3.17}$$

It is important to see that v_{in} *never enters or appears* in the determination of a null driving-point impedance. For instance, in this example with $v_o = 0$, it would have been very tempting to say that the voltage across R_L is v_{in} and proceeded to write equations with v_{in}. Although this would not have been wrong, it would have been a waste of time because v_{in} would have eventually canceled out. □

Example 3.3   Determine $\mathscr{L}'^{(1)}$ and $Z^{(1)}$ for the bridge circuit in Fig. 3.7 in which the output is taken across the bridge resistance R_B.

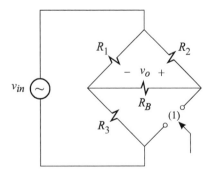

Figure 3.7

The determination of the driving-point impedance with the excitation set to zero, i.e. $v_{in} = 0$, is shown in Fig. 3.8a whence we have:

$$Z^{(1)} = R^{(1)} = R_2 \parallel (R_B + R_1 \parallel R_3) \tag{3.18}$$

The determination of $\mathscr{Z}^{(1)}$ is shown in Fig. 3.8b in which we see that a null in v_o is accompanied with a null in the bridge current so that the test current i_T flows entirely through R_2 and the currents through R_1 and R_3 are the same. Also, since $v = 0$, the voltage across R_2 is the same as the voltage across R_1 and the voltage across R_3 is the same as the test voltage v_T. The voltages and currents with v_o nulled are shown in Fig. 3.8b whence, by the equality of the voltage across R_3 and v_T, we have:

$$v_T = -\frac{i_T R_2}{R_1} R_3 \tag{3.19}$$

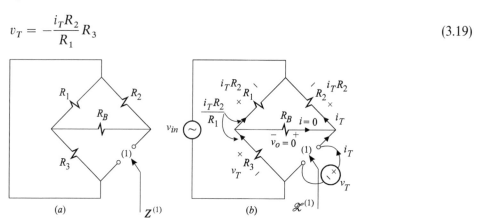

(a) $Z^{(1)}$ (b) $\mathscr{Z}^{(1)}$

Figure 3.8

Hence, the null impedance is negative, real and given by:

$$\mathscr{Z}(1) = \mathscr{R}^{(1)} = -\frac{R_2}{R_1} R_3 \tag{3.20}$$

The significance of a negative null impedance will be discussed in Section 3.3. Once again, observe that v_{in} never needs to enter in the determination of a null impedance and that the expression of a null impedance never contains the element across which the response is nulled.   □

Example 3.4   Determine $Z^{(1)}$ and $\mathscr{Z}^{(1)}$ for the ideal operational amplifier circuit with T-feedback network shown in Fig. 3.9.

The driving-point impedance looking into port (1) with the excitation set to zero is shown in Fig. 3.10a in which we see that v_T must be zero because the voltage drop across each of R_1 and R_2 is zero. It follows that:

$$Z^{(1)} = 0 \tag{3.21}$$

The test current i_T flows entirely through R_3. The fact that this current and the output voltage are undetermined is irrelevant to the determination of $Z^{(1)}$. Note

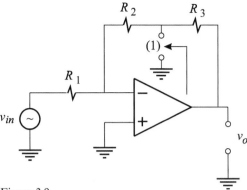

Figure 3.9

that we could have equally well applied a test current source at port (1) and obtained the same result.

The null impedance calculation looking into port (1) with the output nulled is shown in Fig. 3.10b. Because of the virtual ground at the input, the test voltage v_T appears directly across R_2. Also, because of the null in the output voltage, i.e. $v_o = 0$, the test voltage appears across R_3 too. It follows that:

$$i_T = \frac{v_T}{R_2} + \frac{v_T}{R_3} \tag{3.22}$$

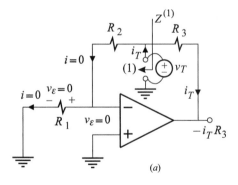

(a)

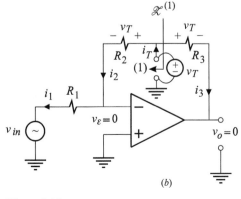

(b)

Figure 3.10

Hence, the null impedance is real and given by:

$$\mathscr{Z}^{(1)} = \mathscr{R}^{(1)} = R_2 || R_3 \tag{3.23}$$

Although the relationship between the input voltage and the test voltage that nulls the output is not important to the determination of $\mathscr{R}^{(1)}$, its determination is left as an exercise (see Problem 3.1).   ☐

Example 3.5   Determine $Z^{(1)}$ and $\mathscr{Z}^{(1)}$ for the ideal operational amplifier circuit shown in Fig. 3.11.

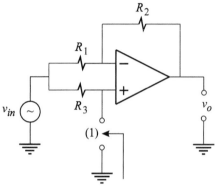

Figure 3.11

With the excitation set to zero as shown in Fig. 3.12a, the driving-point impedance is seen to be real and given by:

$$Z^{(1)} = R^{(1)} = R_3 \tag{3.24}$$

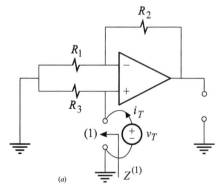

Figure 3.12

The null impedance calculation is shown in Fig. 3.12b. Since $v_\varepsilon = 0$, the test voltage source appears across R_2 when v_o is nulled. It follows that the current through R_2, and hence the current through R_1, are both given by v_T/R_2. The voltage across R_1 is now given by $R_1(v_T/R_2)$ and, since $v_\varepsilon = 0$, this voltage also appears across R_3. Hence, the current through R_3 is given by:

$$-i_T = \frac{R_1(v_T/R_2)}{R_3} \tag{3.25}$$

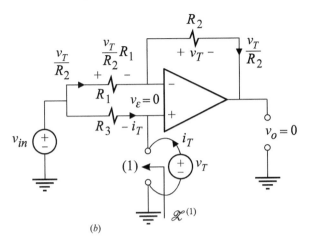

(b)

Figure 3.12 (*cont.*)

Therefore, the null impedance is real and given by:

$$\mathscr{L}^{(1)} = \mathscr{R}^{(1)} = -\frac{R_2 R_3}{R_1} \tag{3.26}$$

The significance of the negative value of $\mathscr{R}^{(1)}$ will be discussed in Example 3.10. $\square$

When studying an impedance function looking into a port as shown in Fig. 3.13, the response in question is the voltage which appears across the current source

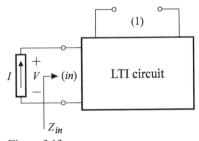

Figure 3.13

(excitation) connected at that port. We will show that, in this case, nulling the voltage across the current source is the same as shorting the current source as shown in Figs. 3.14a and b. The equivalence between Figs. 3.14a and b is a direct consequence of the substitution theorem whereby the short circuit ($R = 0$) in Fig. 3.14b carrying a current I is replaced with a current source carrying the same current I in Fig. 3.14a. A formal proof will be given now to illustrate and review a familiar and special case of the parameters a_{ij} in Eqs. 3.2a and b. In Fig. 3.14a, the two responses are V and v_T and the two excitations are I and i_T so that Eqs. 3.2a and b are written as:

$$\left.\begin{array}{l} V = z_{11}I + z_{12}i_T \\ v_T = z_{21}I + z_{22}i_T \end{array}\right\}$$

(3.27a, b)

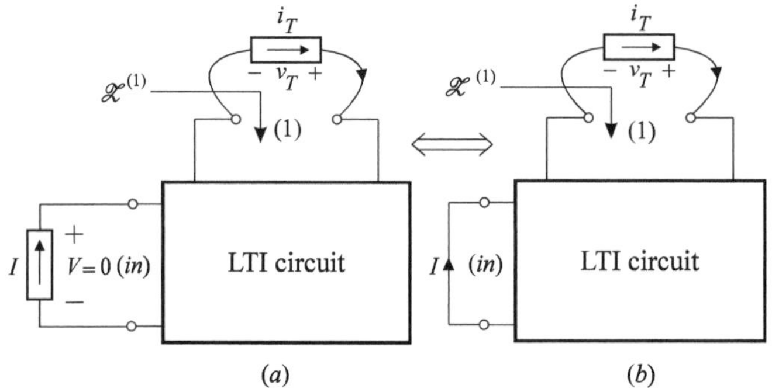

Figure 3.14

These are the familiar z-parameter representations of a two-port linear network, which when used in Figs. 3.14a and b results in the circuits shown in Figs. 3.15a and b, respectively. When V is nulled in Fig. 3.15a, it can be shown that the null impedance looking into port (1) is given by:

$$\mathscr{L}^{(1)} = \frac{z_{11}z_{22} - z_{12}z_{21}}{z_{11}}$$

(3.28)

Equation (3.28) is exactly analogous to Eq. (3.9). When a short is placed across the current source I as shown in Fig. 3.15b, the impedance looking into port (1) is still given by Eq. (3.28):

$$Z^{(1)}_{in \to short} = \frac{z_{11}z_{22} - z_{12}z_{21}}{z_{11}} = \mathscr{L}^{(1)}$$

(3.29)

According to Eq. (3.29), the null impedance looking into port (1) of Fig. 3.14a, with V across the input port nulled, is nothing more than the ordinary driving-point impedance looking into port (1) with port (in) shorted.

Hence, only nulling a voltage across a current source is equivalent to placing a short across that port. Remember that when a voltage across an impedance element is nulled, the current through it vanishes so that placing a short across it would be wrong because that short would carry the short circuit current at that port. Similarly, nulling a current through a voltage source is equivalent to opening the port across which the voltage source is connected.

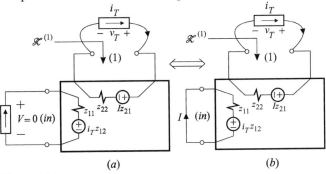

(a)   (b)

Figure 3.15

Example 3.6   In this example, we replace the voltage excitation in the circuit of Example 3.1 with a current excitation i_{in} and study the response v_{in}, or the input impedance function Z_{in}, as shown in Fig. 3.16. Again, we determine $Z^{(1)}$ and $\mathscr{L}^{(1)}$.

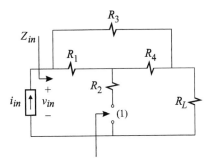

Figure 3.16

With the excitation set to zero, i.e. $i_{in} = 0$, we have the circuit in Fig. 3.17a from which $Z^{(1)}$ or $R^{(1)}$ is given by:

$$Z^{(1)} = R^{(1)} = R_2 + R_4 \parallel (R_1 + R_3) + R_L \tag{3.30}$$

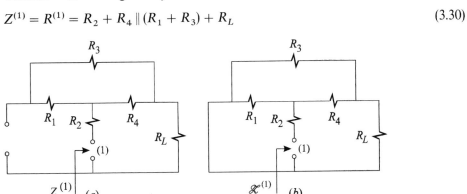

Figure 3.17

As explained earlier, nulling the response v_{in} is the same thing as shorting the input port (*in*). This is shown in Fig. 3.17*b*, which is the same as Fig. 3.4*a* used in the determination of $Z^{(1)}$ for that circuit! Hence we have:

$$\mathcal{L}^{(1)} = \mathcal{R}^{(1)} = R_2 + R_1 \parallel (R_4 + R_L \parallel R_3) \tag{3.31}$$

Note that if we replace the current excitation in the circuit of Example 3.6 back to a voltage source and study the response i_{in}, or the input conductance, then the expression of $Z^{(1)}$ and $\mathcal{L}^{(1)}$ given in Eqs. (3.30) and (3.31) will be interchanged. $\square$

3.3   The EET for impedance elements

Consider the LTI circuit in Fig. 3.18*a* in which the impedance element Z_1 is designated as the extra element. Let the current through Z_1 be I in the presence of the excitation u_{i1}. In Fig. 3.18*b*, Z_1 is replaced with a current source I which, for the *same* u_{i1}, has exactly the *same* value as I in Fig. 3.18*a*. According to the substitution theorem, both circuits in Figs. 3.18*a* and *b* will have the same voltages and currents. The same set of equations in Eqs. (3.2*a*, *b*) applies to the circuit in Fig. 3.18*b*, so that we can write as before:

$$\left. \begin{array}{l} u_{o1} = a_{11}u_{i1} + a_{12}I \\ V = a_{21}u_{i1} + a_{22}I \end{array} \right\} \tag{3.32a, b}$$

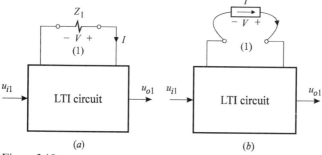

Figure 3.18

The current source I used in these equations does not have an arbitrary value, but is always equal to the current I in Fig. 3.18*a* whose value is given by:

$$I = -\frac{V}{Z_1} \tag{3.32c}$$

Equations (3.32*a*–*c*) can be solved simultaneously by eliminating V and I to yield the transfer function:

$$H \equiv \frac{u_{o1}}{u_{i1}} = a_{11} \frac{1 + \dfrac{a_{11}a_{22} - a_{12}a_{21}}{a_{11}} \dfrac{1}{Z_1}}{1 + \dfrac{a_{22}}{Z_1}} \tag{3.33}$$

The terms a_{11}, a_{22} and $(a_{11}a_{22} - a_{12}a_{21})/a_{11}$ are immediately recognized to be H_o, $Z^{(1)}$ and $\mathscr{Z}^{(1)}$, respectively, from Eqs. (3.4), (3.5) and (3.9b). Upon their substitution in Eq. (3.33), we obtain the desired form of the EET:

$$H = H_o \frac{1 + \dfrac{\mathscr{Z}^{(1)}}{Z_1}}{1 + \dfrac{Z^{(1)}}{Z_1}} \tag{3.34}$$

where:

$$H_o = \left.\frac{u_{o1}}{u_{i1}}\right|_{(1)\to open} = \left.\frac{u_{o1}}{u_{i1}}\right|_{Z_1 \to \infty} \tag{3.35}$$

This is the EET and it can be stated as follows:

Theorem (EET): A transfer function H of an LTI circuit can be determined in three independent steps. First, an element Z_1, connected across port (1), is removed from the circuit ($Z_1 \to \infty$) and a much simpler transfer function, H_o, is determined. The choice of Z_1 is motivated by the greatest simplification which results upon the removal of that element. Second, the ordinary (driving-point) impedance, $Z^{(1)}$, looking into port (1) is determined. Third, the null impedance, $\mathscr{Z}^{(1)}$, looking into port (1) is determined. The transfer function H is given in terms of H_o, $Z^{(1)}$ and $\mathscr{Z}^{(1)}$ according to Eq. (3.34) above. A complete pictorial representation of this theorem is given in Fig. 3.19.

A comparison of Eqs. (3.34) and (3.1) clearly shows that the EET is by far easier to interpret and implement than the bilinear theorem given by Bode in Eq. (3.1), even though both have the same structure. Equation (3.1) suffers from a severe case of maximum entropy because it contains absolutely no recognizable term nor information which can be associated with the physical network save for the element z! Equation (3.34) of the EET on the other hand contains three easily interpretable terms which are directly associated with the physical network.

An alternate and useful form of the EET can be obtained from Eq. (3.34) by the following manipulation:

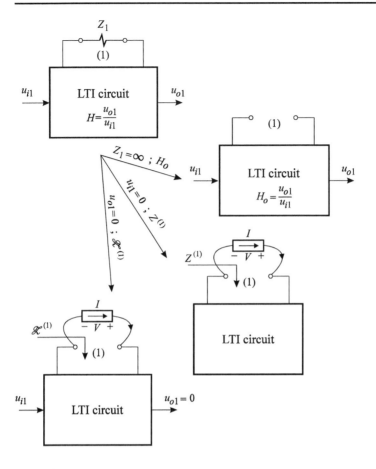

Figure 3.19

$$H = H'_o \frac{1 + \dfrac{Z_1}{\mathscr{L}^{(1)}}}{1 + \dfrac{Z_1}{Z^{(1)}}} \tag{3.36}$$

where H'_o is given by:

$$H'_o = \frac{\mathscr{L}^{(1)}}{Z^{(1)}} H_o \tag{3.37}$$

This form of the EET is shown in Fig. 3.20. It is evident from Eq. (3.36) that H'_o coincides with the value of H when $Z_1 = 0$, or when port (1) is shorted, so that we have:

$$H'_o = \frac{u_{o1}}{u_{i1}} \bigg|_{(1) \to short} \tag{3.38}$$

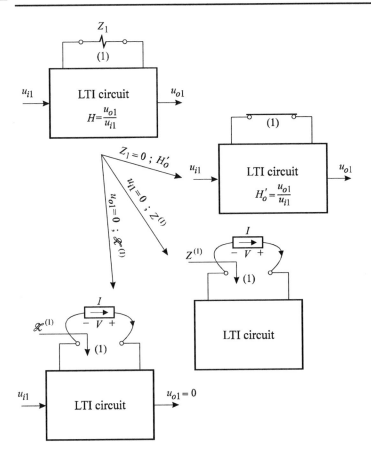

Figure 3.20

Equation (3.37) is actually a corollary of the EET which can be expressed as:

$$\frac{\left.\dfrac{u_{o1}}{u_{i1}}\right|_{(1)\to short}}{\left.\dfrac{u_{o1}}{u_{i1}}\right|_{(1)\to open}} = \frac{\mathscr{L}^{(1)}}{Z^{(1)}} \tag{3.39}$$

It is also possible to derive Eq. (3.39) directly if the extra element Z_1 is replaced with a voltage source V, instead of a current source I, in Fig. 3.18b (see Problem 3.2).

The two forms of the EET give us the flexibility of inserting an extra element either in parallel (Eq. (3.34)) or in series (Eq. (3.36)). This means that we can simplify a circuit either by shorting the port of the extra element and determining H_o' or opening it and determining H_o. Both forms are illustrated in the following examples.

Example 3.7   Consider the circuit in Fig. 3.3 in which a capacitor C_1 is connected across port (1) as shown here in Fig. 3.21. We wish to determine the transfer function:

$$H(s) = \frac{v_o(s)}{v_{in}(s)} \tag{3.40}$$

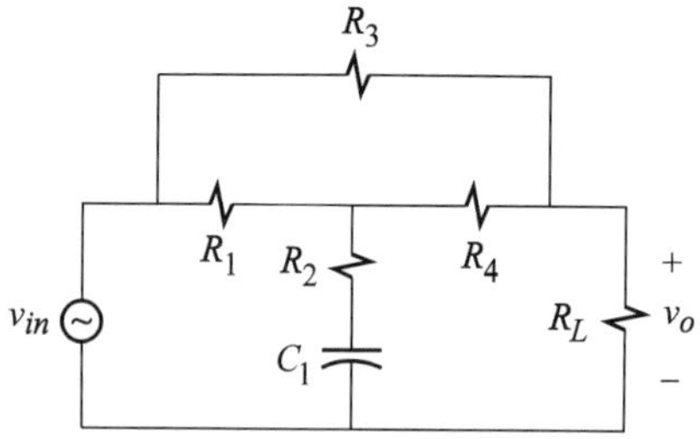

Figure 3.21

We will show now that the EET is the ideal simplification tool for the determination of this transfer function. If we designate the impedance element $Z_1 = 1/sC_1$ as the extra element and remove it from the circuit, i.e. $Z_1 \to \infty$, we obtain the circuit in Fig. 3.22, which is the same as Fig. 3.3, for which we can easily write the transfer function:

$$H_o = \frac{1}{1 + \dfrac{R_3 \,\|\, (R_1 + R_4)}{R_L}} \tag{3.41}$$

Since we let Z_1 be infinite, we realize that H_o is the low-frequency asymptote of the transfer function $H(s)$. According to the first form of the EET given in Eq. (3.34) and explained in Fig. 3.19, $H(s)$ is given by:

$$H = H_o \frac{1 + \dfrac{\mathscr{Z}^{(1)}}{Z_1}}{1 + \dfrac{Z^{(1)}}{Z_1}} \tag{3.42}$$

$$= H_o \frac{1 + sC_1 \mathscr{Z}^{(1)}}{1 + sC_1 Z^{(1)}}$$

in which $Z^{(1)}$ and $\mathscr{Z}^{(1)}$ have been determined in Figs. 3.4a and b and given by Eqs. (3.10) and (3.12). Hence, $H(s)$ can be written in pole-zero form:

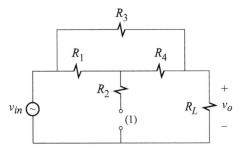

Figure 3.22

$$H = H_o \frac{1 + \dfrac{s}{\omega_{z1}}}{1 + \dfrac{s}{\omega_{p1}}}$$ (3.43)

where:

$$\omega_{z1} = \frac{1}{C_1 \mathscr{R}^{(1)}} = \frac{1}{C_1 \left[R_2 + \dfrac{R_1}{1 + R_3/R_4} \middle\| R_4 \right]}$$ (3.44a)

$$\omega_{p1} = \frac{1}{C_1 R^{(1)}} = \frac{1}{C_1 [R_2 + R_1 \| (R_4 + R_L \| R_3)]}$$ (3.44b)

An asymptotic magnitude plot of this transfer function is shown in Fig. 3.23 in which we see that the pole is always at a lower frequency than the zero. The relative

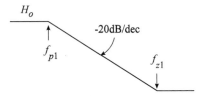

Figure 3.23

position of the pole with respect to the zero can be ascertained by comparing the low-entropy expressions of $\mathscr{R}^{(1)}$ and $R^{(1)}$ and realizing that $\mathscr{R}^{(1)} < R^{(1)}$ for any choice of the values of the resistors in the circuit. ☐

Example 3.8 In this example the capacitor in Example 3.7 is replaced with an inductor L_1 and the output is taken across R_3 as shown in Fig. 3.24. We wish to determine the transfer function:

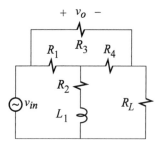

Figure 3.24

$$H(s) = \frac{v_o(s)}{v_{in}(s)} \tag{3.45}$$

The inductor represents an impedance element $Z_1 = sL_1$ and the best way to simplify this circuit is to take this impedance out, i.e. $Z_1 \to \infty$. This results in the circuit shown in Fig. 3.25 for which we can write:

$$H_o = \frac{1}{1 + \dfrac{R_L}{R_3 \parallel (R_1 + R_4)}} \tag{3.46}$$

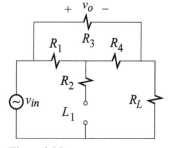

Figure 3.25

The EET can now be applied to yield the complete transfer function:

$$H = H_o \frac{1 + \dfrac{\mathscr{Z}^{(1)}}{Z_1}}{1 + \dfrac{Z^{(1)}}{Z_1}} \tag{3.47}$$

$$= H_o \frac{1 + \dfrac{\mathscr{Z}^{(1)}}{sL_1}}{1 + \dfrac{Z^{(1)}}{sL_1}}$$

in which $Z^{(1)}$ and $\mathcal{Z}^{(1)}$ are given by Eqs. (3.14) and (3.17). Now we can write $H(s)$ in inverted pole-zero form:

$$H = H_o \frac{1 + \dfrac{\omega_{z1}}{s}}{1 + \dfrac{\omega_{p1}}{s}} \tag{3.48}$$

in which:

$$\omega_{z1} = \frac{\mathcal{R}^{(1)}}{L_1} = \frac{R_2 + [R_1 \| (R_4 + R_L)]\left[1 + \dfrac{R_L}{R_1 + R_4}\right]}{L_1} \tag{3.49a}$$

$$\omega_{p1} = \frac{R^{(1)}}{L_1} = \frac{R_2 + R_1 \| (R_4 + R_L \| R_3)}{L_1} \tag{3.49b}$$

The asymptotic magnitude plot of this transfer function is shown in Fig. 3.26 in which H_o is seen to be the high-frequency asymptote. Comparing the expression of $\mathcal{R}^{(1)}$ and $R^{(1)}$ and realizing that $\mathcal{R}^{(1)} > R^{(1)}$ for any choice of the values of the resistors in the circuit, we can ascertain that the zero always comes after the pole.

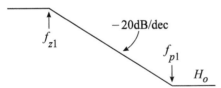

Figure 3.26

Note that in both of these examples we could have taken the extra impedance element as a short rather than an open when determining H_o. The resultant circuit would have been a bridge circuit for which H_o would have been more complicated to determine. □

Example 3.9   In this example we will apply the results derived in Example 3.3 towards the determination of the transfer function of the bridge circuit in Fig. 3.27. We have a choice of taking the capacitor out either as a short or an open. Both choices will result in a simple resistive circuit for which H_o can be derived by inspection. In Fig. 3.28 we take C_1 out as an open, form the Thevenin equivalent of v_{in}, R_3 and R_1, and immediately write:

$$H_o = \frac{1}{1 + R_3/R_1} \frac{1}{1 + (R_2 + R_1 \| R_3)/R_B} \tag{3.50}$$

Application of the first form of the EET yields the complete transfer function:

$$H = H_o \frac{1 + \dfrac{\mathscr{Z}^{(1)}}{Z_1}}{1 + \dfrac{Z^{(1)}}{Z_1}} \tag{3.51}$$

$$= H_o \frac{1 + sC_1 \mathscr{Z}^{(1)}}{1 + sC_1 Z^{(1)}}$$

Substituting for $Z^{(1)}$ and $\mathscr{Z}^{(1)}$ as determined in Eqs. (3.18) and (3.20), respectively, we obtain:

$$H = H_o \frac{1 - \dfrac{s}{\omega_{z1}}}{1 + \dfrac{s}{\omega_{p1}}} \tag{3.52}$$

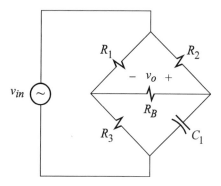

Figure 3.27

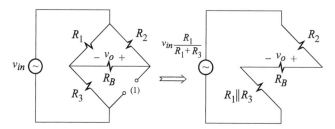

Figure 3.28

where:

$$\omega_{z1} = -\frac{1}{C_1 \mathscr{R}^{(1)}} = \frac{1}{C_1 R_2} \frac{R_1}{R_3} \tag{3.53a}$$

$$\omega_{p1} = \frac{1}{C_1 R^{(1)}} = \frac{1}{C_1 R_2 \parallel (R_B + R_1 \parallel R_3)} \tag{3.53b}$$

An asymptotic magnitude and phase plot of this transfer function is shown in Fig. 3.29 in which it has been assumed that the zero occurs before the pole. For this circuit, the pole may occur before or after the zero. In fact, for a given C_1, it is possible to choose the values of the resistors such that the pole and zero occur at the same frequency resulting in a transfer function which has a magnitude

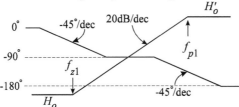

Figure 3.29

response independent of frequency and a phase response with $-90°/dec$ centered at the pole or zero. Such a network is called a phase shifter, but this bridge circuit is not a good phase shifter because it requires simultaneous adjustments of more than one resistor (see Problem 3.4) to achieve variable phase shift and its magnitude response depends on the values of resistors. A far better circuit will be given in Example 3.10. □

The right-half plane (RHP) zero in the transfer function in Eq. (3.52) is due to the negative null impedance looking into port (1). A physical interpretation of the RHP zero is given by examining the circuit at very low and very high frequencies. At very low frequencies, the capacitor port is essentially an open circuit and the response is in phase with the excitation. At very high frequencies, the capacitor is essentially a short circuit as shown in Fig. 3.30 and the response is 180° out of phase with the excitation. Hence, we see that there are two frequency-dependent paths from the excitation to the response: a noninverting path, which is dominant at low frequencies as shown in Fig. 3.28, and an inverting path, which is dominant at high frequencies as shown in Fig. 3.30. The gain of the high-frequency path can be obtained from the transfer function in Eq. (3.52) by letting $s \to \infty$:

$$H'_o \equiv \lim_{s \to \infty} H(s) = -H_o \frac{\omega_{p1}}{\omega_{z1}} \tag{3.54}$$

in which the negative sign of the RHP zero accounts for the fact that the low- and high-frequency gains are of opposite polarity. Of course, H'_o could also have been determined from the circuit in Fig. 3.31, but the point here is to relate the negative sign of the RHP zero with the inverting gain at high frequencies. Equation (3.54) is the same as the corollary in Eq. (3.37).

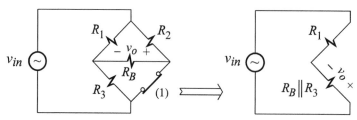

Figure 3.30

Example 3.10 In this example, we will use the results derived in Example 3.4 to determine the transfer function of an ideal operational amplifier with T-feedback as shown in Fig. 3.31. It is only natural to take C_1 out of the circuit as an open as shown in Fig. 3.32 and determine H_o:

$$H_o = -\frac{R_2 + R_3}{R_1}$$

(3.55)

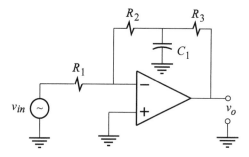

Figure 3.31

The other choice of taking C_1 as a short is ruled out because it yields an infinite value of H_o. Application of the first form of the EET yields the complete transfer function:

$$H = H_o \frac{1 + \dfrac{\mathcal{Z}^{(1)}}{Z_1}}{1 + \dfrac{Z^{(1)}}{Z_1}}$$

(3.56)

$$= H_o \frac{1 + sC_1 \mathcal{Z}^{(1)}}{1 + sC_1 Z^{(1)}}$$

Substituting for $Z^{(1)}$ and $\mathcal{Z}^{(1)}$ from Eqs. (3.21) and (3.23), we get:

$$H(s) = H_o(1 + s/\omega_1)$$

(3.57)

where:

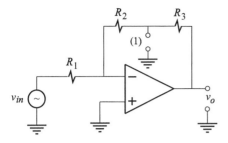

Figure 3.32

$$\omega_1 = \frac{1}{C_1 R_3 \| R_2} \tag{3.58}$$

As expected, the gain increases indefinitely as frequency increases.   □

Example 3.11   In this example, we will use the results derived in Example 3.5 to determine the transfer function of the circuit in Fig. 3.33. Note that there are two paths from the input signal to the output: an inverting path beginning through R_1, with a gain of $-R_2/R_1$, and a noninverting path through R_3, with a gain of $1 + R_2/R_1$. At low frequencies, the noninverting path dominates the inverting path so that the net gain is $1 + R_2/R_1 - R_2/R_1 = 1$. At high frequencies the gain of the noninverting path diminishes to zero while the gain of the inverting path remains at $-R_2/R_1$ and dominates the noninverting path. This causes the net gain at high frequencies to be $-R_2/R_1$. Hence, as explained earlier, we expect a

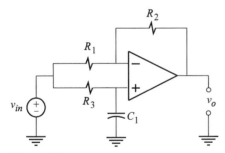

Figure 3.33

RHP zero in the transfer function. If we designate the capacitor as the extra element, then we have a choice of applying the first or the second form of the EET. Let us take the capacitor out of the circuit as a short, i.e. $Z_1 = 1/sC_1 \to 0$, as shown in Fig. 3.34 and determine:

$$H'_o = -\frac{R_2}{R_1} \tag{3.59}$$

Application of the second form of the EET yields:

$$H = H'_o \frac{1 + \dfrac{Z_1}{\mathscr{Z}^{(1)}}}{1 + \dfrac{Z_1}{Z^{(1)}}} \tag{3.60}$$

$$= H'_o \frac{1 + 1/sC_1\mathscr{Z}^{(1)}}{1 + 1/sC_1Z^{(1)}}$$

Substituting for $Z^{(1)}$ and $\mathscr{Z}^{(1)}$ from Eqs. (3.24) and (3.26), we get the transfer function in inverted pole and zero form:

$$H(s) = H'_o \frac{1 - \omega_{z1}/s}{1 + \omega_{p1}/s} \tag{3.61}$$

where:

$$\omega_{z1} = \frac{1}{C_1R_3}\frac{R_1}{R_2} \tag{3.62a}$$

$$\omega_{p1} = \frac{1}{C_1R_3} \tag{3.62b}$$

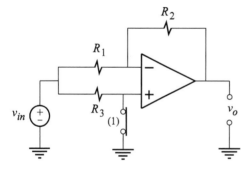

Figure 3.34

Equation (3.61) can also be written in normal pole-zero notation as:

$$H(s) = \frac{1 - s/\omega_{z1}}{1 + s/\omega_{p1}} \tag{3.63}$$

This form of $H(s)$ corresponds to the first form of the EET in which $H_o = 1$. As expected we have an RHP zero which accounts for the inversion in the sign of the gain at very high frequencies.

This circuit can be made into a phase shifter by letting $\omega_{z1} = \omega_{p1} = \omega_0$ so that we get:

$$H(s) = \frac{1 - s/\omega_0}{1 + s/\omega_0} \tag{3.64}$$

According to Eqs. (3.62a) and (3.62b) this requires that $R_2 = R_1$ so that ω_0 is given by:

$$\omega_0 = \frac{1}{C_1 R_3} \tag{3.65}$$

Hence, by adjusting only R_3, we can shift the phase of the output signal anywhere from 0 to $-180°$ without affecting its amplitude. If the positions of R_3 and C_3 are interchanged, a FET can be used as a voltage controlled resistor in place of R_3. In this case the phase can be adjusted from -180 to $-360°$ (see Problem 3.5).   □

Example 3.12   In this example, we will use the results derived in Example 3.6 to determine the input impedance of the circuit in Fig. 3.35a. The best way to simplify this circuit is to take out the impedance element $Z_1 = sL_1$ as an open as shown in Fig. 3.35b whence we have:

$$R_0 = R_3 \parallel (R_1 + R_4) + R_L \tag{3.66}$$

Application of the first form of the EET yields:

$$Z_{in} = R_0 \frac{1 + \mathscr{Z}^{(1)}/Z_1}{1 + Z^{(1)}/Z_1} \tag{3.67}$$

$$= R_0 \frac{1 + \mathscr{Z}^{(1)}/sL_1}{1 + Z^{(1)}/sL_1}$$

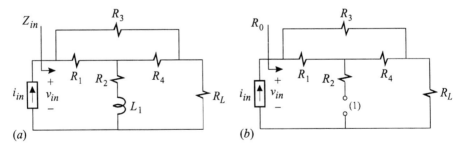

(a)   (b)

Figure 3.35

Substituting for the $Z^{(1)}$ and $\mathscr{Z}^{(1)}$ from Eqs. (3.30) and (3.31) yields:

$$Z_{in}(s) = R_0 \frac{1 + \dfrac{\omega_{z1}}{s}}{1 + \dfrac{\omega_{p1}}{s}} \tag{3.68}$$

where:

$$\omega_{z1} = \frac{R_2 + R_1 \,\|\, (R_4 + R_L \,\|\, R_3)}{L_1} \qquad (3.69a)$$

$$\omega_{p1} = \frac{R_2 + R_4 \,\|\, (R_1 + R_3) + R_L}{L_1} \qquad (3.69b)$$

The expression of $Z_{in}(s)$ in Eq. (3.68) is in inverted pole-zero form which can also be written in normal pole-zero form as:

$$Z_{in}(s) = R_0' \frac{1 + \dfrac{s}{\omega_{z1}}}{1 + \dfrac{s}{\omega_{p1}}} \qquad (3.70)$$

In this expression, it can be seen that R_0' is the input resistance of a resistive bridge circuit in which R_2 is taken as the extra element (see Problem 3.6).   □

3.4   The EET for dependent sources

So far, the designated extra element in the EET has been an impedance element. In this section, the EET will be extended to dependent sources. This extension is quite natural once an impedance element is recognized to be a special case of a dependent source. This is shown in Fig. 3.36a in which Z is replaced by a dependent current source whose value depends on the voltage across its terminals multiplied by $1/Z$. In Fig. 3.36b, the same impedance is replaced by a voltage source whose

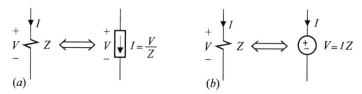

(a)   (b)

Figure 3.36

value depends on the current through it multiplied by Z. These two cases can be easily extended to an arbitrary dependent source whose value depends on the voltage or current elsewhere in the circuit rather across or through its own terminals. With this in mind, consider the LTI circuit in Fig. 3.37 in which u_{i1} is the only independent excitation and u_{o1} is a response. A dependent generator u_{i2}, which can be either a voltage source or a current source, is connected across port (1) and its value depends on some other response u_{o2} with a gain A, i.e. $u_{i2} = Au_{o2}$. Our objective now is to express the transfer function $H = u_{o1}/u_{i1}$ as some bilinear transformation of A. This can be done quite easily by writing the system equations as before:

$$u_{o1} = a_{11}u_{i1} + a_{12}u_{i2}$$

$$u_{o2} = a_{21}u_{i1} + a_{22}u_{i2}$$ (3.71a, b, c)

$$u_{i2} = Au_{o2}$$

The simultaneous solution of these equations yields:

$$\frac{u_{o1}}{u_{i1}} = a_{11} \frac{1 + \left(-\dfrac{a_{11}a_{22} - a_{12}a_{21}}{a_{11}} \right) A}{1 + (-a_{22})A}$$ (3.72)

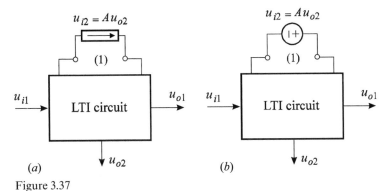

Figure 3.37

In this equation, a_{11} is the transfer function with $A = 0$ and is determined by opening the dependent current source as shown in Fig. 3.38a or shorting the dependent voltage source as shown in Fig. 3.38b. Hence, a_{11} can be written as:

$$a_{11} = \left. \frac{u_{o1}}{u_{i1}} \right|_{A=0} \equiv H_o$$ (3.73)

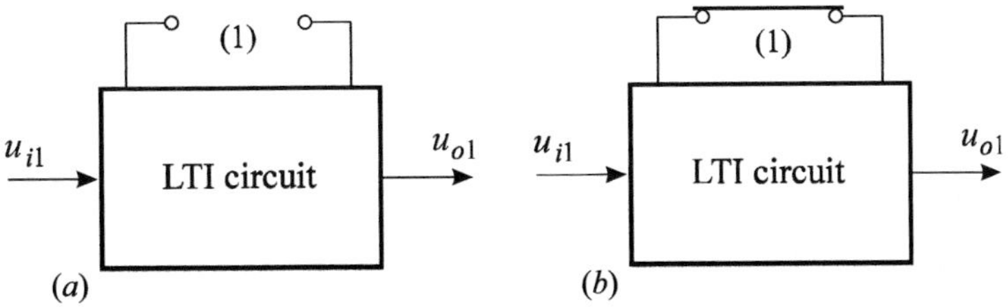

Figure 3.38   Determination of H_o by letting A = 0

If we set the excitation u_{i1} to zero in Eq. (3.71b), we see that $-a_{22}$ is the inverse gain which relates u_{o2} to an *independent* source $-u_{i2}$ connected at port (1) with $u_{i1} = 0$:

$$-a_{22} = \left. \frac{u_{o2}}{-u_{i2}} \right|_{u_{i1}=0} \equiv \overline{A^{(1)}}$$ (3.74)

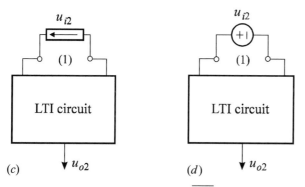

Figure 3.38 (*cont.*)   Determination of $\overline{\mathscr{A}^{(1)}}$ with $u_{i1} = 0$

This is shown in Fig. 3.38c when u_{i2} is an independent current source and Fig. 3.38d when u_{i2} is an independent voltage source. In these figures we see that $\overline{A^{(1)}}$ is nothing more than a transfer function which relates u_{o2} to an *independent* source u_{i2}; which is of the same type as the dependent source (current or voltage), has a polarity opposite to that of the dependent source, and is connected at the same place as the dependent source. Note that in this figure the excitation u_{i1} is set to zero.

If in Eqs. (3.71a, b) we let u_{i2} be an *independent* source *in addition* to u_{i1}, we can then null the response u_{o1} and obtain:

$$-\frac{a_{11}a_{22} - a_{12}a_{21}}{a_{11}} = \frac{u_{o2}}{-u_{i2}}\bigg|_{u_{o1}=0} \equiv \overline{\overline{\mathscr{A}^{(1)}}} \tag{3.75}$$

This is shown in Fig. 3.38e when u_{i2} is a current source and in Fig. 3.38f when u_{i2} is a voltage source. In this figure we see that $\overline{\overline{\mathscr{A}^{(1)}}}$ is very similar to $\overline{A^{(1)}}$ except that it is determined with the response null ($u_{o1} = 0$) rather than the excitation set to zero ($u_{i1} = 0$).

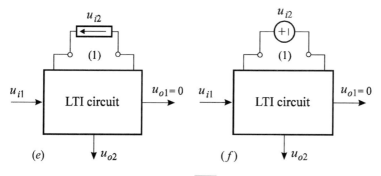

Figure 3.38 (*cont.*)   Determination of $\overline{\overline{\mathscr{A}^{(1)}}}$ with $u_{o1} = 0$

Substituting Eqs. (3.73), (3.74) and (3.75) in (3.72) we obtain the first form of the EET for a dependent source:

$$H = H_o \frac{1 + A \mathscr{A}^{(1)}}{1 + A \overline{A^{(1)}}}$$

(3.76)

To summarize then, the three components in the EET for a dependent source are the transfer function H_o with the dependent source set to zero, the inverse gain $\overline{A^{(1)}}$, and the null inverse gain $\mathscr{A}^{(1)}$. Note that the units of the inverse gains are the reciprocal of the units of A of the dependent source.

Example 3.13 Determine the output impedance, R_{out}, of the amplifier circuit in Fig. 3.39 using the nonideal model of the operational amplifier shown. The output impedance R_{out} is a transfer function which relates the response V to a current source, I, connected at the output, i.e. $R_{out} = V/I$.

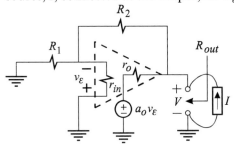

Figure 3.39

According to the EET in Eq. (3.76), we let $a_o = 0$ and obtain from Fig. 3.40a:

$$R_0 = r_o \parallel (R_2 + R_1 \parallel r_{in})$$

(3.77)

To determine the null inverse gain $\bar{\alpha}$, we replace the dependent voltage generator of the operational amplifier with an independent voltage generator, v_T, pointing in the opposite direction and determine the gain $\bar{\alpha} = v_\varepsilon / v_T$ with the response V nulled. As explained earlier, nulling the response of an impedance function looking into a port is the same as shorting that port. This is shown in Fig. 3.40b whence we have:

$$\overline{\alpha^{(1)}} = 0$$

(3.78)

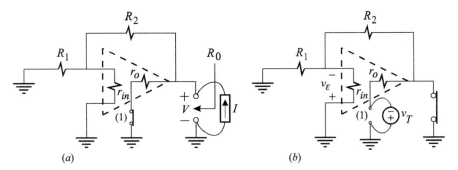

(a)   (b)

Figure 3.40

To determine the inverse gain with the excitation removed, we simply set $I = 0$, or open the output port, as shown in Fig. 3.40c whence we have:

$$\overline{a^{(1)}} = \frac{1}{1 + \dfrac{r_o + R_2}{r_{in} \parallel R_1}}$$

(3.79)

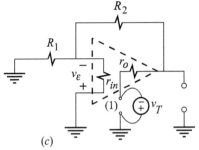

(c)

Figure 3.40 (cont.)

Substitution of Eqs. (3.77), (3.78) and (3.79) in the EET given by Eq. (3.76) yields:

$$R_o = R_0 \frac{1 + \overline{\alpha^{(1)}} a_o}{1 + \overline{a^{(1)}} a_o}$$

(3.80)

$$= \frac{r_o \parallel (R_2 + R_1 \parallel r_{in})}{1 + \dfrac{a_o}{1 + (r_o + R_2)/r_{in} \parallel R_1}}$$

This result is in low-entropy form and can be simplified readily to:

$$R_o = r_o \frac{1 + R_2/R_1}{a_o}$$

(3.81)

in which we have assumed $r_o \ll (R_2, R_1), a_o \gg 1$ and $r_{in} \gg R_1$.   □

Example 3.14   Determine the input impedance of the noninverting amplifier in Fig. 3.41 using the nonideal model of the operational amplifier shown.

The input impedance with $a_o = 0$ is shown in Fig. 3.42a where we see:

$$R_0 = r_{in} + R_1 \parallel (R_2 + r_o)$$

(3.82)

The inverse gain with the excitation removed, or the input port open, is shown in Fig. 3.42b where we see:

$$\overline{a^{(1)}} = 0$$

(3.83)

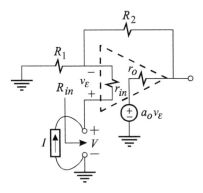

Figure 3.41

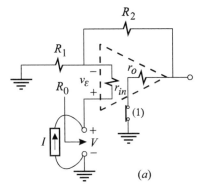

(a)

Figure 3.42

The null inverse gain with the response null, or the input port short, is shown in Fig. 3.42c where we see:

$$\overline{\alpha^{(1)}} = \frac{1}{1 + \dfrac{R_2 + r_o}{r_{in} \| R_1}}$$

(3.84)

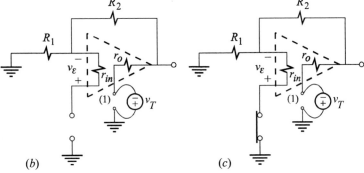

(b)   (c)

Figure 3.42 (cont.)

Substituting these in the EET, we obtain:

$$R_{in} = R_0 \frac{1 + \overline{\alpha^{(1)}} a_o}{1 + \overline{a^{(1)}} a_o} \tag{3.85}$$

$$= [r_{in} + R_1 \parallel (R_2 + r_o)] \left[1 + \frac{a_o}{1 + (R_2 + r_o)/r_{in} \parallel R_1} \right]$$

This expression is in low-entropy form and can be approximated readily as:

$$R_{in} \approx r_{in} \frac{a_o}{1 + R_2/R_1} \tag{3.86}$$

in which we have assumed $r_o \ll R_2$ and $r_{in} \gg R_1, R_2$. □

In some cases, notably in feedback amplifiers in which A is part of the open-loop gain, it is more convenient to determine a transfer function under the assumption that $A \to \infty$. In this case the EET in Eq. (3.76) can be written as:

$$H = H'_o \frac{1 + 1/A\overline{\mathscr{A}^{(1)}}}{1 + 1/A\overline{A^{(1)}}} \tag{3.87}$$

in which

$$H'_o = \frac{u_{o1}}{u_{i1}} \bigg|_{A \to \infty} \tag{3.88}$$

and is shown in Figs. 3.43a and b for a dependent voltage source and current source, respectively. In a feedback system, if A is part of the open-loop gain or the main plant, then H'_o is the ideal closed-loop gain determined by the feedback network alone. Note that in Figs. 3.43a and b, we have $u_{o2} = 0$ because any finite value of the dependent source $u_{i2} = Au_{o2}$ with an infinite gain A must require that u_{o2} be zero. This reminds us of ideal operational amplifier circuits in which the differential voltage between the inverting and the noninverting terminals is taken to be zero in the same way as $u_{o2} = 0$ in Fig. 3.43a and b. Also, in this figure, one is not concerned with the infinite gain A of the dependent source $u_{i2} = Au_{o2}$, in the same way one is never concerned with the infinite gain of the dependent voltage or current source at the output of an ideal operational amplifier. In determining H'_o for such circuits, one only utilizes the fact that $u_{o2} = 0$ and is never concerned with the dependent source with infinite gain.

The form of the EET in Eq. (3.87) is ideal for studying transfer functions of nonideal operational amplifier circuits with finite gain a_o. In this form, one can immediately see the dominant constituents of the transfer function in H'_o and minor deviations from it in a bilinear function of a_o multiplying it. This form of the EET can also be used to formulate the closed-loop response of a general feedback amplifier circuit as will be discussed in Chapter 5.

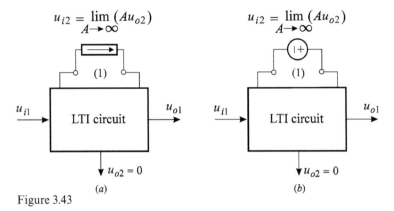

$$u_{i2} = \lim_{A \to \infty} (Au_{o2})$$

$$u_{i2} = \lim_{A \to \infty} (Au_{o2})$$

Figure 3.43

Example 3.15   Determine the transfer function of the inverting operational amplifier circuit shown in Fig. 3.44 taking into account the finite input and output impedance and the finite gain of the amplifier.

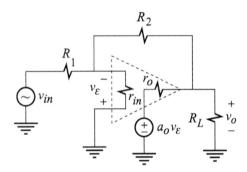

Figure 3.44

Since the gain a_o of the operational amplifier is typically a very large number, we can assume that it is infinite and obtain the circuit in Fig. 3.45a in which $v_\varepsilon = 0$. In this figure, v_ε corresponds to u_{o2} of the general case shown in Fig. 3.43. With $v_\varepsilon = 0$, the current through r_{in} is zero so that the current through R_1, given by v_{in}/R_1, flows entirely through the feedback resistor R_2. The output voltage is now given by $v_o = -R_2(v_{in}/R_1)$ and we have the well-known relation:

$$A_o = -\frac{R_2}{R_1} \tag{3.89}$$

The inverse gain $\overline{a^{(1)}}$ is determined from Fig. 3.45b in which the excitation v_{in} is set to zero and the dependent source $a_o v_\varepsilon$ is replaced with an independent voltage source v_T pointing in the opposite direction. Taking a Thevenin equivalent of v_T, r_o and R_L, yields immediately:

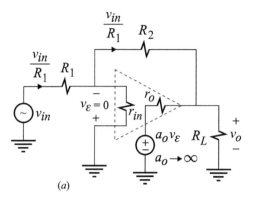

(a)

Figure 3.45

$$\overline{a^{(1)}} \equiv \left.\frac{v_\varepsilon}{V_T}\right|_{v_{in}=0} = \frac{1}{1 + \dfrac{r_o}{R_L}} \; \frac{1}{1 + \dfrac{R_2 + r_o \| R_L}{R_1 \| r_{in}}} \tag{3.90}$$

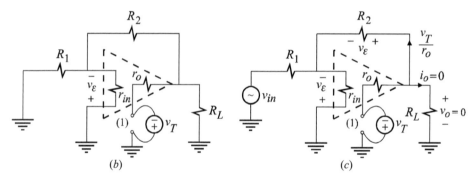

(b)   (c)

Figure 3.45 (*cont.*)

The null inverse gain $\overline{\alpha^{(1)}}$ is determined from Fig. 3.45c in which the response has been nulled, i.e. $v_o = 0$. With $v_o = 0$, the current through R_L is zero, the current through r_o is v_T/r_o, and the voltage across R_2 is v_ε. It follows that:

$$\overline{\alpha^{(1)}} \equiv \left.\frac{v_\varepsilon}{v_T}\right|_{v_o \equiv 0} = -\frac{R_2}{r_o} \tag{3.91}$$

The nonideal gain is now given by substituting the above in the EET:

$$A = A_o \frac{1 + 1/a_o \overline{\alpha^{(1)}}}{1 + 1/a_o \overline{a^{(1)}}} \tag{3.92}$$

$$= -\frac{R_2}{R_1}\frac{1 - \dfrac{1}{a_o}\dfrac{r_o}{R_2}}{1 + \dfrac{1}{a_o}\left(1 + \dfrac{r_o}{R_L}\right)\left(1 + \dfrac{R_2 + r_o\,\|\,R_L}{R_1\,\|\,r_{in}}\right)}$$

This is the expression of the gain given in Problem 2.7. $\square$

Example 3.16 Determine the transfer function of the noninverting operational amplifier circuit shown in Fig. 3.46 taking into account the finite input and output impedance and the finite gain of the amplifier.

As in the previous example, if we let $a_o \to \infty$ we obtain the gain of the ideal noninverting amplifier:

$$A_o = 1 + \frac{R_2}{R_1} \tag{3.93}$$

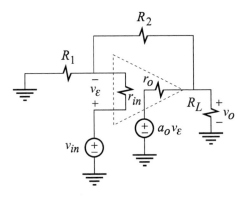

Figure 3.46

The null inverse gain is determined in Fig. 3.47a in which the dependent voltage generator is replaced with an independent one pointing in the opposite direction. With $v_o = 0$, the current through r_o is v_T/r_o, the voltage across R_2 and R_1 is $(v_T/r_o)R_2$ and, finally, the current through R_1 is $[(v_T/r_o)R_2]/R_1$. Since the sum of the currents in R_1 and R_2 flows through r_{in}, the voltage across r_{in} is:

$$v_e = (1 + R_2/r_1)(v_T/r_o)r_{in} \tag{3.94}$$

It follows that:

$$\overline{\alpha^{(1)}} = \frac{1 + R_2/R_1}{r_o}r_{in} \tag{3.95}$$

The determination of the inverse gain with $v_{in} = 0$ is shown in Fig. 3.47b which is the same as Fig. 3.45b so that $\overline{a^{(1)}}$ is still given by Eq. (3.90).

Substituting Eqs. (3.93), (3.95) and (3.90) in the EET given by Eq. (3.87) we obtain:

$$A = A_o \frac{1 + 1/a_o \overline{\alpha^{(1)}}}{1 + 1/a_o \overline{a^{(1)}}}$$

(3.96)

$$= \left[1 + \frac{R_2}{R_1}\right] \frac{1 + \dfrac{r_o/r_{in}}{a_o(1 + R_2/R_1)}}{1 + \dfrac{1}{a_o}\left(1 + \dfrac{r_o}{R_L}\right)\left(1 + \dfrac{R_2 + r_o \| R_L}{R_1 \| r_{in}}\right)}$$

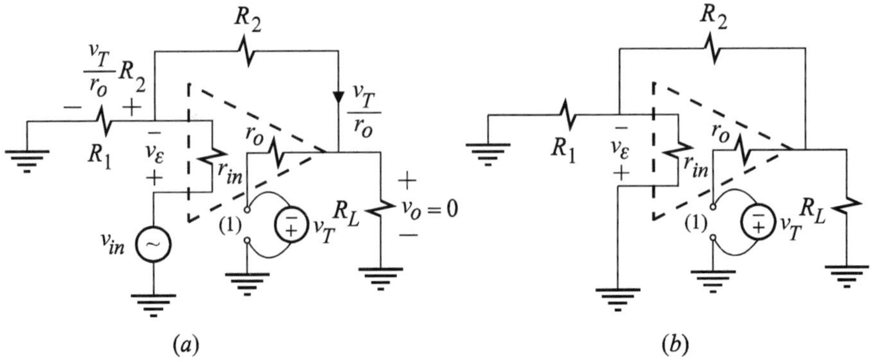

(a)   (b)

Figure 3.47

From the expressions of the gain of the inverting and the noninverting nonideal operational amplifiers obtained above in Eqs. (3.92) and (3.96), we can see why the simple expressions of the ideal closed-loop gain, $-R_2/R_1$ and $1 + R_2/R_1$, are excellent approximations of the real closed-loop gain. ☐

3.5   Review

The extra element theorem (EET) is a tool which simplifies the determination of a transfer function, H, of an electrical circuit by breaking up the analysis into three separate, independent and simpler parts. In the first part, an impedance element or a dependent source of the network, connected across a certain port, is designated as an extra element and set to zero or infinity. The choice of the extra element is mostly motivated by the greatest and most obvious simplification of the network which yields a simpler transfer function H_o. In the second part, the excitation of the transfer function applied to the network is set to zero and a second transfer function is determined at the port of the extra element with the extra element removed. If the extra element is an impedance, this transfer function is the impedance looking into that port. If the extra element is a controlled source, this transfer function is the ratio of the response of the controlling variable to an independent source connected at that port which is of the same type but opposite polarity as the dependent source. Either transfer function is simpler than the original one, H, because it is determined on a circuit which does not contain the extra element. In the third part, the excitation of the transfer function is retained and an additional excitation is applied at the port of the extra element with the extra element removed. With two excitations, the response of the transfer function, H, is nulled and a third "transfer function," called null transfer function, is determined as in the second part. The desired transfer function is then assembled from these three separate transfer functions and the extra element using the bilinear form. Thus, when the reactive element in a first-order circuit is designated as the extra element, the EET reduces its analysis to the analysis of three resistive circuits.

Problems

3.1 Nulling a response. Determine the relationship between v_T and v_{in} in Example 3.4 which nulls the output voltage. A knowledge of this relationship is *not* required for the determination of a null driving-point impedance or a null inverse gain. Null response calculations are always performed by assuming that the response has been nulled and following that null on the circuit diagram all the way back to v_T or i_T.

3.2 The EET. Derive the second form of the EET in Eq. (3.36) beginning with Fig. 3.18b in which Z is replaced with a voltage V carrying a current I.

3.3 Differential amplifier with adjustable gain. The differential amplifier in Fig. 3.48a is a subtractor with gain R_2/R_1. Its output is given by:

$$v_o = \frac{R_2}{R_1}(v_2 - v_1) \tag{3.97}$$

This circuit is not suitable if variable gain is required because it requires simultaneous adjustment of two resistors. An extra degree of freedom is added to this circuit by splitting each R_2 into R_{2A} and R_{2B} thus providing a new port across which R_{adj} can be connected as shown in Fig. 3.48b. Show that the gain of this circuit is given by:

$$v_o = \frac{R_{2A} + R_{2B}}{R_1}\left(1 + \frac{2R_{2A} \| R_{2B}}{R_{adj}}\right)(v_2 - v_1) \tag{3.98}$$

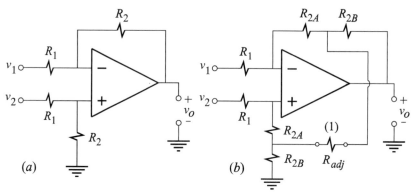

Figure 3.48

If $R_{2A} = R_{2B} = R_2$, then we have:

$$v_o = \frac{2R_2}{R_1}\left(1 + \frac{R_2}{R_{adj}}\right)(v_2 - v_1) \tag{3.99}$$

Hints: (*i*) Since the circuit is symmetrical, you can consider the gain from either input to the output as shown in Fig. 3.48c in which R_{adj} is designated as the extra element and removed from the circuit. The gain of this circuit is simply:

$$A_o = -\frac{R_{2A} + R_{2B}}{R_1} \tag{3.100}$$

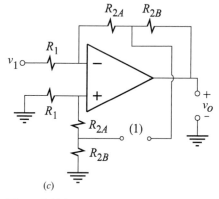

(c)

Figure 3.48 (*cont.*)

(*ii*) Determine the ordinary driving-point impedance, $R^{(1)}$, looking into port (1) as shown in Fig. 3.48d.

(*iii*) Determine the null driving-point impedance, $\mathscr{R}^{(1)}$, looking into port (1) as shown in Fig. 3.48e. Apply the first form of the EET in Eq. (3.34).

Note that:

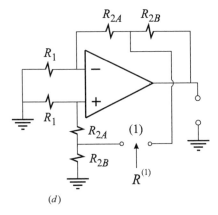

(d)

Figure 3.48 (cont.)

$$i_T = \frac{v_y}{R_{2B} \| R_{2A}} - \frac{v_z}{R_{2A}}$$

$$i_T = -\frac{v_x}{R_{2B} \| R_{2A}} + \frac{v_z}{R_{2A}} \Bigg\}$$ (3.101a–c)

$$v_T = v_y - v_x$$

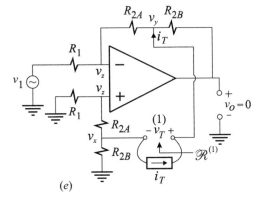

(e)

Figure 3.48 (cont.)

3.4 Bridge phase shifter. For the bridge circuit in Example 3.9, determine the relationship between the resistors so that the magnitudes of the pole and zero are numerically equal. Determine the complete transfer function with this condition satisfied. Note how the gain varies as the pole and zero move together to vary the phase at a given frequency.

3.5 Adjustable phase shifter. The circuit in Example 3.11 can be easily converted into an adjustable phase shifter by interchanging C_3 and R_3 and using a FET to emulate R_3 as shown in Fig. 3.49. Verify that the results of Example 3.11 hold for this circuit as well.

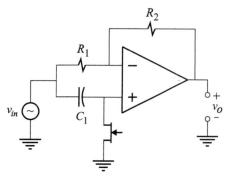

Figure 3.49

3.6 Input resistance of a bridge circuit. Verify that the expression of R'_o in Eq. (3.70) is the input impedance of the bridge circuit with R_2 taken as the extra element.

3.7 The effect of the feedback resistor on the gain of the common-emitter amplifier. Verify Eq. (2.47) in which R_f is treated as the extra element.

3.8 The effect of the collector-to-emitter resistance on the gain of the common-emitter amplifier.

(a) Using the EET, determine the effect of the collector-to-emitter resistance, r_{ce}, on the voltage gain of the common-emitter in Figs. 3.50a and b and show that it is given by:

$$A = -\frac{R_C}{R_E}\frac{\alpha}{1+\dfrac{(R_s+r_\pi)/R_E}{1+\beta}}\frac{1-\dfrac{R_E}{r_{ce}\beta}}{1+\dfrac{R_C}{r_{ce}}\dfrac{1+\dfrac{R_E}{(R_s+r_\pi)\|R_C}}{1+\dfrac{R_E(1+\beta)}{(R_s+r_\pi)}}} \qquad (3.102)$$

Hint: (i) Let r_{ce} be the extra element and let $r_{ce} \to \infty$ as shown in Fig. 3.50c and determine the voltage gain.

(ii) Determine next the null driving-point impedance looking into port (c) as shown in Fig. 3.50d.

(iii) Determine the ordinary driving-point impedance looking into port (c) as shown in Fig. 3.50e. To determine $R^{(c)}$, you can apply the EET for impedance functions with R_E as the extra element as shown in Figs. 3.50f and g. This is an example of nested EET.

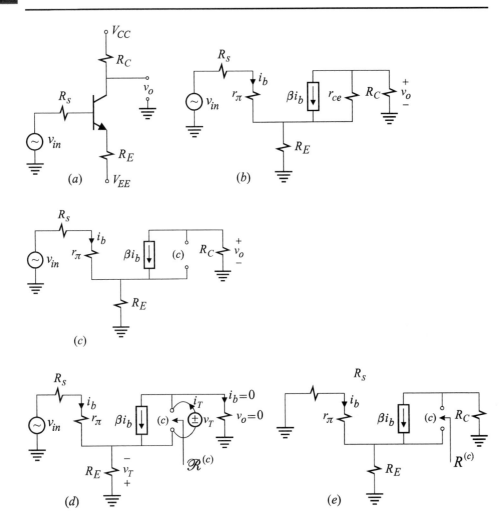

Figure 3.50

(b) Note that as $\beta \to \infty$, the gain $A \to -R_C/R_E$. The reason for this is that R_E feeds back a voltage v_f proportional to the output voltage, which when $\beta \to \infty$, becomes:

$$v_f = i_e R_E = i_c R_E = -\frac{v_o}{R_C} R_E \qquad (3.103)$$

because $i_b \to 0$ and the collector and emitter currents become identical. Hence, with $\beta \to \infty$ the foreward gain becomes infinite and the closed-loop gain A is determined entirely by the feedback network and is given by the reciprocal of the feedback gain v_f/v_o. Determine another expression of the voltage gain treating β as the extra element and show that it is given by:

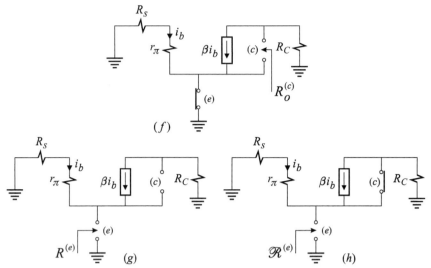

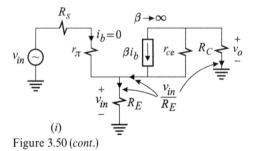

Figure 3.50 (cont.)

$$A = -\frac{R_C}{R_E} \frac{1 - \dfrac{R_E}{r_{ce}\beta}}{1 + \dfrac{1}{\beta}\left(1 + \dfrac{R_C}{r_{ce}}\right)\left[1 + \dfrac{R_s + r_\pi}{R_E \| (r_{ce} + R_C)}\right]} \tag{3.104}$$

Hints: (i) Let β be the extra element and let $\beta \to \infty$ as shown in Fig. 3.50i and determine the voltage gain.

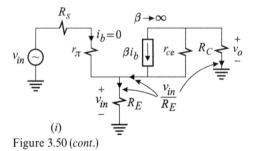

(i)
Figure 3.50 (cont.)

(ii) Determine the inverse current $\bar{\mathscr{B}}^{(c)}$ with the response nulled as shown in Fig. 3.50j.

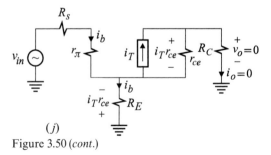

(j)
Figure 3.50 (cont.)

(iii) Determine the inverse current gain $\bar{B}^{(c)}$ with the excitation removed as shown in Fig. 3.50k and apply the EET for dependent sources.

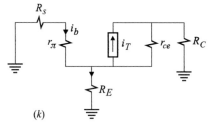

(k)

Figure 3.50 (*cont.*)

(c) Knowing that $R_E/\beta r_{ce} \ll 1$ always, show that if we assume $R_C \gg (R_s + r_\pi)$, the expression of A given in part (a) can be approximated as:

$$A = -\frac{R_C}{R_E} \frac{\alpha}{1 + \dfrac{(R_s + r_\pi)/R_E}{1+\beta}} \frac{1}{1 + \dfrac{R_C}{r_{ce}} \dfrac{1 + \dfrac{R_E}{R_s + r_\pi}}{1 + \dfrac{R_E(\beta+1)}{R_s + r_\pi}}} \qquad (3.105)$$

(d) We would like to know how large R_C can get before the effect of r_{ce} cannot be ignored. This is particularly important if the load is a current source. Using the expression above, show that a quantitative criterion for ignoring r_{ce}, is given by:

$$r_{ce} > R_C \frac{1 + \dfrac{R_E}{R_s + r_\pi}}{1 + \dfrac{R_E(\beta+1)}{R_s + r_\pi}} \qquad (3.106)$$

Show that if β is large enough so that the gain is given by $A \approx -R_c/R_E$, then the criterion of ignoring r_{ce} becomes:

$$r_{ce} > \frac{R_C}{\beta}\left(1 + \frac{R_s + r_\pi}{R_E}\right) \qquad (3.107)$$

This shows that, in the presence of R_E and with large β, r_{ce} can be comparable to R_C and still have no effect on the gain.

3.9 Inductive behavior of the output impedance of the emitter-follower. Show that the output imedpance of the emitter-follower shown in Figs. 3.51a and b is given by:

$$Z_o = R_o \frac{1 + s/\omega_1}{1 + s/\omega_2} \qquad (3.108)$$

where:

$$R_o = R_L \| r_{ce} \left\| \frac{R_s + r_\pi}{1 + g_m r_\pi} \right. \qquad (3.109)$$

$$\omega_1 = \frac{1}{C_\pi r_\pi \parallel R_s} \qquad (3.110)$$

$$\omega_2 = \frac{1}{C_\pi r_\pi \left\| \dfrac{R_s + R_L \parallel r_{ce}}{1 + g_m R_L \parallel r_{ce}} \right.} \qquad (3.111)$$

In this derivation, C_μ has been ignored because, as we shall see later, its contribution with a small R_s occurs at very high frequencies. Note that if we let $R_L \to \infty$, we obtain the impedance looking directly into the emitter. Determine R_o, ω_1 and ω_2 with $R_L \to \infty$ and show that if $R_s \gg 1/g_m$, then $\omega_1 \ll \omega_2$ and the output impedance increases at 20 dB/dec between ω_1 and ω_2, like an inductive impedance. Show that the effective inductance is $L_e = C_\pi R_s/g_m$.

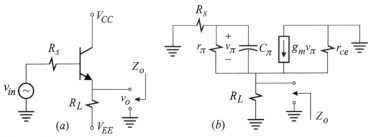

Figure 3.51

It is important to consider this inductive behavior when designing an emitter-follower with a capacitative load in order to prevent the output voltage from ringing or preventing the system from becoming unstable altogether.

REFERENCES

1. R. D. Middlebrook, "Null double injection and the extra element theorem", *IEEE Transactions on Education*, Vol. 32, No. 3, Aug. 1989, pp. 167–180.
2. H. W. Bode, *Network Analysis and Feedback Amplifier Design*, Princeton, N.J., Van Nostrand, 1945.
3. K. S. Yeung, "An open- and short-circuit technique for analyzing electronic circuits", *IEEE Transactions on Education*, Vol. E-30, No. 1, Feb. 1987, pp. 55–56.
4. E. V. Sorensen, "General relations governing the exact sensitivity of linear networks", *Proceedings of the IEE*, Vol. 114, No. 9, Sept. 1967, pp. 1209–1212.
5. S. R. Parker, E. Peskin and P. M. Chirilian, "Application of a bilinear theorem to network sensitivity", *IEEE Transactions on Circuit Theory*, Sept. 1965, pp. 448–450.
6. R. A. Rohrer, "Circuit Partitioning Simplified", *IEEE Transactions on Circuits and Systems*, Vol. 35, No. 1, Jan. 1988, pp. 2–5.
7. S. Sabharwal, "The N-element theorem: Part I and II" Internal technical notes of graduate students, T72 and T77, EE Department, Caltech.
8. R. D. Middlebrook, V. Vorpérian and J. Lindal, "The *N* Extra Element Theorem", *IEEE Transactions on Circuits and Systems – I: Fundamental Theory and Application*, Vol. 45, No. 9, Sept. 1998, pp. 919–935.

4 The *N*-extra element theorem
Divide and conquer

4.1 Introduction

The *N*-extra element theorem (NEET) is an extension of the EET to *N* elements consisting of impedance elements and dependent sources connected across an *N*-port network as shown in Fig. 4.1. In this figure, $A_a u_a, A_b u_b, \ldots, A_x u_x$ are arbitrary dependent voltage and current sources controlled by internal voltages or currents, $u_a, u_b, \ldots, u_x$, with gains $A_a, A_b, \ldots, A_x$, respectively. The most useful application of the NEET is when these elements are either inductors or capacitors rather than arbitrary impedance branches because, upon their removal, one can determine the complete *N*th-order transfer function from the remaining purely resistive network.

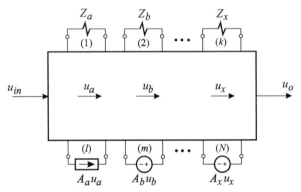

Figure 4.1

In general, to solve a complex circuit effectively, the NEET must be applied piecemeal. For instance, in a circuit containing several reactive elements, resistors and dependent sources, the reactive elements are removed first. If the remaining circuit, which consists of resistors and dependent sources, is fairly complex, then the NEET can be applied by designating certain resistors and dependent sources as extra elements. The contribution of the reactive elements is then determined by a separate application of the NEET to the reactive elements (see Problem 3.8 for example).

Although the NEET is a relatively straight-forward and intuitive procedure, the best way to develop it is to start with the 2-EET for impedance elements. The most general form of the NEET and a proof will be given in Sections 4.4 and 4.5, respectively.

4.2   The 2-EET for impedance elements

Consider the LTI network in Fig. 4.2 in which the impedance elements Z_1 and Z_2 are connected across ports (1) and (2), respectively. We would like to determine a certain transfer function of this network:

$$H(s) = \frac{u_o(s)}{u_{in}(s)} \tag{4.1}$$

Since each impedance element can be removed from the circuit either by shorting it or opening it, there are four possible ways of removing both elements. In Fig. 4.3, we show Z_1 and Z_2 removed as open circuits or, equivalently, port (1) and port (2) opened.

Several important definitions are required for the development of NEET.

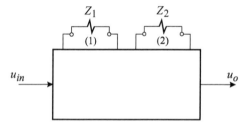

Figure 4.2

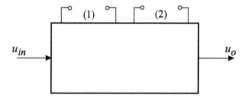

Figure 4.3

Reference network: The network in Fig. 4.3 is called the reference network which is obtained from the original network in Fig. 4.2 by opening ports (1) and (2). There are four reference networks which can be obtained from the original network in Fig. 4.2. The remaining three reference networks are shown in Figs. 4.4*a–c*.

Reference state of a port: The open or short state of a port in the reference network is called its reference state. In Fig. 4.3, the reference state of port (1) is open and the reference state of port (2) is open. In Fig. 4.4*a*, the reference state of port (1)

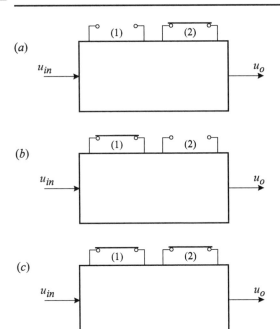

Figure 4.4

is open and the reference state of port (2) is short. In Fig. 4.4*b*, the reference state of port (1) is short and the reference state of port (2) is open. In Fig. 4.4*c*, the reference state of both ports is short.

Opposite state of a port: The opposite state of a port is defined as the opposite of its reference state. Hence in Fig. 4.3, the opposite state of port (1) is a short, while in Fig. 4.4*c* the opposite state of port (1) is an open.

Reference transfer function: The transfer function of the reference network in Fig. 4.3 is called the *reference transfer function* and is written as:

$$H_o = \frac{u_o}{u_{in}}\bigg|_{\substack{(1)\to open \\ (2)\to open}} \tag{4.2}$$

According to the 2-EET, the transfer function in Eq. (4.1) is given in terms of H_o, Z_1 and Z_2 according to the following *structure*, which is a generalization of the structure of the EET:

$$H(s) = H_o \frac{1 + \dfrac{\overset{*}{}}{Z_1} + \dfrac{\overset{*}{}}{Z_2} + \dfrac{\overset{*}{}\ \overset{*}{}}{Z_1\ Z_2}}{1 + \dfrac{\overset{*}{}}{Z_1} + \dfrac{\overset{*}{}}{Z_2} + \dfrac{\overset{*}{}\ \overset{*}{}}{Z_1\ Z_2}} \tag{4.3}$$

The asterixes in Eq. (4.3) correspond to certain port impedances of the reference network which are independent of the extra elements Z_1 and Z_2 and are discussed later. At the moment, we would like to discuss two important features of the structure of Eq. (4.3), namely, the *placement* and the *combination* of the extra elements Z_1 and Z_2 in the numerator and the denominator.

1. **Placement of Z_1 and Z_2.**   In Eq. (4.3), Z_1 and Z_2 are placed in the denominator of the small fractions which occur in the numerator and the denominator of the transfer function so that when they both become infinite, $H(s)$ reduces to the reference transfer function, H_o, i.e. $Z_1, Z_2 \to \infty \Rightarrow H(s) \to H_o$. We can also see that when Z_1 becomes infinite, Eq. (4.3) reduces to the EET for Z_2 and, similarly, when Z_2 becomes infinite, (4.3) reduces to the EET for Z_1. It follows that, for the reference network in Fig. 4.4c, Z_1 and Z_2 are placed in the numerator of the small fractions in the 2-EET:

$$H(s) = H'_o \frac{1 + \dfrac{Z_1}{*} + \dfrac{Z_2}{*} + \dfrac{Z_1}{*}\dfrac{Z_2}{*}}{1 + \dfrac{Z_1}{*} + \dfrac{Z_2}{*} + \dfrac{Z_1}{*}\dfrac{Z_2}{*}} \tag{4.4}$$

in which H'_o is the reference transfer function of the network in Fig. 4.4c and defined as:

$$H'_o = \left. \frac{u_o}{u_{in}} \right|_{\substack{(1)\to short \\ (2)\to short}} \tag{4.5}$$

It is clear then, that in order to recover the transfer function, the placement of Z_1 and Z_2 in the 2-EET for the reference network in Fig. 4.4a, is:

$$H(s) = H''_o \frac{1 + \dfrac{*}{Z_1} + \dfrac{Z_2}{*} + \dfrac{*}{Z_1}\dfrac{Z_2}{*}}{1 + \dfrac{*}{Z_1} + \dfrac{Z_2}{*} + \dfrac{*}{Z_1}\dfrac{Z_2}{*}} \tag{4.6}$$

in which:

$$H''_o = \left. \frac{u_o}{u_{in}} \right|_{\substack{(1)\to open \\ (2)\to short}} \tag{4.7}$$

For the reference circuit in Fig. 4.4b, Z_1 and Z_2 are interchanged in Eq. (4.6).

2. **The combinations of Z_1 and Z_2.**   There are *two* combinations in which Z_1 and Z_2 appear in the structure of the 2-EET in Eqs. (4.3), (4.4) or (4.6). In the first combination, one impedance element is taken at a time to generate the second and third terms in the numerator and the denominator. In the second combination, two impedance elements are taken at a time to generate the last term. The number of terms each combination produces is given by:

$$\text{one at a time} \Rightarrow \binom{2}{1} = 2 \text{ terms}$$

$$\text{two at a time} \Rightarrow \binom{2}{2} = 1 \text{ term} \tag{4.8}$$

which is the familiar combinatoric formula. Now we begin to see the trend for more extra elements. For example, for three extra elements, there would be three combinations (one at a time, two at a time and three at a time) with each combination producing the following number of terms:

$$\left. \begin{array}{l} \text{one at a time} \Rightarrow \binom{3}{1} = 3 \text{ terms} \\[2em] \text{two at a time} \Rightarrow \binom{3}{2} = 3 \text{ terms} \\[2em] \text{three at a time} \Rightarrow \binom{3}{3} = 1 \text{ term} \end{array} \right\} \tag{4.9}$$

Hence, in the 3-EET there are seven terms added to unity in the numerator and the denominator, respectively.

Next, we determine the port impedances of the reference network which correspond to the asterixes in the structure of the 2-EET. As in the case of the EET, we shall determine two types of port impedances: *ordinary* driving-point impedances and *null* driving-point impedances. For the 2-EET, the four ordinary driving-point impedances, $Z_{(j)}^{(i)}$ are defined as:

$Z^{(1)} \equiv$ Ordinary driving-point impedance looking in port (1)
    with port (2) in its *reference* state.

$Z^{(2)} \equiv$ Ordinary driving-point impedance looking in port (2)
    with port (1) in its *reference* state.

$Z_{(2)}^{(1)} \equiv$ Ordinary driving-point impedance looking in port (1)
    with port (2) in its *opposite* state.

$Z_{(1)}^{(2)} \equiv$ Ordinary driving-point impedance looking in port (2)
    with port (1) in its *opposite* state.

Since these are ordinary driving point impedances, they are determined with the excitation set to zero, i.e. $u_{in} = 0$. Observe that, in the notation above, a subscript is used only to designate a port which is in its opposite state. Hence, if a port does not appear in the subscript, it is considered to be in its reference state.

The null driving-point impedances are defined exactly in the same way as the ordinary driving-point impedances and are determined under double injection with the response nulled, i.e. $u_o = 0$. The following notation is used:

$\mathcal{Z}_{(j)}^{(i)} \equiv$ Null driving-point impedance looking in port (i)
     with port (j) in its *opposite* state.

Each of these port impedances is worked out in the following examples.

Example 4.1   Let the reference network in Fig. 4.3, repeated here in Fig. 4.5*a*, be purely resistive. The ordinary driving-point resistances $R^{(1)}$, $R^{(2)}$, $R_{(2)}^{(1)}$ and $R_{(1)}^{(2)}$ are shown in Figs. 4.5*b–e*, respectively.

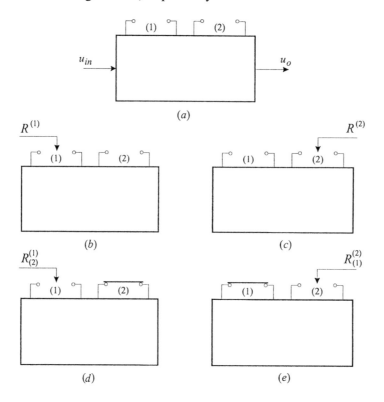

Figure 4.5

Since these are ordinary driving-point resistances, the excitation of the transfer function is set to zero, i.e. $u_{in} = 0$. In Fig. 4.5*b*, $R^{(1)}$ is determined with port (2) open, which correspnds to its reference state in Fig. 4.5*a*. In Fig. 4.5*c*, $R^{(2)}$ is determined with port (1) open, which is its reference state in Fig. 4.5*a*. In Fig. 4.5*d*, $R_{(2)}^{(1)}$ is determined with port (2) short, which is the opposite of its reference state in Fig. 4.5*a*. In Fig. 4.5*e*, $R_{(1)}^{(2)}$ is determined with port (1) short, which is the opposite of its reference state in Fig. 4.5*a*.   □

Example 4.2   Let the reference network in Fig. 4.4*c*, repeated here in Fig. 4.6*a*, be purely resistive. The ordinary driving-point resistances $R^{(1)}$, $R^{(2)}$, $R^{(1)}_{(2)}$ and $R^{(2)}_{(1)}$ for this network are shown in Figs. 4.6b–e, respectively. We can see that the state of the other port for each driving-point resistance of this reference network is exactly opposite that of the corresponding driving-point resistance in Example 4.1. The reason for this is that the reference state of both ports in Fig. 4.6a is opposite to those in Fig. 4.5a.

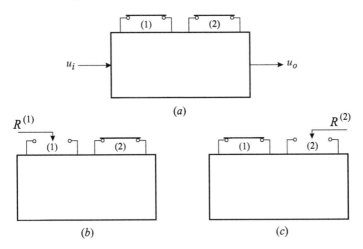

Figure 4.6

Note that the reference state of port (1), whether short or open, has nothing to do with $R^{(1)}$, because $R^{(1)}$ is determined looking into port (1) regardless of its reference state. The same applies to $R^{(2)}$. In other words, for the reference network in Fig. 4.6a, $R^{(1)}$ and $R^{(2)}$ should not be confused with zero simply because their reference state is a short. □

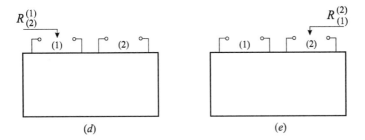

Figure 4.6 (*cont.*)

Example 4.3   For the reference network in Fig. 4.4a, repeated here in Fig. 4.7a, the null driving-point resistances $\mathscr{R}^{(1)}$, $\mathscr{R}^{(2)}$, $\mathscr{R}^{(1)}_{(2)}$ and $\mathscr{R}^{(2)}_{(1)}$ are shown in Figs. 4.7b–e, respectively.

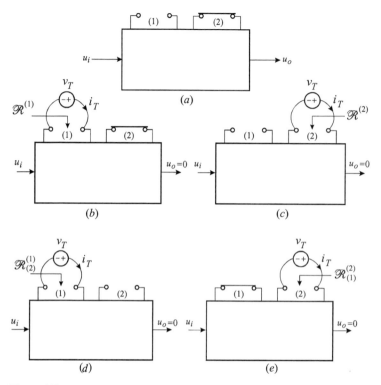

Figure 4.7

As explained in Chapter 3, these null driving-point resistances are determined by nulling the response of the transfer function, u_o, with the simultaneous application of two excitations: the excitation of the transfer function, u_i, and the test source, v_T (or i_T), connected at the port of interest (null double injection.)   □

All the asterixes in the four different forms of the 2-EET are now replaced with the null and ordinary driving-point impedances. As in the case of the EET, the null driving-point impedances go into the numerator of the transfer function while the ordinary driving point impedances go into its denominator. For the first form of the 2-EET in Eq. (4.3), whose reference circuit is shown in Fig. 4.3, these port impedances are substituted as:

$$H(s) = H_o \frac{1 + \dfrac{\mathscr{Z}^{(1)}}{Z_1} + \dfrac{\mathscr{Z}^{(2)}}{Z_2} + \dfrac{\mathscr{Z}^{(1)}}{Z_1}\dfrac{\mathscr{Z}^{(2)}_{(1)}}{Z_2}}{1 + \dfrac{Z^{(1)}}{Z_1} + \dfrac{Z^{(2)}}{Z_2} + \dfrac{Z^{(1)}}{Z_1}\dfrac{Z^{(2)}_{(1)}}{Z_2}} \tag{4.10}$$

Hence, in order to determine $H(s)$, we must perform seven separate and independent calculations on the reference network. Although this may seem like a much more difficult task to perform than determining $H(s)$ directly, the following examples will prove to the contrary. Notice that only three of the four ordinary and null driving-point impedances are used in Eq. (4.10). This means that we have a choice in picking the order of Z_1 and Z_2 when forming the product in the last term in the numerator and the denominator. Hence, since the order is immaterial, the last term in the numerator can be written as:

$$\frac{\mathscr{L}^{(2)}}{Z_2} \frac{\mathscr{L}^{(1)}_{(2)}}{Z_1} \tag{4.11}$$

and the last term in the denominator can be written as:

$$\frac{Z^{(2)}}{Z_2} \frac{Z^{(1)}_{(2)}}{Z_1} \tag{4.12}$$

A comparison of these with their corresponding terms in the numerator and denominator of Eq. (4.10) yields the following equalities:

$$\mathscr{L}^{(1)}\mathscr{L}^{(2)}_{(1)} = \mathscr{L}^{(2)}\mathscr{L}^{(1)}_{(2)} \tag{4.13}$$

$$Z^{(1)}Z^{(2)}_{(1)} = Z^{(2)}Z^{(1)}_{(2)} \tag{4.14}$$

Equation (4.14) can be rewritten as:

$$\frac{Z^{(1)}}{Z^{(1)}_{(2)}} = \frac{Z^{(2)}}{Z^{(2)}_{(1)}} \tag{4.15}$$

Equation (4.15) is a well-known theorem of two-port networks which states that the ratio of the impedance looking into port (1) with port (2) open to the impedance looking into port (1) with port (2) short is equal to the ratio of the impedance looking into port (2) with port (1) open to the impedance looking into port (2) with port (1) short (see Problems 4.1 and 4.2).

For the second form of the 2-EET in Eq. (4.4), whose reference circuit is shown in Fig. 4.4c, these port impedances are substituted as:

$$H(s) = H'_o \frac{1 + \dfrac{Z_1}{\mathscr{L}^{(1)}} + \dfrac{Z_2}{\mathscr{L}^{(2)}} + \dfrac{Z_1}{\mathscr{L}^{(1)}} \dfrac{Z_2}{\mathscr{L}^{(2)}_{(1)}}}{1 + \dfrac{Z_1}{Z^{(1)}} + \dfrac{Z_2}{Z^{(2)}} + \dfrac{Z_1}{Z^{(1)}} \dfrac{Z_2}{Z^{(2)}_{(1)}}} \tag{4.16}$$

For the remaining two forms of the 2-EET which correspond to the reference networks in Figs. 4.4a and b we have, respectively:

$$H(s) = H''_o \frac{1 + \dfrac{\mathscr{Y}^{(1)}}{Z_1} + \dfrac{Z_2}{\mathscr{Y}^{(2)}} + \dfrac{\mathscr{Y}^{(1)}}{Z_1}\dfrac{Z_2}{\mathscr{Y}^{(2)}_{(1)}}}{1 + \dfrac{Z^{(1)}}{Z_1} + \dfrac{Z_2}{Z^{(2)}} + \dfrac{Z^{(1)}}{Z_1}\dfrac{Z_2}{Z^{(2)}_{(1)}}} \qquad (4.17)$$

$$H(s) = H'''_o \frac{1 + \dfrac{Z_1}{\mathscr{Y}^{(1)}} + \dfrac{\mathscr{Y}^{(2)}}{Z_2} + \dfrac{Z_1}{\mathscr{Y}^{(1)}}\dfrac{\mathscr{Y}^{(2)}_{(1)}}{Z_2}}{1 + \dfrac{Z_1}{Z^{(1)}} + \dfrac{Z^{(2)}}{Z_2} + \dfrac{Z_1}{Z^{(1)}}\dfrac{Z^{(2)}_{(1)}}{Z_2}} \qquad (4.18)$$

As mentioned earlier, one of the most useful applications of the NEET is in the analysis of reactive circuits. In the following examples the 2-EET is applied to solve second-order networks.

Example 4.4   Determine the transfer function of the passive second-order filter in Fig. 4.8.

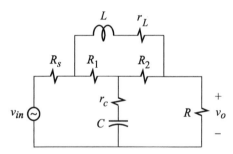

Figure 4.8

If L and C are designated as the two extra elements, the reference network can be formed by considering this circuit at zero frequency in which the capacitor is

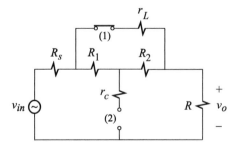

Figure 4.9

replaced with an open and the inductor is replaced with a short as shown in Fig. 4.9. The transfer function of the reference circuit in Fig. 4.9 is given by:

$$H_o \equiv \frac{v_o}{v_{in}} = \frac{1}{1 + \dfrac{R_s + r_L \| (R_1 + R_2)}{R}} \tag{4.19}$$

According to the 2-EET, the transfer function of the circuit in Fig. 4.8 is given (see Eq. (4.18)) by:

$$H(s) = H_o \frac{1 + \dfrac{sL}{\mathscr{R}^{(1)}} + \dfrac{\mathscr{R}^{(2)}}{1/sC} + \dfrac{sL}{\mathscr{R}^{(1)}} \dfrac{\mathscr{R}^{(2)}_{(1)}}{1/sC}}{1 + \dfrac{sL}{R^{(1)}} + \dfrac{R^{(2)}}{1/sC} + \dfrac{sL}{R^{(1)}} \dfrac{R^{(2)}_{(1)}}{1/sC}}$$

$$= H_o \frac{1 + s\left(\dfrac{L}{\mathscr{R}^{(1)}} + C\mathscr{R}^{(2)}\right) + s^2 LC \dfrac{\mathscr{R}^{(2)}_{(1)}}{\mathscr{R}^{(1)}}}{1 + s\left(\dfrac{L}{R^{(1)}} + CR^{(2)}\right) + s^2 LC \dfrac{R^{(2)}_{(1)}}{R^{(1)}}} \tag{4.20a, b}$$

The ordinary driving-point impedances are determined first. For these, the excitation of the transfer function is set to zero by replacing v_{in} with a short circuit as shown in Figs. 4.10a, b and 11. For the circuit in Fig. 4.10a, $R^{(1)}$ is readily given by:

$$R^{(1)} = r_L + (R_1 + R_2) \| (R_s + R) \tag{4.21}$$

In Fig. 4.10b, $R^{(2)}$ is equal to r_c plus the impedance of a bridge circuit in which r_L is the bridge element. The impedance of a bridge circuit was derived in Chapter 1 using the EET, with the bridge resistance as the designated extra element. Hence, according to Eq. (1.5), $R^{(1)}$ is given by:

$$R^{(2)} = r_c + (R_1 + R_s) \| (R_2 + R) \frac{1 + \dfrac{R_1 \| R_s + R_2 \| R}{r_L}}{1 + \dfrac{(R_1 + R_2) \| (R_s + R)}{r_L}} \tag{4.22}$$

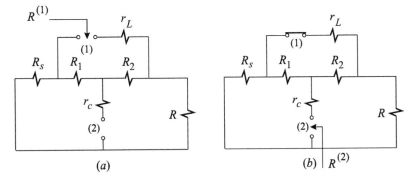

(a)   *(b)* $R^{(2)}$

Figure 4.10

In Fig. 4.11, port (1) is changed to its opposite state for the determination of $R^{(2)}_{(1)}$, which is readily given by:

$$R^{(2)}_{(1)} = r_c + (R_1 + R_s) \| (R_2 + R)$$ (4.23)

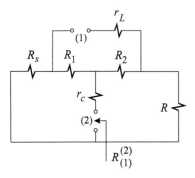

Figure 4.11

In Fig. 4.12, $\mathcal{R}^{(1)}$ is determined using null double injection. The null in the output voltage is accompanied by a null in the output current so that the test current, i_T, flows entirely through R_2 and R_1. It follows that:

$$\mathcal{R}^{(1)} = r_L + R_2 + R_1$$ (4.24)

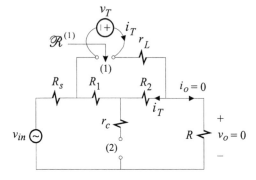

Figure 4.12

Since $\mathcal{R}^{(2)} = r_c + \mathcal{R}^{(2')}$, we determine $\mathcal{R}^{(2')}$ as shown in Fig. 4.13a, whence we see that when the output voltage is nulled, the test voltage, v_T, appears entirely across R_2. The current through R_2, which is now given by v_T/R_2, flows entirely through r_L because of the null in the output current. It follows that the voltage across R_1 is equal to $v_T + (v_T/R_2)r_L$ and the test current is given by:

$$\begin{aligned} i_T &= \frac{v_T}{R_2} + \frac{v_T + (v_T/R_2)r_L}{R_1} \\ &= v_T\left(\frac{1}{R_2} + \frac{1 + r_L/R_2}{R_1}\right) \end{aligned} \Bigg\}$$ (4.25a, b)

Recognizing the parallel combination in Eq. (4.25b), we can write $\mathscr{R}^{(2)}$ as:

$$\mathscr{R}^{(2)} = r_c + R_2 \left\| \frac{R_1}{1 + r_L/R_2} \right. \tag{4.26}$$

To obtain $\mathscr{R}^{(2)}_{(1)}$, port (1) must be in its opposite state, or opened, as shown in Fig. 4.13b whence $\mathscr{R}^{(2')}_{(1)}$ is seen to be zero. It is also very useful to see that opening port (1) is the same as letting $r_L \to \infty$, so that $\mathscr{R}^{(2)}_{(1)}$ can be obtained directly from the expression of $\mathscr{R}^{(2)}$ by letting $r_L \to \infty$. Either way, we have:

$$\mathscr{R}^{(2)}_{(1)} = r_c \tag{4.27}$$

Substituting the results for the ordinary and null driving-point impedances in Eq. (4.20b), we obtain the complete transfer function:

$$H(s) = H_o \frac{1 + b_1 s + b_2 s^2}{1 + a_1 s + a_2 s^2} \tag{4.28}$$

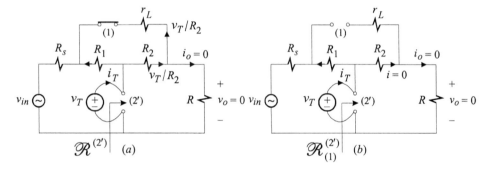

Figure 4.13

in which:

$$H_o = \frac{1}{1 + [R_s + r_L \| (R_1 + R_2)]/R} \tag{4.29}$$

$$a_1 = \frac{L}{r_L + (R_1 + R_2) \| (R_s + R)}$$

$$+ C \left[r_c + (R_1 + R_s) \| (R_2 + R) \frac{1 + \dfrac{R_1 \| R_s + R_2 \| R}{r_L}}{1 + \dfrac{(R_1 + R_2) \| (R_s + R)}{r_L}} \right] \tag{4.30}$$

$$a_2 = LC \frac{r_c + (R_1 + R_s) \| (R_2 + R)}{r_L + (R_1 + R_2) \| (R_s + R)} \tag{4.31}$$

$$b_1 = \frac{L}{r_L + R_2 + R_1} + C\left(r_c + R_2 \left\|\frac{R_1}{1 + r_L/R_2}\right)\right.$$

(4.32)

$$b_2 = CL\frac{r_c}{r_L + R_2 + R_1}$$

(4.33)

Although these are rather impressive, low-entropy equations, the quadratic factors in the numerator and the denominator cannot be left as they are. If the elements are chosen such that the resonance determined by L and C has a high Q, then the quadratic must be written as $1 + s(\omega_o/Q) + (s/\omega_o)^2$, otherwise it must be written in factored form as $(1 + s/\omega_1)(1 + s/\omega_2)$ in which $-\omega_1$ and $-\omega_2$ are the real roots. Approximate factoring of quadratics and higher-order frequency polynomials will be discussed in later chapters. In the next two examples, numerical values will be used to illustrate both cases. ☐

Example 4.5   The transfer function for the component values shown in Fig. 4.14 is given by:

$$H(s) = 1.00\frac{1 + 2.557 \times 10^{-9}\, s}{1 + 1.268 \times 10^{-8}\, s + 4.999 \times 10^{-15}\, s^2}$$

(4.34)

A magnitude and phase plot of this transfer function is shown in solid lines in Fig. 4.15. The disparate values of the components shown may look a little odd to you. In fact, they look odd to the engineer who designed the circuit too! His original intent was to design a simple low-pass filter stage with a dc gain of unity and a high-frequency attenuation of 0.2 as shown by the dashed lines in Fig. 4.15. When the circuit was laid out on a printed circuit board without his supervision, or without paying attention to his instructions, two parasitic elements, R_2 and C, were picked up by leaving a very large trace area between R_1 and R. This trace, together with the ground plane, formed a parasitic capacitance, modeled by $C \approx 100\,\text{pF}$, and a distributed resistance, approximately modeled by $R_2 \approx 0.52\,\Omega$. We now determine the transfer function in Eq. (4.34) analytically and obtain expressions for the unintentional resonant frequency and its Q-factor.

The resonance at $\omega_o = (2\pi)2.25 \times 10^6$ rad/s is due to the complex poles in the denominator, which is written as:

$$1 + \frac{1}{Q}\left(\frac{s}{\omega_o}\right) + \left(\frac{s}{\omega_o}\right)^2$$

(4.35)

A comparison of the coefficients of s^2 in Eqs. (4.28) and (4.35) yields the expression of the resonant frequency:

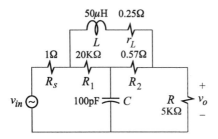

Figure 4.14

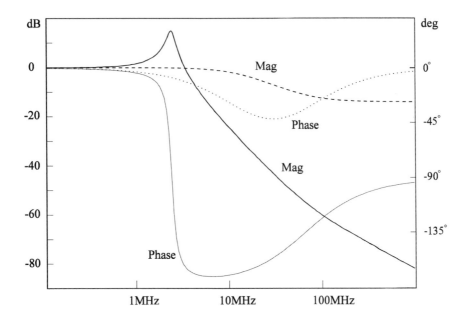

Figure 4.15

$$\omega_o = \frac{1}{\sqrt{a_2}} = \frac{1}{\sqrt{LC}} \sqrt{\frac{r_L + (R_1 + R_2) \| (R_s + R)}{r_c + (R_1 + R_s) \| (R_2 + R)}} \tag{4.36}$$

Comparison of the coefficients of s in Eqs. (4.28) and (4.35) yields:

$$\frac{1}{Q} = a_1 \omega_o \tag{4.37}$$

Squaring both sides of Eq. (4.36), we obtain:

$$\frac{1}{r_L + (R_1 + R_2) \| (R_s + R)} = \frac{1}{\omega_o^2 LC} \frac{1}{r_c + (R_1 + R_s) \| (R_2 + R)} \tag{4.38}$$

Substituting Eq. (4.38) in the expression of a_1 in Eq. (4.30) and using the result in

Eq. (4.37) we get:

$$\frac{1}{Q} = \frac{1}{\omega_o C} \frac{1}{r_c + (R_1 + R_s) \| (R_2 + R)}$$

$$+ \frac{1}{\omega_o L}\left[r_c + \frac{r_L + R_1 \| R_s + R_2 \| R}{1 + r_c/(R_1 + R_s) \| (R_2 + R)}\right] \qquad (4.39)$$

It is always convenient to express Q in terms of parallel and series combinations of other Q-factors which can be defined appropriately. In Eq. (4.39) the following Q-factors can be defined:

$$\left. \begin{array}{l} Q_c = \omega_o C[r_c + (R_1 + R_s) \| (R_2 + R)] \\[2ex] Q_L = \omega_o L\left[r_c + \dfrac{r_L + R_1 \| R_s + R_2 \| R}{1 + r_c/(R_1 + R_s) \| (R_2 + R)}\right]^{-1} \end{array} \right\} \qquad (4.40a, b)$$

It follows that:

$$\frac{1}{Q} = \frac{1}{Q_c} + \frac{1}{Q_L} \qquad (4.41)$$

This can be recognized as a parallel combination:

$$Q = Q_c \| Q_L \qquad (4.42)$$

As you may have noticed, it is possible to derive other expressions of Q by performing different substitutions. For instance, if Eq. (4.38) is not substituted in Eq. (4.30) before applying (4.37), the two Q-factors in the parallel combination in Eq. (4.41) would have different expressions and the one associated with $\omega_o C$ would have a longer expression. Hence, the reason for substituting Eq. (4.38) was solely for simplification purposes.

If r_c and r_L are parasitic resistances which are much smaller than the other resistances in the circuit, then ω_o, Q_c and Q_L can be approximated:

$$\left. \begin{array}{l} \omega_o \approx \dfrac{1}{\sqrt{LC}}\sqrt{\dfrac{(R_1 + R_2) \| (R_s + R)}{(R_1 + R_s) \| (R_2 + R)}} \\[3ex] Q_c \approx \omega_o C[(R_1 + R_s) \| (R_2 + R)] \\[3ex] Q_L \approx \dfrac{\omega_o L}{R_1 \| R_s + R_2 \| R} \end{array} \right\} \qquad (4.43a\text{--}c)$$

For the values in Fig. 4.14, these can be further approximated:

$$\left.\begin{aligned}
\omega_o &\approx \frac{1}{\sqrt{LC}}\sqrt{\frac{R_2 \parallel R}{R_s \parallel R}} \approx \frac{1}{\sqrt{LC}} \\[2mm]
Q_c &\approx \omega_o C(R_1 \parallel R) \\[2mm]
Q_L &\approx \frac{\omega_o L}{R_s + R_2}
\end{aligned}\right\} \tag{4.43d–f}$$

These expressions yield $\omega_o = (2\pi)2.55 \times 10^6$ rad/s, $Q_c = 5.65$ and $Q_L = 450$. Since $Q = Q_c \parallel Q_L$, we have $Q \approx Q_c = 15$ dB, which agree well with Fig. 4.15.

In the numerator we have a single zero which can be written as:

$$1 + \frac{s}{\omega_z} \tag{4.44}$$

in which:

$$\left.\begin{aligned}
\omega_z &= \frac{1}{b_1} = \frac{1}{\dfrac{L}{R_2 + R_1} + CR_2 \parallel R_1} \\[4mm]
&\approx \frac{R_1}{L}
\end{aligned}\right\} \tag{4.45a, b}$$

The complete transfer function is now written as:

$$H(s) = \frac{1 + \dfrac{s}{\omega_z}}{1 + \dfrac{1}{Q}\left(\dfrac{s}{\omega_o}\right) + \left(\dfrac{s}{\omega_o}\right)^2} \tag{4.46}$$

The numerical value of the zero is $\omega_z = (2\pi)63.5 \times 10^6$ rad/s, which is in agreement with Fig. 4.15.   □

Example 4.6   When the components of the circuit in the previous example are chosen as shown in Fig. 4.16, it behaves as a low-pass filter with a real input impedance $R_{in} = R$ at all frequencies. Such a filter is known as an equalizer and is used with a source which needs to be terminated with its internal impedance, R, at all frequencies. Using the results of the previous two examples, the transfer function v_o/v_{in} is given by:

$$\frac{v_o}{v_{in}} = \frac{1}{2} \frac{1 + \dfrac{s}{2}\left(\dfrac{L}{R} + RC\right)}{1 + s\left(\dfrac{L}{R} + RC\right) + s^2 LC} \tag{4.47}$$

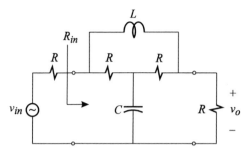

Figure 4.16

If we let:

$$\frac{L}{R} = RC \Rightarrow R = \sqrt{\frac{L}{C}} \tag{4.48}$$

then Eq. (4.47) reduces to a low-pass transfer function:

$$\frac{v_o}{v_{in}} = \frac{1}{2} \frac{1 + sRC}{1 + s2RC + (sRC)^2} \tag{4.49}$$

$$= \frac{1}{2} \frac{1}{1 + sRC}$$

This transfer function and the input resistance are shown in Fig. 4.17.

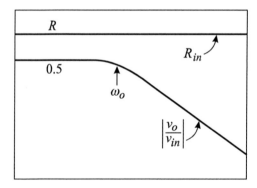

Figure 4.17

The input resistance is real at all frequencies and equal to R (see Problem 4.3). $\square$

Example 4.7   Determine the voltage gain of the common-emitter amplifier in Fig. 4.18*a* using the high-frequency small-signal model of the transistor shown in Fig. 4.18*b*. The parasitic base resistance, $r_{bb'}$, is absorbed in the source resistance R_s.

Two good candidates for extra elements are r_μ and r_o because they usually have very high values and are often ignored when the load is a resistor element rather than a current source.

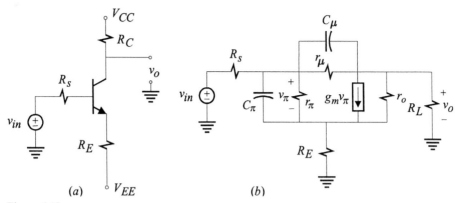

(a)   (b)

Figure 4.18

With these elements taken out as infinite, we obtain the reference circuit shown in Fig. 4.19 whose gain is readily given by:

$$A'_o = -\frac{\beta R_L}{R_s + r_\pi + (1 + \beta)R_E} \tag{4.50}$$

$$= -\frac{R_L}{\dfrac{R_E + r_e}{\alpha} + \dfrac{R_s}{\beta}}$$

in which $\beta = g_m r_\pi$ and $\alpha = \beta/(1 + \beta)$. For design purposes, when $\beta \gg 1$ Eq. (4.50) can be approximated as $A'_o \approx -R_L/R_E$.

The gain A_o, according to the 2-EET, is given by:

$$A_o = A'_o \frac{1 + \dfrac{\mathcal{R}^{(1)}}{r_\mu} + \dfrac{\mathcal{R}^{(2)}}{r_o} + \dfrac{\mathcal{R}^{(1)}}{r_\mu}\dfrac{\mathcal{R}^{(2)}_{(1)}}{r_o}}{1 + \dfrac{R^{(1)}}{r_\mu} + \dfrac{R^{(2)}}{r_o} + \dfrac{R^{(1)}}{r_\mu}\dfrac{R^{(2)}_{(1)}}{r_o}} \tag{4.51}$$

Each of the port impedances in Eq. (4.51) is determined next.

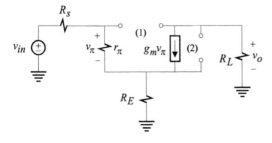

Figure 4.19

$\underline{\mathcal{R}^{(1)}}$: This is shown in Fig. 4.20 in which we see that a null in the output voltage causes a null in the output current which in turn causes (*i*) the test current i_T to be equal to $-\beta i_b$ and (*ii*) the sum of the voltage drops across R_E and r_n to be equal to v_T. Hence we have:

$$\left.\begin{aligned}v_T &= i_b r_\pi + (1 + \beta)i_b R_E\\ i_T &= -\beta i_b\end{aligned}\right\}$$

$$(4.52a, b)$$

These two yield:

$$\left.\begin{aligned}\mathscr{R}^{(1)} &= \frac{r_\pi + (1 + \beta)R_E}{\beta}\\[2mm] &= -\frac{r_e + R_E}{\alpha}\end{aligned}\right\}$$

$$(4.53a, b)$$

in which $r_e = r_\pi/(1 + \beta) = r_\pi(\alpha/\beta)$.

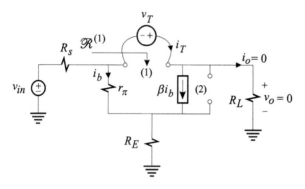

Figure 4.20

$\mathscr{R}^{(2)}$: This is shown in Fig. 4.21. As a result of the null in the output voltage, and hence in the output current, the test current i_T is equal to βi_b and the voltage across R_E, which is now given by $i_b R_E$, is equal to the negative of the test voltage v_T. Hence:

$$\left.\begin{aligned}i_T &= \beta i_b\\ v_T &= -i_b R_E\end{aligned}\right\}$$

$$(4.54a, b)$$

These two yield:

$$\mathscr{R}^{(2)} = -\frac{R_E}{\beta} \qquad\qquad (4.55)$$

$\mathscr{R}^{(1)}_{(2)}$: This is shown in Fig. 4.22. With port (2) shorted and the output voltage nulled, the voltage across R_E and R_L is zero and the test current flows entirely through r_π so that we have:

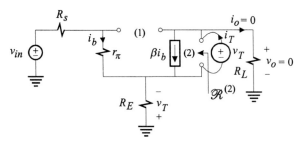

Figure 4.21

$$\mathcal{R}^{(1)}_{(2)} = r_\pi \tag{4.56}$$

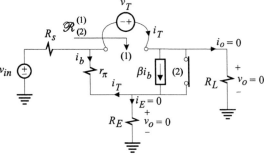

Figure 4.22

$R^{(1)}$: This is shown in Fig. 4.23. We shall now perform a few basic steps in our heads. First reflect r_π into the emitter circuit as $r_e = r_\pi/(1 + \beta)$ and let it combine with the emitter-resistor R_E. Next, designate $R_E + r_e$ as the extra element and apply the EET to the impedance function $R^{(1)}$ to obtain:

$$
\left.
\begin{aligned}
R^{(1)} &= R_0^{(1)} \dfrac{1 + \dfrac{\mathcal{R}^{(E)}}{R_E + r_e}}{1 + \dfrac{R^{(E)}}{R_E + r_e}} \\[3mm]
&= (R_s + R_L)\dfrac{1 + \dfrac{R_s \parallel R_L}{R_E + r_e}}{1 + \dfrac{R_s}{(1 + \beta)(R_E + r_e)}}
\end{aligned}
\right\} \tag{4.57a, b}
$$

The components of Eq. (4.57b) are derived as follows. When $R_E + r_e$ is taken as an open at port (E), the emitter, the base and the collector currents simultaneously become zero and the impedance seen by v_T at port (1) simply becomes $R_s + R_L$. The null impedance $\mathcal{R}^{(E)}$ looking back into the circuit from the emitter port is obtained by shorting port (1) and is given by $R_s \parallel R_L$. The ordinary impedance $R^{(E)}$

looking back from the emitter port is obtained by opening port (1) and is equal to the impedance in the base divided by $1 + \beta$, i.e. $R_s/(1 + \beta)$.

By factoring out $R_s \parallel R_L$ and $R_s/(1 + \beta)$ from the numerator and the denominator, respectively, we can write Eq. (4.57b) as:

$$R^{(1)} = R_L \frac{R_s \parallel (1 + \beta)(R_E + r_e)}{R_s \parallel R_L \parallel (R_E + r_e)} \tag{4.57c}$$

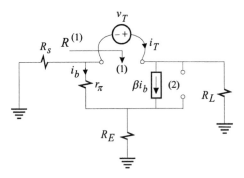

Figure 4.23

$R^{(2)}$: This is shown in Fig. 4.24. If R_E is designated as the extra element and taken out as a short, then application of the EET to the impedance function $R^{(2)}$ yields:

$$R^{(2)} = R_L \frac{1 + \dfrac{R_E}{R_L \parallel (R_s + r_\pi)}}{1 + \dfrac{R_E(1 + \beta)}{R_s + r_n}}$$

$$= R_L \frac{R_E \left\| \dfrac{R_s + r_\pi}{1 + \beta} \right.}{R_E \parallel R_L \parallel (R_s + r_\pi)} \tag{4.58a, b}$$

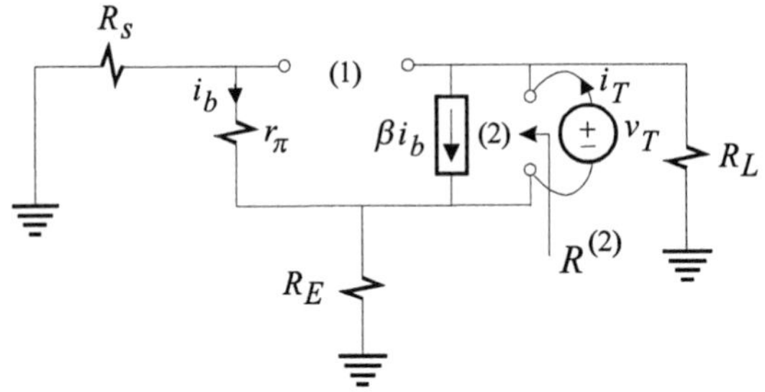

Figure 4.24

The derivation of the components in Eq. (4.58a) is very similar to the derivation of the components of (4.57b).

$R_{(1)}^{(2)}$: This is shown in Fig. 4.25 in which the parallel combination of r_π and the dependent current generator $i_b\beta$ acts as an effective resistance $r_e = r_\pi/(1 + \beta)$

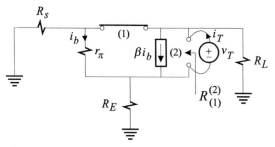

Figure 4.25

(which can also be easily recognized as the dynamic resistance of the diode formed by the base–emitter junction when the base and the collector are shorted.) It follows that:

$$R_{(1)}^{(2)} = r_e \parallel (R_E + R_s \parallel R_L) \tag{4.59}$$

Before substituting these results in Eq. (4.51), we simplify the product $R^{(1)}R_{(1)}^{(2)}$:

$$
\left.
\begin{aligned}
R^{(1)}R_{(1)}^{(2)} &= R_L R_s \parallel (1 + \beta)(R_E + r_e)\frac{r_e \parallel (R_E + R_s \parallel R_L)}{R_s \parallel R_L \parallel (R_E + r_e)} \\[2mm]
&= R_L \frac{R_s(1 + \beta)(R_E + r_e)}{R_s + (1 + \beta)(R_E + r_e)} \frac{r_e(R_E + R_s \parallel R_L)}{(R_s \parallel R_L)(R_E + r_e)} \\[2mm]
&= R_L r_\pi \frac{1 + \dfrac{R_E}{R_s \parallel R_L}}{1 + \dfrac{(1 + \beta)(R_E + r_e)}{R_s}}
\end{aligned}
\right\} \tag{4.60a–c}
$$

The expression of A_o in Eq. (4.51) is now given by:

$$A_o = -\frac{R_L}{\dfrac{R_E + r_e}{\alpha} + \dfrac{R_s}{\beta}}$$

$$
\times \frac{1 - \dfrac{1}{\alpha}\dfrac{r_e + R_E}{r_\mu} - \dfrac{1}{\beta}\dfrac{R_E}{r_o} - \dfrac{1}{\alpha}\dfrac{R_E}{r_o}\dfrac{r_e}{r_\mu}}{1 + \dfrac{R_L}{r_\mu}\dfrac{R_s \parallel [(R_E + r_e)(1 + \beta)]}{R_s \parallel R_L \parallel (R_E + r_e)} + \dfrac{R_L}{r_o}\dfrac{R_E\left\|\dfrac{R_s + r_\pi}{1 + \beta}\right.}{R_E \parallel R_L \parallel (R_s + r_\pi)} + \dfrac{R_L}{r_\mu}\dfrac{r_\pi}{r_o}\dfrac{1 + \dfrac{R_E}{R_s \parallel R_L}}{1 + \dfrac{(1 + \beta)(R_E + r_e)}{R_s}}}
$$

$$\tag{4.61}$$

Since $R_\mu, r_o \gg R_E, r_e$ the numerator always simplifies to unity. The denominator simplifies in one of two different ways depending upon the value of R_L. If the amplifier is current loaded so that $R_L \gg R_s$, R_E and $\beta \gg 1$, then A_o simplifies as:

$$A_o \approx -\frac{R_L}{R_E + r_e}$$

$$\times \frac{1}{1 + \dfrac{R_L}{r_\mu} \dfrac{1 + \dfrac{R_s}{R_E + r_e}}{1 + \dfrac{R_s}{(R_E + r_e)\beta}} + \dfrac{R_L}{r_o} \dfrac{1 + \dfrac{R_E}{R_s + r_\pi}}{1 + \dfrac{\beta R_E}{R_s + r_\pi}} + \dfrac{R_L}{r_\mu} \dfrac{r_\pi}{r_o} \dfrac{1 + \dfrac{R_E}{R_s}}{1 + \dfrac{\beta(R_E + r_e)}{R_s}}}$$

$$(4.62)$$

If R_L is a resistor, then in almost all cases $R_L \ll r_o, r_\mu$ so that the second denominator in Eq. (4.62) can be approximated as unity. $\qquad \square$

4.3   The 2-EET for dependent sources

Sometimes, it is desirable to determine the dependence of a transfer function on two dependent generators explicitly in the form of the 2-EET, as may be the case in a two-stage amplifier with feedback. In this section we shall discuss the form of the 2-EET for the arbitrary linear network shown in Fig. 4.26 in which two dependent sources, having gains A_1 and A_2, are applied at ports (1) and (2), respectively. These sources can be voltage or current sources controlled by internal currents or voltages, u_1 and u_2. As usual, a dependent generator can either be set to zero or to infinity, the latter being suited for feedback amplifiers.

If we let $A_1 = A_2 = 0$, then we obtain the reference circuit in Fig. 4.27, which has

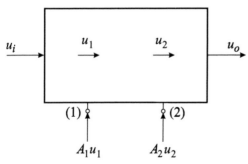

Figure 4.26

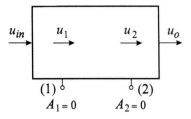

Figure 4.27

a gain H_o. According to the 2-EET, the complete transfer function can be expressed in terms of A_1 and A_2:

$$H = H_o \frac{1 + A_1 \mathcal{\bar{A}}^{(1)} + A_2 \mathcal{\bar{A}}^{(2)} + A_1 A_2 \mathcal{\bar{A}}^{(1)} \mathcal{\bar{A}}^{(2)}_{(1)}}{1 + A_1 \bar{A}^{(1)} + A_2 \bar{A}^{(2)} + A_1 A_2 \bar{A}^{(1)} \bar{A}^{(2)}_{(1)}}$$

(4.63)

in which the gains with bars over them are defined as:

$\mathcal{\bar{A}}^{(k)}_{(j)} \equiv$ The negative inverse gain, with respect to A_k, from port (k) to the controlling voltage or current, u_k, with the gain A_j in its opposite state and the response null. This is given by:

$$\mathcal{\bar{A}}^{(k)}_{(j)} = -\left. \frac{u_k}{u_T} \right|_{\substack{u_o = 0 \\ A_j \to opp.\,state}}$$

(4.64)

in which u_T is an independent test source of the same type as the dependent source (current or voltage) connected at port (k). The four possible cases of $\mathcal{\bar{A}}^{(k)}_{(j)}$ for two extra elements are shown in Fig. 4.28.

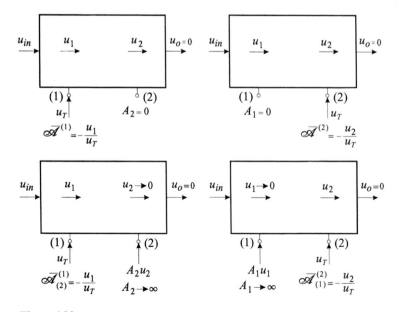

Figure 4.28

$\overline{A}_{(j)}^{(k)} \equiv$ The negative inverse gain, with respect to A_k, from port (k) to the controlling voltage or current, u_k, with the gain A_j in its opposite state and the excitation set to zero. This is given by:

$$\overline{A}_{(j)}^{(k)} = -\frac{u_k}{u_T}\bigg|_{\substack{u_{in}=0 \\ A_j \to opp.\ state}} \tag{4.65}$$

in which u_T is an independent test source of the same type as the dependent source connected at port (k). The four possible cases of $\overline{A}_{(j)}^{(k)}$ are shown in Fig. 4.29.

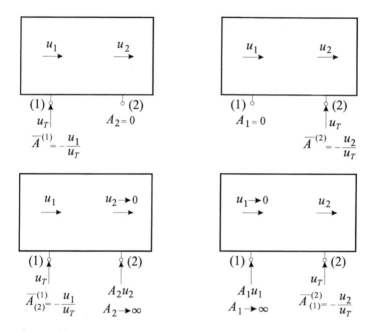

Figure 4.29

If we let both gains be infinite, then the 2-EET takes the following form:

$$H = H'_o \frac{1 + \dfrac{1}{A_1 \mathscr{A}^{(1)}} + \dfrac{1}{A_2 \mathscr{A}^{(2)}} + \dfrac{1}{A_1 A_2 \mathscr{A}^{(1)} \mathscr{A}_{(1)}^{(2)}}}{1 + \dfrac{1}{A_1 \overline{A}^{(1)}} + \dfrac{1}{A_2 \overline{A}^{(2)}} + \dfrac{1}{A_1 A_2 \overline{A}^{(1)} \overline{A}_{(1)}^{(2)}}} \tag{4.66}$$

in which H'_o is shown in Fig. 4.30 and is given by:

$$H'_o = \frac{u_o}{u_{in}}\bigg|_{\substack{A_1 \to \infty \\ A_2 \to \infty}} \tag{4.67}$$

The other two forms of the 2-EET are left as an exercise.

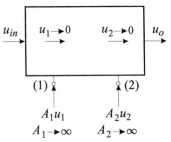

$$(1) \qquad\qquad (2)$$
$$A_1 u_1 \qquad\qquad A_2 u_2$$
$$A_1 \to \infty \qquad A_2 \to \infty$$

Figure 4.30

Example 4.8   Determine the voltage of the current feedback pair in Fig. 4.31 by applying the 2-EET to β_1 and β_2.

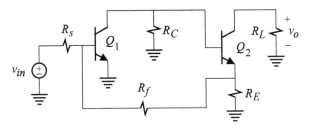

Figure 4.31

According to the 2-EET, the gain is given by:

$$A = A_o \frac{1 + \dfrac{1}{\beta_1 \bar{\mathscr{B}}^{(1)}} + \dfrac{1}{\beta_2 \bar{\mathscr{B}}^{(2)}} + \dfrac{1}{\beta_1 \beta_2 \bar{\mathscr{B}}^{(1)} \bar{\mathscr{B}}^{(2)}_{(1)}}}{1 + \dfrac{1}{\beta_1 \bar{B}^{(1)}} + \dfrac{1}{\beta_2 \bar{B}^{(2)}} + \dfrac{1}{\beta_1 \beta_2 \bar{B}^{(1)} \bar{B}^{(2)}_{(1)}}} \qquad (4.68)$$

Each component in Eq. (4.68) is determined next.

A_o: If we let $\beta_1, \beta_2 \to \infty$ then $i_{b_1} = i_{b_2} = 0$ as shown in Fig. 4.32. Since $i_{b_1} = 0$, the voltage drop across r_π is zero and the base of Q_1 is at virtual ground. It follows that the current through R_f is the same as the current through R_s, which is given by v_{in}/R_s. Because of the virtual ground at the base of Q_1, the voltage drop across R_f, which is given by $(v_{in}/R_s)R_f$, is equal to the voltage R_E so that the current through R_E is given by $(v_{in}/R_s)R_f/R_E$. Since $i_{b_2} = 0$, the collector current of Q_2, or the output current, is equal to its emitter current which is given by the sum of the currents in R_f and R_E:

$$i_o = \frac{v_{in}}{R_s} + \frac{v_{in}}{R_s} \frac{R_f}{R_E} \qquad (4.69)$$

It follows that:

$$A_o = \frac{R_L}{R_s}\left(1 + \frac{R_f}{R_E}\right) \tag{4.70}$$

$\mathscr{B}^{(1)}$: This is the null inverse gain with respect to β_1 with $\beta_2 \to \infty$ and the response nulled, i.e. $v_o = 0$. It is determined by replacing β_1 with an independent test current source, i_T, pointing in the opposite direction and solving for the

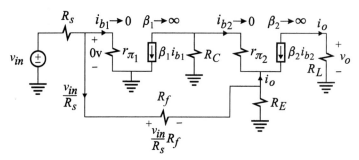

Figure 4.32

current i_{b_1} as shown in Fig. 4.33. Since the response is nulled, the output current and hence the collector, emitter and base currents of Q_2 are all zero. It follows that the voltage across R_C, which is given by $i_T R_C$, is the same as the voltage across R_E. The voltage at the base of Q_1 is given by the sum of the voltage drops across R_f and R_E so that i_{b_1} is given by:

$$i_{b_1} = \frac{1}{r_{\pi_1}}\left(i_T R_C + \frac{i_T R_C}{R_E}R_f\right) \tag{4.71}$$

It follows that:

$$\mathscr{B}^{(1)} = \frac{i_{b_1}}{i_T} = \frac{R_C}{r_{\pi_1}}\left(1 + \frac{R_f}{R_E}\right) \tag{4.72}$$

$\mathscr{B}^{(2)}$: This is the null inverse gain with respect to β_2 with $\beta_1 \to \infty$ and the response nulled, i.e. $v_o = 0$. It is determined by replacing β_2 with an independent test current source, i_T, pointing in the opposite direction and solving for the current i_{b_2} as shown in Fig. 4.34. Since the output voltage is nulled, the output current and $i_T = 0$. Since i_{b_2} remains finite because of v_{in}, it follows that:

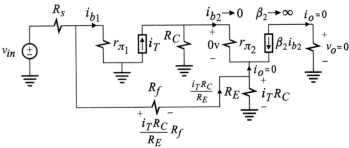

Figure 4.33

$$\mathcal{B}^{(2)} = \frac{i_{b2}}{i_T} \to \infty \tag{4.73}$$

$\mathcal{B}^{(2)}_{(1)}$: This is very similar to $\mathcal{B}^{(2)}$ except that in this case $\beta_1 = 0$. It can be seen in Fig. 4.34, however, that whether β_1 is zero or infinite, $i_T = 0$ and i_{b2} is finite so that:

$$\mathcal{B}^{(2)}_{(1)} = \frac{i_{b2}}{i_T} \to \infty \tag{4.74}$$

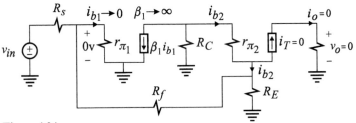

Figure 4.34

$\bar{B}^{(1)}$: This is the inverse gain with respect to β_1 with $\beta_2 \to \infty$ and with the excitation set to zero, i.e. $v_{in} = 0$. This is shown in Fig. 4.35. Since $\beta_2 \to \infty$, the base current of Q_2 and hence v_{π_2} are both zero. It follows that the voltage drop across R_C, which is given by $i_T R_C$, is equal to the voltage across R_E. The current i_{b_1} due to $i_T R_C$ is obtained after a current division between R_s and r_{π_1} of the current in R_f:

$$i_{b_1} = \frac{i_T R_C}{R_f + R_s \| r_{\pi_1}} \frac{R_s}{R_s + r_{\pi_1}} \tag{4.75}$$

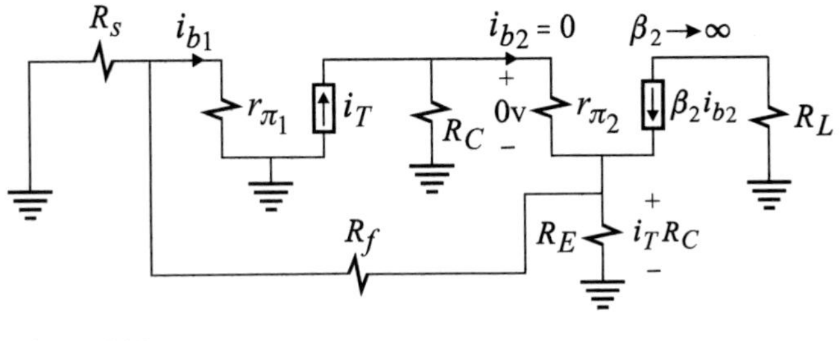

Figure 4.35

It follows that:

$$\bar{B}^{(1)} = \frac{R_C}{r_{\pi_1}} \frac{1}{1 + R_f/R_s \| r_{\pi_1}}$$ (4.76)

$\bar{B}^{(2)}$: This is inverse gain with respect to β_2 with $\beta_1 \rightarrow \infty$ and $v_{in} = 0$ as shown in Fig. 4.36. Since $i_{b_1} = 0$, the voltage at the base of Q_1 is zero so that there is no current flow through r_{π_1} and R_s, nor hence through R_f. It follows that the voltage across R_E and hence the emitter current of Q_2 are both zero, and that $i_T = i_{b_2}$ so that:

$$\bar{B}^{(2)} = 1$$ (4.77)

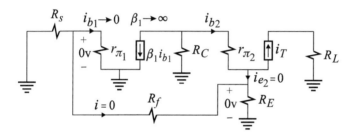

Figure 4.36

$\bar{B}^{(1)}_{(2)}$: Since $\bar{B}^{(2)}$ turned out so simple, it is preferable to determine $\bar{B}^{(1)}_{(2)}$ rather than $\bar{B}^{(2)}_{(1)}$ for the product term of the 2-EET. This is the inverse gain with respect to β_1 with $\beta_2 = 0$ and $V_{in} = 0$ as shown in Fig. 4.37. After two successive current divisions between R_C and $r_{\pi_2} + R_E \| (R_f + R_s \| r_{\pi_1})$ and R_s and r_{π_1} we obtain:

$$\bar{B}^{(1)}_{(2)} = \frac{1}{1 + \dfrac{r_{\pi_2} + R_E \| (R_f + R_s \| r_{\pi_1})}{R_C}} \frac{1}{1 + \dfrac{r_{\pi_1}}{R_s}}$$ (4.78)

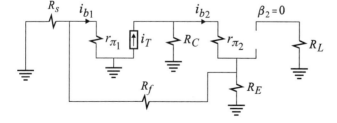

Figure 4.37

Substituting these results in Eq. (4.68) we obtain the gain A:

$$A_o = \frac{R_L}{R_s} \frac{1 + \dfrac{R_f}{R_E} + \dfrac{r_{\pi_1}}{\beta_1 R_C}}{\dfrac{1}{\alpha_2} + \dfrac{r_{\pi_1}}{\beta_1 R_C}\left(1 + \dfrac{R_f}{R_s \| r_{\pi_1}}\right) + \dfrac{1 + \dfrac{r_{\pi_1}}{R_s}}{\beta_1 \beta_2}\left[1 + \dfrac{r_{\pi_2} + R_E \| (R_f + R_s \| r_{\pi_1})}{R_c}\right]}$$

(4.79)

in which we have made use of $1/\alpha_2 = 1 + 1/\beta_2$. ▫

4.4   The NEET

The NEET[1,2] just for impedance elements has 2^N forms. If in addition to imped-ance elements, we include the four types of dependent sources, then the number of forms the NEET takes becomes unmanageable. In order to write a single expres-sion for the NEET for all possible cases, we need to generalize the concept of a linear element.

The *general linear element* α is defined as an element that relates any voltage and any current of a linear network through a simple dependency relation α given by:

$$\alpha = \frac{u_y}{u_x} \qquad (4.80)$$

in which:

$u_x \equiv$ A voltage or a current in a linear network.

$u_y \equiv$ A dependent voltage or a current source with gain α and controlled by u_x.

It is easily seen that α can be any of four types of dependent sources, as shown in Fig. 4.38a–d. In Fig. 4.38e we show how an ordinary resistive element R (or in general an impedance element Z) can be modeled by a current-controlled voltage source (CCVS) which depends on the current passing through it. Hence, for a resistor, $u_y = v_y$ and $u_x = i_x$ in Eq. (4.80) so that we have:

$$\alpha = R = \frac{v_y}{i_x} \qquad (4.81)$$

In Fig. 4.38f, we show the same resistor modeled as a conductive element (or in general an admittance element) using a voltage-controlled current source (VCCS) which depends on the voltage across it. In this case $u_y = i_y$ and $u_x = v_x$ in Eq. (4.80) so that we have:

$$\alpha = G = \frac{i_y}{v_x} \qquad (4.82)$$

Next we define the *reference element function* as:

$$e(\alpha) = \begin{cases} \alpha; & \text{if } \alpha = 0 \text{ in the reference circuit.} \\ 1/\alpha; & \text{if } \alpha \to \infty \text{ in the reference circuit.} \end{cases} \tag{4.83}$$

Hence, the value of the reference element function is always zero in the reference circuit. For convenience, the argument, α, will be dropped from $e(\alpha)$. We define next the *reference null inverse gain function*:

$$\mathscr{E}^{(i)}_{(j,k,\ldots,m)} = \begin{cases} \mathscr{A}^{(i)}_{(j,k,\ldots,m)}; & \text{if } \alpha_i = 0 \text{ in the reference circuit} \\ 1/\mathscr{A}^{(i)}_{(j,k,\ldots,m)}; & \text{if } \alpha_i \to \infty \text{ in the reference circuit} \end{cases} \tag{4.84}$$

in which:

$$\mathscr{A}^{(i)}_{j,k,\ldots,m)} \equiv \text{null inverse gain with respect to } \alpha_i \text{ while} \\ \alpha_j, \alpha_k, \ldots, \alpha_m \text{ are in their opposite state} \tag{4.85}$$

Similarly, we define the *reference inverse gain function*:

$$E^{(i)}_{(j,k,\ldots,m)} = \begin{cases} \overline{A}^{(i)}_{(j,k,\ldots,m)}; & \text{if } \alpha_i = 0 \text{ in the reference circuit} \\ 1/\overline{A}^{(i)}_{j,k,\ldots,m)}; & \text{if } \alpha_i \to \infty \text{ in the reference circuit} \end{cases} \tag{4.86}$$

in which:

$$\overline{A}^{(i)}_{j,k,\ldots,m)} \equiv \text{inverse gain with respect to } \alpha_i \text{ while} \\ \alpha_j, \alpha_k, \ldots, \alpha_m \text{ are in their opposite state} \tag{4.87}$$

Finally, we define the following index permutation symbol:

$$\varepsilon_{ijk\ldots,m} = \begin{cases} 1, & \text{if } i < j < k < \cdots < m \\ 0, & \text{otherwise} \end{cases} \tag{4.88}$$

Note that $\varepsilon_i = 1$ for $i = 1, 2, \ldots, N$. The NEET can now be written as:

$$H = H_o \frac{1 + e_i\varepsilon_i\mathscr{E}^{(i)} + e_ie_j\varepsilon_{ij}\mathscr{E}^{(i)}\mathscr{E}^{(j)}_{(i)} + e_ie_je_k\varepsilon_{ijk}\mathscr{E}^{(i)}\mathscr{E}^{(j)}_{(i)}\mathscr{E}^{(k)}_{(i,j)} \cdots + e_1e_2\cdots e_N\mathscr{E}^{(1)}\mathscr{E}^{(2)}_{(1)}\cdots\mathscr{E}^{(N)}_{(1,2,\ldots,N-1)}}{1 + e_i\varepsilon_iE^{(i)} + e_ie_j\varepsilon_{ij}E^{(i)}E^{(j)}_{(i)} + e_ie_je_k\varepsilon_{ijk}E^{(i)}E^{(j)}_{(i)}E^{(k)}_{(i,j)} \cdots + e_1e_2\cdots e_NE^{(1)}E^{(2)}_{(1)}\cdots E^{(N)}_{(1,2,\ldots,N-1)}} \tag{4.89}$$

in which summation over repeated indices of $e_ie_j\cdots e_m\varepsilon_{ij\ldots m}$ is assumed. According to Eq. (4.88) and the summation rule, there are $\binom{N}{1}$ terms in the summation $e_i\varepsilon_i$, $\binom{N}{2}$ terms in the summation $e_ie_j\varepsilon_{ij}$, etc. Since the order in which the elements, or the ports, are taken is not important, we have the following $k!$ redundancy relations for the null inverse gains (and the inverse gains with $\mathscr{E} \to E$) determined at k arbitrary ports, $n_1, n_2, \ldots, n_k$:

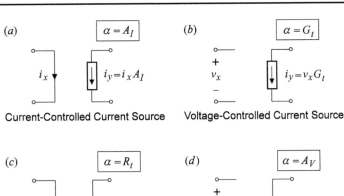

(a) $\boxed{\alpha = A_I}$

i_x $i_y = i_x A_I$

Current-Controlled Current Source

(b) $\boxed{\alpha = G_t}$

$+$
v_x $i_y = v_x G_t$
$-$

Voltage-Controlled Current Source

(c) $\boxed{\alpha = R_t}$

i_x $v_y = i_x R_t$

Current-Controlled Voltage Source

(d) $\boxed{\alpha = A_V}$

$+$
v_x $v_y = v_x A_V$
$-$

Voltage-Controlled Voltage Source

(e) $\boxed{\alpha = R}$

$\dfrac{v_y}{i_x} = R$ $i_x \quad + \quad v_y$ $\Longleftrightarrow$ $v_y = i_x R$

Resistor as a CCVS

(f) $\boxed{\alpha = G}$

$\dfrac{i_y}{v_x} = G$ $i_y \quad + \quad v_x$ $\Longleftrightarrow$ $v_x \quad i_y = v_x G$

Conductor as a VCCS

Figure 4.38

$$\mathscr{E}^{(n_1)} \mathscr{E}^{(n_2)}_{(n_1)} \cdots \mathscr{E}^{(n_k)}_{(n_1, n_2, \ldots, n_{k-1})} = \mathscr{E}^{(n_2)} \mathscr{E}^{(n_1)}_{(n_2)} \cdots \mathscr{E}^{(n_k)}_{(n_1, n_2, \ldots, n_{k-1})} \qquad (4.90)$$

$$= \mathscr{E}^{(n_3)} \mathscr{E}^{(n_2)}_{(n_3)} \cdots \mathscr{E}^{(n_k)}_{(n_1, n_2, \ldots, n_{k-1})}$$

$$= \quad \cdots$$

$$= \mathscr{E}^{(n_k)} \mathscr{E}^{(n_{k-1})}_{(n_k)} \cdots \mathscr{E}^{(n_1)}_{(n_{k-1}, n_{k-2}, \ldots, n_2)}$$

These redundancy relations are generalizations of the open-short theorem in Eqs. (4.13) and (4.14) discussed earlier. Quite often an indeterminacy of the type infinity-times-zero can occur when determining the product of these gains for a given order. Such an indeterminacy often can be removed simply by changing the order in which the ports or the elements are taken in the product (see Problem 4.5).

Equation (4.89) is the most general and compact form of the NEET and, as such, it is of limited analytical use simply because it is *very* general. To this end, all we need to know and appreciate is the intuitive procedure of the NEET and the structure of its components.

Example 4.9 Determine the denominator of the transfer function of the notch filter discussed in Examples 2.9 and 2.10 and shown here in Fig. 4.39. To obtain a notch response, the capacitors must all be equal and $R_3 = 6(R_1 + R_2)$.

The transfer function is of the form:

$$A(s) = A_o \frac{N(s)}{D(s)}$$

(4.91)

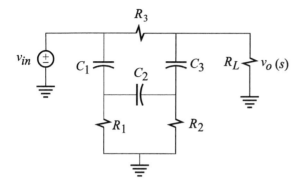

Figure 4.39

in which A_o is the low-frequency gain. The numerator $N(s)$ was determined using the method discussed in Example 2.9 and will be redetermined here in the next example using the 3-EET.

We designate the capacitors as the three extra elements and remove them to obtain the reference circuit in Fig. 4.40a whence we have:

$$A_o = \frac{1}{1 + R_3/R_L}$$

(4.92)

The denominator is determined by setting the excitation of the transfer function to zero, i.e. $v_{in} = 0$, as shown in Fig. 4.40b. Note in Fig. 4.40a that once v_{in} is shorted, R_L always appears in parallel with R_3 as shown in Fig. 4.40b. According to the 3-EET the denominator is given by:

$$D(s) = 1 + s[C_1 R^{(1)} + C_2 R^{(2)} + C_3 R^{(3)}]$$

(4.93)

$$+ s^2 [C_1 C_2 R^{(1)} R_{(1)}^{(2)} + C_1 C_3 R^{(1)} R_{(1)}^{(3)} + C_2 C_3 R^{(2)} R_{(2)}^{(3)}]$$

$$+ s^3 C_1 C_2 C_3 R^{(1)} R_{(1)}^{(2)} R_{(1,2)}^{(3)}$$

In what follows, we shall see how the entire denominator in Eq. (4.93) is determined simply by inspecting the circuit in Fig. 4.40b.

The impedance looking into any one port while the other two are in their reference states (open) is easily determined from Fig. 4.40b as follows:

$$R^{(1)} = R_1$$

(4.94)

$$R^{(2)} = R_2 + R_3 \| R_L \tag{4.95}$$

$$R^{(3)} = R_1 + R_2 \tag{4.96}$$

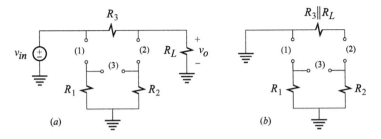

Figure 4.40

In the coefficient of s^2, we need to determine three additional port impedances. The impedance $R^{(2)}_{(1)}$ is determined looking into port (2) with port (1) shorted, or in its opposite state. In Fig. 4.40b this is seen to be the same as $R^{(2)}$ so that we have:

$$R^{(2)}_{(1)} = R^{(2)} = R_2 + R_3 \| R_L \tag{4.97}$$

The impedance $R^{(3)}_{(1)}$ is determined by looking into port (3) with port (1) shorted and is given by:

$$R^{(3)}_{(1)} = R_2 \tag{4.98}$$

The impedance $R^{(3)}_{(2)}$ is determined by looking into port (3) with port (2) shorted and is given by:

$$R^{(3)}_{(2)} = R_1 + R_2 \| R_3 \| R_L \tag{4.99}$$

Finally, in the coefficient of s^3 we need to determine $R^{(3)}_{(1,2)}$ which is the impedance looking into port (3) with ports (1) and (2) shorted. This is given by:

$$R^{(3)}_{(1,2)} = R_2 \| R_3 \| R_L \tag{4.100}$$

Substituting Eqs. (4.94)–(4.100) in (4.93) we obtain:

$$D(s) = 1 + a_1 s + a_2 s^2 + a_3 s^3 \tag{4.101}$$

in which:

$$\left.\begin{array}{l} a_1 = R_1(C_1 + C_2) + R_2(C_2 + C_3) + R_3 \| R_L C_2 \\[6pt] a_2 = R_1 R_2 (C_1 C_2 + C_1 C_3 + C_2 C_3) \\ \qquad + C_2 R_3 \| R_L (C_1 R_1 + C_3 R_1 + C_3 R_2) \\[6pt] a_3 = C_1 C_2 C_3 R_1 R_2 (R_3 \| R_L) \end{array}\right\} \tag{4.102a–c}$$

With $C_1 = C_2 = C_3 = C$, Eqs. (4.102a, b) become:

$$a_1 = [2(R_1 + R_2) + R_3 \parallel R_L]C$$

$$a_2 = [3R_1 R_2 + R_3 \parallel R_L(2R_1 + R_2)]C^2 \Bigg\}$$ (4.103a–c)

$$a_3 = R_1 R_2 (R_3 \parallel R_L)C^3$$

Now that we have determined $D(s)$ in low-entropy form, we must ascertain whether its roots correspond to three real poles or a pole and a quadratic. Since the roots of $D(s)$ correspond to the eigenvalues of the structure obtained by reducing the excitation to zero, we conclude that $D(s)$ must have three real roots because shorting v_{in} in this circuit yields a passive, third-order RC network. If the roots are all negative and well separated, say by a factor of three or more, then $D(s)$ can be factored analytically to an excellent approximation:

$$D(s) \approx (1 + a_1 s)\left(1 + \frac{a_2}{a_1} s\right)\left(1 + \frac{a_3}{a_2} s\right)$$ (4.104)

This approximation can be justified simply by expanding Eq. (4.104):

$$D(s) \approx 1 + s\left(a_1 + \frac{a_2}{a_1} + \frac{a_3}{a_2}\right)$$ (4.105)

$$+ s^2 a_2 \left[\frac{a_3}{a_2}\left(a_1 + \frac{a_2}{a_1}\right) + 1\right] + s^3 a_3$$

If the roots, as given by Eq. (4.104), are well separated, then we have:

$$a_1 > \frac{a_2}{a_1} > \frac{a_3}{a_2}$$ (4.106)

According to these inequalities, the coefficient of s in Eq. (4.105) is dominated by a_1 and the coefficient of s^2 is dominated by a_2 so that $D(s)$ in Eq. (4.104) or (4.105) is approximately the same as $D(s)$ in Eq. (4.101). The factored form of $D(s)$ in Eq. (4.104) is expressed in terms of poles:

$$D(s) = \left(1 + \frac{s}{\omega_1}\right)\left(1 + \frac{s}{\omega_2}\right)\left(1 + \frac{s}{\omega_3}\right)$$ (4.107)

in which:

$$\omega_1 \approx \frac{1}{a_1} = \frac{1}{C(\frac{1}{3}R_3 + R_3 \parallel R_L)}$$ (4.108)

$$\omega_2 \approx \frac{a_1}{a_2} = \frac{1}{CR_2} \frac{\frac{1}{3}R_3 + R_3 \parallel R_L}{3R_1 + R_3 \parallel R_L\left(1 + \frac{2R_1}{R_2}\right)}$$ (4.109)

$$\omega_3 \approx \frac{a_2}{a_3} = \frac{1}{CR_1 \parallel R_3 \parallel (\frac{1}{2}R_2) \parallel R_L} \tag{4.110}$$

When R_1 and R_2 are set up as a potentiometer, so that the circuit becomes a tunable notch filter, as shown in Fig. 2.16, we have:

$$\left.\begin{array}{l} R_1 = kR_3/6 \\ R_2 = (1-k)R_3/6 \end{array}\right\} \tag{4.111a, b}$$

in which recall that $R_1 + R_2 = R_3/6$. Substituting Eqs. (4.111) in (4.109) and (4.110) we obtain:

$$\omega_2 \approx \frac{4}{R_3C} \frac{1}{1-k} \frac{1 + 3\dfrac{R_3 \parallel R_L}{R_3}}{k + 2\dfrac{1+k}{1-k} \dfrac{R_3 \parallel R_L}{R_3}} \tag{4.112}$$

$$\omega_3 \approx \frac{1}{R_3C}\left(1 + \frac{6}{k} + \frac{12}{1-k} + \frac{R_3}{R_L}\right) \tag{4.113}$$

The transfer function in factored pole-zero form is now given by:

$$\frac{v_o(s)}{v_{in}(s)} = A_o \frac{\left(1 + \dfrac{s^2}{\omega_o^2}\right)\left(1 + \dfrac{s}{\omega_z}\right)}{\left(1 + \dfrac{s}{\omega_1}\right)\left(1 + \dfrac{s}{\omega_2}\right)\left(1 + \dfrac{s}{\omega_3}\right)} \tag{4.114}$$

in which ω_z and ω_o are given by Eqs. (2.55a) and (2.60) in Chapter 2 and are written here in terms of k:

$$\omega_o = \frac{1}{CR_3\sqrt{3k(1-k)}} \tag{4.115}$$

$$\omega_z = \frac{3}{R_3C} \tag{4.116}$$

Using the following numerical values, we shall compare the exact and the approximately factored transfer functions. For a 1.84-kHz notch filter feeding a load of $30\,k\Omega$ we have:

$R_1 = R_2 = 5\,k\Omega$

$R_3 = 60\,k\Omega$

$R_L = 30\,k\Omega$

$C = 0.01\,\mu F$

The numerical values of A_o, ω_z and ω_o are given by:

$$A_o = \frac{1}{3}, \omega_z = (2\pi)796 \,\text{rad/s}, \omega_o = (2\pi)1.838 \,\text{rad/s}$$

The numerical values of the coefficients a_i in the denominator in Eq. (4.101) are given by:

$$a_1 = 4 \times 10^{-4}, a_2 = 3.75 \times 10^{-8}, a_3 = 5 \times 10^{-13}$$

Note that these values satisfy the inequality in Eq. (4.106). The numerical values of the approximate poles in Eqs. (4.108)–(4.110) are given by:

$$\omega_1 \approx (2\pi)389 \,\text{rad/s}, \omega_2 \approx (2\pi)1698 \,\text{rad/s}, \omega_3 \approx (2\pi)11.94 \times 10^3 \,\text{rad/s}$$

The approximate and exact transfer functions are plotted in Fig. 4.41 and are seen to be in close agreement.

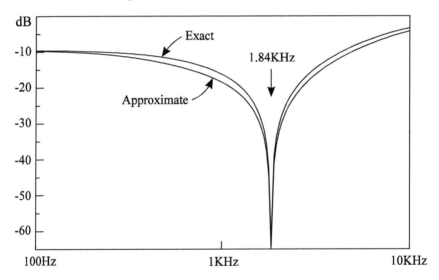

Figure 4.41

It is important to realize that the value of obtaining the analytical factors is not in trying to obtain a highly precise numerical result but rather in providing a good insight into the denominator.   □

Example 4.10   In this example we shall determine the numerator of the transfer in Section 4.3 using the 3-EET. As we shall find out, sometimes it is easier to determine the numerator using the technique described in Chapter 2 and it is a good idea to know both techniques equally well. The null driving-point impedances are determined next.

$\underline{\mathscr{R}^{(1)}}$: $\mathscr{R}^{(1)}$ is shown in Fig. 4.42*a* whence we see that a null in the output voltage causes a null in the output current and hence a null in i_3 so that the total voltage drop across R_3 and R_L is zero and the test voltage source appears across R_1. It follows that:

$$\mathcal{R}^{(1)} = R_1 \tag{4.117}$$

$\mathcal{R}^{(2)}$: This is shown in Fig. 4.42b whence we see that a null in the output voltage causes the test voltage source to appear directly across R_2. It follows that:

$$\mathcal{R}^{(2)} = R_2 \tag{4.118}$$

$\mathcal{R}^{(3)}$: This is shown in Fig. 4.42c whence we see that a null in the output voltage does not affect *this* calculation and that:

$$\mathcal{R}^{(3)} = R_1 + R_2 \tag{4.119}$$

It is interesting to observe that the only way to null the response in this case is by adjusting the excitation to zero. This circuit and the one in Fig. 4.42a represent a special case in which the test source has no influence on the response and the only way to null the response is by adjusting the excitation to zero.

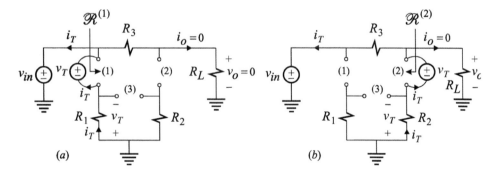

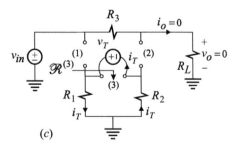

Figure 4.42

$\mathcal{R}^{(2)}_{(1)}$: This is shown in Fig. 4.43a whence we see that a null in the output voltage causes the test voltage source to appear directly across R_2. It follows that:

$$\mathcal{R}^{(2)}_{(1)} = R_2 \tag{4.120}$$

$\mathscr{R}^{(3)}_{(1)}$: This is shown in Fig. 4.43b in which we see that a null in the output voltage implies a zero voltage drop across R_3, which in turn implies that the voltage drop across R_1 is zero. Observe that this is in contrast with $\mathscr{R}^{(3)}$ above in which a null in the output did not result in a null across R_1. The path of the test current, i_T, is shown in Fig. 4.43b whence we have:

$$\mathscr{R}^{(3)}_{(1)} = R_2 \tag{4.121}$$

$\mathscr{R}^{(3)}_{(2)}$: This is shown in Fig. 4.43c in which we see that the null in the output voltage is copied on R_2 because of the short across port (2). The path of the test current, i_T, is shown in Fig. 4.43c whence we have:

$$\mathscr{R}^{(3)}_{(2)} = R_1 \tag{4.122}$$

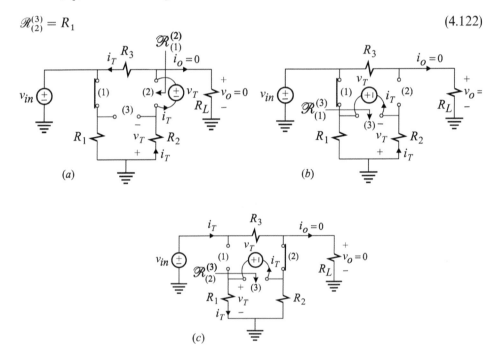

(a)   (b)   (c)

Figure 4.43

$\mathscr{R}^{(3)}_{(1,2)}$: This is shown in Fig. 4.44 in which by carefully tracing the path of the test current, i_T, we obtain:

$$\mathscr{R}^{(3)}_{(1,2)} = R_3 \tag{4.123}$$

Observe that one can easily get confused by v_T appearing across R_1 and conclude that R_3 should be in parallel with R_1 in Eq. (4.123). What is actually happening here is that the output is nulled by setting $v_T = v_{in}$ so that the current through R_1 is due to v_{in} and the current through R_3 is due to v_T. Both of these currents share a common path provided by the short across port (1).

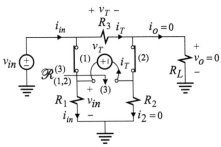

Figure 4.44

Substituting the above in the expression of the numerator given by the 3-EET, we get:

$$N(s) = 1 + s2(R_1 + R_2)C + s^2 3R_1R_2C^2 + s^3 R_1R_2R_3C^3 \tag{4.124}$$

If R_3 is chosen equal to $6(R_1 + R_2)$, then the numerator factors exactly as discussed in Example 4.2.   □

4.5   A proof of the NEET

In this section we shall give a direct proof of the most general form of the NEET in Eq. (4.89) rather than an indirect proof using induction.[2] For the arbitrary linear time invariant (LTI) network in Fig. 4.45, we can express the output signal, u_o, as a linear combination of the primary excitation, or input signal, u_{in}, and N other excitations, u_{yi}, applied to the system:

$$u_o = a_0 u_{in} + c^T u_y \tag{4.125}$$

in which $u_y^T = [u_{y1}, u_{y2}, \ldots, u_{yN}]$. The internal signals, u_{xi}, can also be expressed as a linear combination of u_{in} and u_{yi}:

$$u_x = b u_{in} + A u_y \tag{4.126}$$

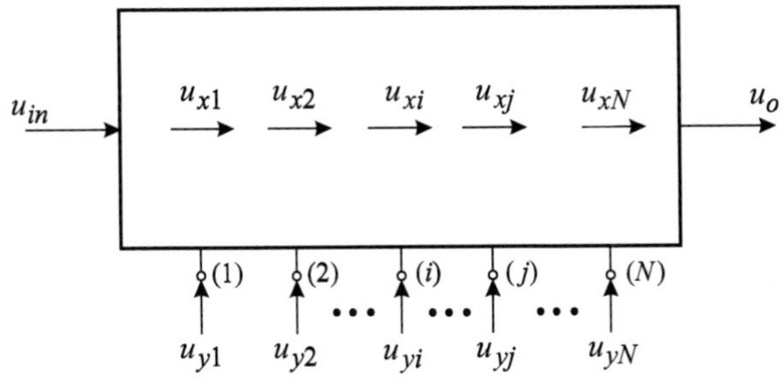

Figure 4.45

N arbitrary linear elements can be formed, if each excitation in the vector $\boldsymbol{u}_y$ is made to depend linearly on a corresponding internal signal in the vector $\boldsymbol{u}_x$:

$$u_y = \alpha u_x \tag{4.127}$$

in which α is a diagonal matrix whose elements, α_i, are the designated extra N generalized linear elements whose reference values are zero.

Substituting Eq. (4.127) in (4.125) and (4.126) we get:

$$\left.\begin{array}{r} u_o a_0^{-1} - a_0^{-1} c^T \alpha u_x = u_{in} \\ [I - \alpha A] u_x = b u_{in} \end{array}\right\} \tag{4.128a, b}$$

Equations (4.128a) and (4.128b) can be written in partitioned matrix form:

$$\begin{bmatrix} a_0^{-1} & -a_0^{-1} c^T \alpha \\ 0 & I - \alpha A \end{bmatrix} \begin{bmatrix} u_o \\ u_x \end{bmatrix} = \begin{bmatrix} 1 \\ b \end{bmatrix} u_{in} \tag{4.129}$$

According to Cramer's rule, the input–output transfer function H is given by:

$$H = \frac{u_o}{u_{in}} = \frac{\begin{vmatrix} 1 & -a_0^{-1} c^T \alpha \\ b & I - \alpha A \end{vmatrix}}{\begin{vmatrix} a_0^{-1} & -a_0^{-1} c^T \alpha \\ 0 & I - \alpha A \end{vmatrix}} \tag{4.130}$$

The determinant in the denominator can be easily expanded:

$$\begin{vmatrix} a_0^{-1} & -a_0^{-1} c^T \alpha \\ 0 & I - \alpha A \end{vmatrix} = a_0^{-1} |I - \alpha A| \tag{4.131}$$

In the determinant of the numerator we left-multiply the first row by b and subtract it from the second row to obtain:

$$\begin{vmatrix} 1 & -a_0^{-1} c^T \alpha \\ b & I - \alpha A \end{vmatrix} = \begin{vmatrix} 1 & -a_0^{-1} c^T \alpha \\ 0 & I - \alpha A + a_0^{-1} bc^T \alpha \end{vmatrix} = |I - \alpha(A - a_0^{-1} bc^T)| \tag{4.132}$$

(In factoring α as shown in the last step, we have used the fact that α is a diagonal matrix and hence it commutes with the matrix bc^T.) Substituting (4.131) and (4.132) in Eq. (4.130) we obtain the NEET in determinant form:

$$H = a_0 \frac{|I - \alpha(A - a_0^{-1} bc^T)|}{|I - \alpha A|} \tag{4.133}$$

We can see from this equation that if we set $\alpha = 0$, the transfer function reduces to a_0 which is the gain of the reference network in Fig. 4.46, i.e. $H_o = a_0$. Equation (4.133) can now be written as:

$$H = H_o \frac{|I - \alpha \mathcal{A}|}{|I - \alpha A|} \tag{4.134}$$

in which:

$$\mathcal{A} = A - a_0^{-1} bc^T \tag{4.135}$$

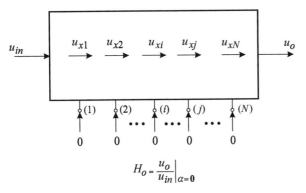

$$H_o = \left.\frac{u_o}{u_{in}}\right|_{\alpha=0}$$

Figure 4.46

If the determinants in the numerator and the denominator of Eq. (4.134) are carried out and the results interpreted in terms of the null and ordinary inverse gains with respect to α, the expression of NEET is obtained. This is shown next.

The matrix $[\boldsymbol{I} - \alpha\boldsymbol{A}]$ is given by:

$$[\boldsymbol{I} - \alpha\boldsymbol{A}] = (-1) \begin{bmatrix} \alpha_1 a_{11} - 1 & \alpha_1 a_{12} & \cdots & \alpha_1 a_{1N} \\ \alpha_2 a_{21} & \alpha_2 a_{22} - 1 & \cdots & \alpha_2 a_{2N} \\ \cdots & & & \\ \alpha_N a_{N1} & \alpha_N a_{N2} & \cdots & \alpha_N a_{NN} {}^{-1} \end{bmatrix} \tag{4.136}$$

Using the summation rule over repeated indices and the index permutation symbol $\varepsilon_{ijk\ldots m}$, $|\boldsymbol{I} - \alpha\boldsymbol{A}|$ can be expressed as:

$$|\boldsymbol{I} - \alpha\boldsymbol{A}| = 1 - \alpha_i \varepsilon_{ij}\Delta(i,j) - \alpha_i \alpha_j \alpha_k \varepsilon_{ijk}\Delta(i,j,k) - \cdots$$

$$- \alpha_1 \alpha_2 \cdots \alpha_N \Delta(1,2,\ldots,N) \tag{4.137}$$

in which

$\Delta(i,j,\ldots,m) \equiv$ Determinant of matrix obtained from matrix A by deleting all rows and columns different from $i,j,\ldots,m$. For example, the first three such determinants are:

$$\left.\begin{aligned} \Delta(i) &= a_{ii} \\[2mm] \Delta(i,j) &= \begin{vmatrix} a_{ii} & a_{ij} \\ a_{ji} & a_{jj} \end{vmatrix} \\[2mm] \Delta(i,j,k) &= \begin{vmatrix} a_{ii} & a_{ij} & a_{ik} \\ a_{ji} & a_{jj} & a_{jk} \\ a_{ki} & a_{kj} & a_{kk} \end{vmatrix} \end{aligned}\right\} \tag{4.138a–c}$$

These determinants will now be interpreted in terms of the ordinary inverse gains with respect to α with $u_{in} = 0$. According to Eq. (4.126), if we set $u_{in} = 0$, we obtain:

$$u_{xi} = a_{i1}u_{y1} + a_{i2}u_{y2} \cdots + a_{ii}u_{yi} \cdots + a_{iN}u_{yN} \tag{4.139}$$

Let u_{yi} in this equation be the only *independent* excitation while all the other u_{yn} are allowed to depend on u_{xn} through α_n as given by Eq. (4.127), i.e. $u_{yn} = \alpha_n u_{xn}$. Now, if we let all $\alpha_n = 0$, then all the dependent u_{yn} will vanish and according to Eq. (4.139) we get:

$$u_{xi} = a_{ii}u_{yi} \tag{4.140}$$

This can be written as:

$$-a_{ii} = \left.\frac{u_{xi}}{-u_{yi}}\right|_{u_{in}=0} \tag{4.141}$$

The right-hand side of Eq. (4.141) is immediately recognized to be the negative inverse gain with respect to α_i from port (i) to the controlling internal signal, u_{xi}, with $u_{in} = 0$ and all other elements, α_n, in their reference state. In fact, the right-hand side of Eq. (4.141) is an operational way of determining $-a_{ii}$ which, using the notation developed in the previous section, can be written as:

$$\bar{A}^{(i)} = -a_{ii} \tag{4.142}$$

The operation corresponding to $\bar{A}^{(i)}$ is shown in Fig. 4.47.

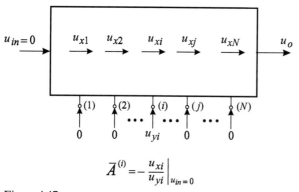

Figure 4.47

In Fig. 4.48, let u_{yj} be an independent excitation applied at port (j) and let all other α_n be in their reference states, i.e. $\alpha_n = 0$, except for α_i which will be in its opposite state, i.e. $\alpha_i \to \infty$. Now, except for u_{yj} (which is the independent excitation) and u_{yi}, all other $u_{yn} = 0$. Also note that $u_{xi} = 0$ because $u_{xi} = u_{yi}/\alpha_i$ and $\alpha_i \to \infty$ (this reminds us of infinite-gain opamp circuits in which the differential input voltage of the opamp is zero). According to Eq. (4.126), we have:

$$\left.\begin{array}{l} u_{xi} = a_{ij}u_{yj} + a_{ii}u_{yi} = 0 \\ u_{xj} = a_{jj}u_{yj} + a_{ji}u_{yi} \end{array}\right\} \tag{4.143a, b}$$

Simultaneous solution of these two yields:

$$\left.\frac{u_{xj}}{-u_{yj}}\right|_{\alpha_i \to \infty} = \frac{a_{jj}a_{ii} - a_{ji}a_{ij}}{a_{ii}}$$ (4.144)

The left-hand side of Eq. (4.144) is seen to be the negative inverse gain from port (j) to the controlling internal signal u_{xj} with $u_{in} = 0$ and the extra element α_i in its opposite state which, in terms of the notation adopted in Section 4.3, is simply $\bar{A}_{(i)}^{(j)}$. Equation (4.144) can be written as:

$$\left.(-a_{ii})\frac{u_{xj}}{-u_{yj}}\right|_{\substack{\alpha_i \to \infty \\ u_{in} = 0}} = -(a_{jj}a_{ii} - a_{ji}a_{ij})$$ (4.145)

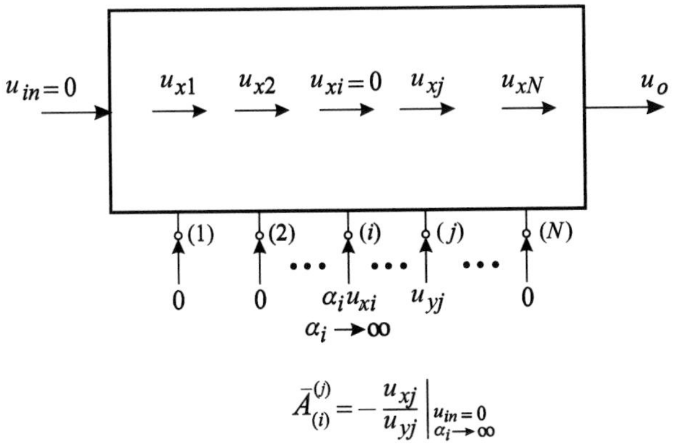

$$\bar{A}_{(i)}^{(j)} = -\left.\frac{u_{xj}}{u_{yj}}\right|_{\substack{u_{in} = 0 \\ \alpha_i \to \infty}}$$

Figure 4.48

which can be expressed in terms of $\bar{A}^{(i)}$ and $\bar{A}_{(i)}^{(j)}$ as:

$$\bar{A}^{(i)}\bar{A}_{(i)}^{(j)} = -\Delta(i,j)$$ (4.146)

Following the same procedure, we can show in general that:

$$\bar{A}^{(i)}\bar{A}_{(i)}^{(j)}\bar{A}_{(i,j)}^{(k)} \cdots \bar{A}_{(i,j,k,\ldots,m-1)}^{(m)} = -\Delta(i,j,k,\ldots,m)$$ (4.147)

The determinant in Eq. (4.137) can now be written as:

$$|I - \alpha A| = 1 + \alpha_i \varepsilon_i \bar{A}^{(i)} + \alpha_i \alpha_j \varepsilon_{ij} \bar{A}^{(i)} \bar{A}_{(i)}^{(j)} + \alpha_i \alpha_j \alpha_k \varepsilon_{ijk} \bar{A}^{(i)} \bar{A}_{(i)}^{(j)} \bar{A}_{(i,j)}^{(k)} + \cdots$$

$$+ \alpha_1 \alpha_2 \cdots \alpha_N \bar{A}^{(1)} \bar{A}_{(1)}^{(2)} \cdots \bar{A}_{(1,2,\ldots,N-1)}^{(N)}$$ (4.148)

This proves the denominator of the NEET in Eq. (4.89) when the reference values of all the extra elements are zero, i.e. $e_i = \alpha_i$ and $\mathcal{E}_{(i,j,k,\ldots)}^{(m)} = \bar{A}_{(i,j,k,\ldots)}^{(m)}$. When the reference value of some of the extra elements is taken as infinite, it can be shown that these elements and their corresponding inverse gains will appear as reciprocals in Eq. (4.148). For these elements, the reference element function and the reference inverse gain function in Eq. (4.89) are $e_i = 1/\alpha_i$ and $\mathcal{E}_{(i,j,k,\ldots)}^{(m)} = 1/\bar{A}_{(i,j,k,\ldots)}^{(m)}$.

The determinant in the numerator of Eq. (4.134) has the same expansion in Eq. (4.137) in which A is replaced with $\mathscr{A}$, which is given by:

$$
\mathscr{A} = \begin{bmatrix}
a_{11} - a_0^{-1}b_1c_1 & a_{12} - a_0^{-1}b_1c_2 & \cdots & a_{1N} - a_0^{-1}b_1c_N \\
a_{21} - a_0^{-1}b_2c_1 & a_{22} - a_0^{-1}b_2c_2 & \cdots & a_{2N} - a_0^{-1}b_2c_N \\
\cdots & \cdots & \cdots & \cdots \\
a_{N1} - a_0^{-1}b_Nc_1 & a_{N2} - a_0^{-1}b_Nc_2 & & a_{NN} - a_0^{-1}b_Nc_N
\end{bmatrix}
\tag{4.149}
$$

Applying the expansion in Eq. (4.137) to $|I - \alpha\mathscr{A}|$, we obtain:

$$
|I - \alpha\mathscr{A}| = 1 - \alpha_i\varepsilon_i\Delta'(i) - \alpha_i\alpha_j\varepsilon_{ij}\Delta'(i,j) - \alpha_i\alpha_j\alpha_k\varepsilon_{ijk}\Delta'(i,j,k) - \cdots
$$

$$
- \alpha_1\alpha_2\cdots\alpha_N\Delta'(1,2,\ldots,N)
\tag{4.150}
$$

in which

$\Delta'(i,j,\ldots,m) \equiv$ Determinant of the matrix obtained from matrix $\mathscr{A}$ by deleting all rows and columns different from $i,j,\ldots,m$. For example, the first three such determinants are:

$$
\left.
\begin{aligned}
\Delta'(i) &= a_{ii} - a_0^{-1}b_ic_i \\[2mm]
\Delta'(i,j) &= \begin{vmatrix}
a_{ii} - a_0^{-1}b_ic_i & a_{ij} - a_0^{-1}b_ic_j \\
a_{ji} - a_0^{-1}b_jc_i & a_{jj} - a_0^{-1}b_jc_j
\end{vmatrix} \\[2mm]
\Delta'(i,j,k) &= \begin{vmatrix}
a_{ii} - a_0^{-1}b_ic_i & a_{ij} - a_0^{-1}b_ic_j & a_{jk} - a_0^{-1}b_ic_k \\
a_{ji} - a_0^{-1}b_jc_i & a_{jj} - a_0^{-1}b_jc_j & a_{jk} - a_0^{-1}b_jc_k \\
a_{ki} - a_0^{-1}b_kc_i & a_{kj} - a_0^{-1}b_kc_j & a_{kk} - a_0^{-1}b_kc_k
\end{vmatrix}
\end{aligned}
\right\}
\tag{4.151a–c}
$$

These determinants will now be expressed in terms of null inverse gains. To do so, let u_{yi} be an independent external excitation applied at port (i) in addition to the input excitation u_{in} as shown in Fig. 4.49. Let all other u_{yn} be dependent on u_{xn} through α_n and set all these to their reference value, i.e. $u_{yn} = \alpha_n u_{xn} = 0$. Now, with the simultaneous application of u_{in} and u_{yi}, we can null the response, u_o, and write:

$$
\left.
\begin{aligned}
u_o &= a_0 u_{in} + c_i u_{yi} = 0 \\
u_{xi} &= b_i u_{in} + a_{ii} u_{yi}
\end{aligned}
\right\}
\tag{4.152a, b}
$$

From these, the negative of the null inverse gain from port (i) to the internal controlling signal u_{xi} follows immediately:

$$
\bar{\mathscr{J}}^{(i)} = \left.\frac{u_{xi}}{-u_{yi}}\right|_{u_o = 0} = -(a_{ii} - a_0^{-1}c_ib_i) = -\Delta'(i)
\tag{4.153}
$$

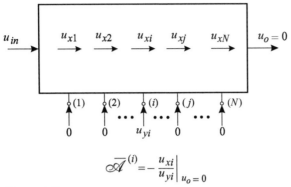

Figure 4.49

Proceeding in the same manner, we can show that:

$$\mathscr{A}^{(i)}\mathscr{A}^{(j)}_{(i)} = -\Delta'(i,j) \tag{4.154}$$

In general, it can be shown that:

$$\mathscr{A}^{(i)}\mathscr{A}^{(j)}_{(i)}\mathscr{A}^{(k)}_{(i,j)}\cdots\mathscr{A}^{(m)}_{(i,j,k,\ldots,m-1)} = -\Delta'(i,j,k,\ldots,m) \tag{4.155}$$

Substituting (4.154) and (4.155) in Eq. (4.150) we obtain the numerator of the NEET in (4.89) for the case when all the reference values of all the extra elements are zero. When the reference value of some of the extra elements is taken as infinite, these elements and their null inverse gain will appear as reciprocals in Eq. (4.148). Using the reference element function and the reference null inverse gain function we obtain the numerator in Eq. (4.89) for the general case.

4.6   Review

The NEET is a generalization of the EET in which N elements of an LTI circuit are removed by setting their values either to zero or to infinity for the purpose of determining a certain transfer function H in that circuit. The structure and the implementation of the NEET is very intuitive and simple. Once the designated N elements are removed, we obtain the reference circuit in which we determine the reference transfer function H_o corresponding to H. The values of the N elements are reinstated by performing two sets of calculations on the reference circuit. The first set of calculations is performed with the response of the transfer function nulled using null double injection and the second set of calculations is performed with the excitation of the transfer function set to zero. Each calculation corresponds to the inverse of the relation defining an extra element. For example, the inverse relation for a resistor connected across a port is the conductance looking into that port and the inverse relation for a transconductance connected at a port is the corresponding transresistance looking into that port. The transfer function H can be expressed in terms of H_o and the two sets of calculations as given by the NEET in Eq. (4.89). A very useful application of the NEET is in the determination of the frequency response of reactive networks in which all the reactive elements are removed and only the remaining purely resistive circuit is analyzed.

Problems

4.1 Open-short theorem for an ordinary impedance function. Using any of the two-port parameter representations of a linear network prove the result in Eq. (4.15). In Fig. 4.50, the z-parameter representation is shown.

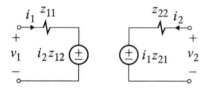

Figure 4.50

Note: Proof of the 2-EET, or the NEET, already constitutes a proof of this theorem (and the one in Problem 4.2) simply because the order in which the ports are taken in that proof is unimportant. This exercise is another proof.

4.2 Open-short theorem for a null impedance function. Repeat Problem 4.1 for a null impedance function for the network in Fig. 4.51.

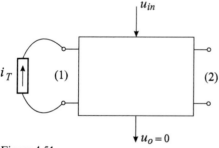

Figure 4.51

Hint: Recognize that a null double injection is equivalent to connecting an infinite gain amplifier with input u_o and output u_{in}.

4.3 Input impedance of an equalizer. Show that the input impedance of the equalizer filter in Fig. 4.16 is real and equal to R.

4.4 The contribution of an ideal $1:n$ transformer to a transfer function of an LTI circuit

(a) Consider the LTI circuit in Fig. 4.52a into which an ideal $1:n$ transformer is inserted (shunt–series) as shown Fig. 4.52b. Show that any transfer function A_o of the circuit in Fig. 4.52a is modified to A in Fig. 4.52b by a biquadratic factor in n:

$$A = A_o \frac{1 + a_1 n + a_2 n^2}{1 + b_1 n + b_2 n^2} \tag{4.156}$$

where:

$$
\left.
\begin{aligned}
a_1 &= \mathcal{N}_v + \mathcal{N}_i \\
a_2 &= \mathcal{N}_v \mathcal{N}_{i(n_v \to \infty)} \\
    &= \mathcal{N}_i \mathcal{N}_{v(n_i \to \infty)} \\
b_1 &= N_i + N_v \\
b_2 &= N_v N_{i(n_v \to \infty)} \\
    &= N_i N_{v(n_i \to \infty)}
\end{aligned}
\right\} \tag{4.157a–d}
$$

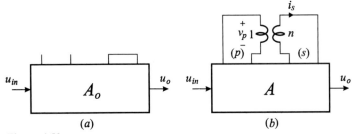

Figure 4.52

In these equations N and $\mathcal{N}$ are the inverse and null inverse gains with respect to the dependent voltage and current sources in the equivalent circuit model of an ideal transformer shown in Fig. 4.52c. For example, N_i is the inverse gain with respect to the dependent current source $n_i i_s$ with $n_v = 0$ and $\mathcal{N}_{v(n_i \to \infty)}$ is the null inverse gain with respect to the dependent voltage source $n_v v_p$ with $n_i \to \infty$. These are shown in Figs. 4.52d and e. Note that n_v and n_i are both equal to n but they are represented by different symbols in order to distinguish between the two sources.

Hint: Use the 2-EET for dependent sources.

Figure 4.52 (cont.)

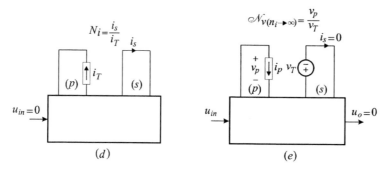

$$N_i = \frac{i_s}{i_T}$$

$$\mathcal{N}_{v(n_i \to \infty)} = \frac{v_p}{v_T}$$

(d)   (e)

Figure 4.52 (*cont.*)

(b) Show that if the same transformer above is inserted shunt–shunt in an LTI with a transfer function A_o as shown in Figs. 4.52f and g, then the transfer function is modified to:

$$A = A_o \frac{\mathcal{R}^{(s)}}{R^{(s)}} \frac{1 + a_1 n + a_2 n^2}{1 + b_1 n + b_2 n^2} \tag{4.158}$$

provided that $A \neq 0$ for $n = 0$ which is the same thing as saying $\mathcal{R}^{(s)} \neq 0$.

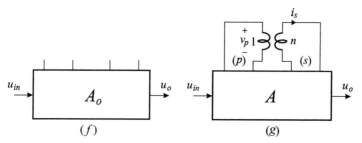

(f)   (g)

Figure 4.52 (*cont.*)

(c) Show that if the same transformer above is inserted series–series in an LTI with a transfer function A_o as shown in Figs. 4.52h and i, then the transfer function is modified to:

$$A = A_o \frac{\mathcal{R}^{(p)}}{R^{(p)}} \frac{1 + a_1 n + a_2 n^2}{1 + b_1 n + b_2 n^2} \tag{4.159}$$

provided that $A \neq 0$ for $n = 0$ which is the same thing as saying $\mathcal{R}^{(p)} \neq 0$.

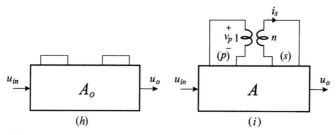

(h)   (i)

Figure 4.52 (*cont.*)

(d) If in a circuit, containing an ideal $1:n$ transformer, setting $n = 0$ or $n \to \infty$ reduces the desired transfer function to zero, you have the option of setting n_v or n_i to zero or infinity independently of each other in order to obtain a finite transfer function A_o and then apply the 2-EET. In this case show that the transfer function can be written as:

$$A = A_o \frac{n}{b_0 + b_1 n + b_2 n^2} \tag{4.160}$$

4.5 Indeterminacy! Determine the voltage gain of the two-stage CE amplifier with emitter-follower feedback in Fig. 4.53a in terms of β_1, β_2 and β_3 using the 3-EET for dependent sources and show that it is given by:

$$\frac{v_o}{v_{in}} = \left(1 + \frac{R_a}{R_b}\right)$$

$$\times \frac{1 + \dfrac{1}{\beta_3}\left(1 + \dfrac{r_{\pi 3} + R_a \| R_b}{R_E}\right) + \dfrac{1 + \beta_1}{\beta_1 \beta_2 \beta_3} \dfrac{1 + \dfrac{r_{\pi 2}}{R_C}}{1 + \dfrac{R_a}{R_b}}}{1 + \dfrac{1}{\beta_3} + \dfrac{1}{\beta_1 \beta_2} \dfrac{r_{\pi 1}}{R_b}\left(1 + \dfrac{R_a + R_b}{R_L}\right)\left(1 + \dfrac{r_{\pi 2}}{R_C}\right)\left\{1 + \dfrac{1}{\beta_3}\left[1 + \dfrac{r_{\pi 3} + R_b \|(R_a + R_L)}{R_E \| r_{\pi 1}}\right]\right\}}$$

$$+ \frac{1}{\beta_2 \beta_3}\left(1 + \frac{R_a + R_b \| r_{\pi 3}}{R_L}\right)\left(1 + \frac{r_{\pi 3}}{R_b}\right)\left(1 + \frac{r_{\pi 2}}{R_C}\right) \tag{4.161}$$

(a)

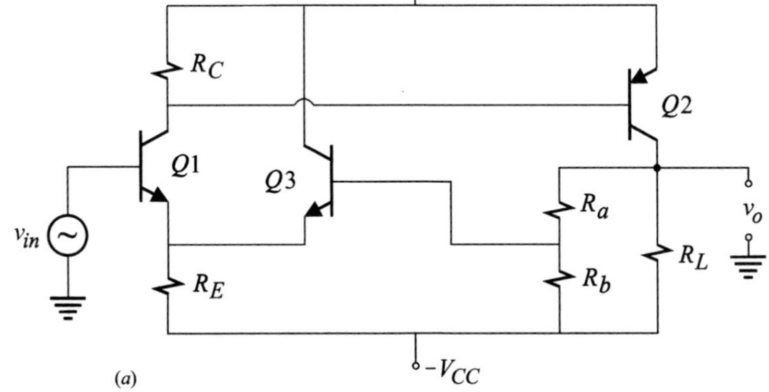

Figure 4.53

The equivalent circuit model is shown in Fig. 4.53b. In this problem you will encounter several indeterminacies. The first two indeterminacies are of the infinity-times-zero type and occur in the products $\bar{\mathcal{B}}^{(2)}\bar{\mathcal{B}}^{(3)}_{(2)}$ and $\bar{B}^{(2)}\bar{B}^{(3)}_{(2)}$. These indeterminacies can be removed by a simple change of order, i.e. by determining the products $\bar{\mathscr{B}}^{(3)}\bar{\mathscr{B}}^{(2)}_{(3)}$ and $\bar{B}^{(3)}\bar{B}^{(2)}_{(3)}$. The third indeterminacy is an infinity times zero which is *irremovable* by a change of order and concerns the product $\bar{B}^{(1)}\bar{B}^{(2)}_{(1)}$ or $\bar{B}^{(2)}\bar{B}^{(1)}_{(2)}$. Such an indeterminacy can be removed by introducing a resistor appropriately in the circuit so that $\bar{B}^{(1)}$ and $\bar{B}^{(2)}_{(1)}$ become finite. Later, the added resistor is allowed to vanish in the product $\bar{B}^{(1)}\bar{B}^{(2)}_{(1)}$.

Hint: When determining $\bar{B}^{(1)}$ and $\bar{B}^{(2)}_{(1)}$, add r_{o1} in parallel with the dependent generator $\beta_1 i_{b1}$ and show that these are given by:

$$\bar{B}^{(1)} = \frac{r_{o1}}{r_{\pi 1}} \tag{4.162}$$

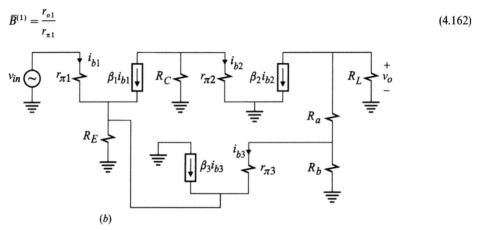

(b)

Figure 4.53 (cont.)

$$\bar{B}^{(2)}_{(1)} = \frac{R_L R_b}{R_L + R_a + R_b} \frac{1}{r_{o1} + R_C \| r_{\pi 2}} \frac{R_C}{R_C + r_{\pi 2}} \tag{4.163}$$

By letting $r_{o1} \to \infty$ in the product $\bar{B}^{(1)}\bar{B}^{(2)}_{(1)}$ show that:

$$\bar{B}^{(1)}\bar{B}^{(2)}_{(1)} = \frac{1}{r_{\pi 1}} \frac{R_L R_b}{R_L + R_a + R_b} \frac{R_C}{R_C + r_{\pi 2}} \tag{4.164}$$

Verify the following null inverse gains:

$$\bar{\mathscr{B}}^{(1)}, \bar{\mathscr{B}}^{(2)}, \bar{\mathscr{B}}^{(1)}_{(3)} \to \infty$$

$$\bar{\mathscr{B}}^{(1)}_{(2,3)}, \bar{\mathscr{B}}^{(1)}_{(2)} \to 1$$

$$\bar{\mathscr{B}}^{(3)}_{(2)} \to 0$$

$$\bar{\mathscr{B}}^{(3)} = \frac{1}{1 + \dfrac{r_{\pi 3} + R_a \| R_b}{R_E}} \tag{4.165a-e}$$

$$\bar{\mathscr{B}}^{(2)}_{(3)} = \frac{R_C}{R_C + r_{\pi 2}} \left(1 + \frac{R_a}{R_b}\right) \left(1 + \frac{r_{\pi 3} + R_a \| R_b}{R_E}\right)$$

Also, verify the following inverse gains:

$$\bar{B}^{(1)}, \bar{B}^{(2)} \to \infty$$

$$\bar{B}^{(3)} = 1$$

$$\bar{B}^{(1)}_{(2)}, \bar{B}^{(2)}_{(1)}, \bar{B}^{(3)}_{(2)} \to 0$$

$$\bar{B}^{(2)}_{(3)} = \cfrac{1}{1 + \cfrac{R_a + R_b \parallel r_{\pi 3}}{R_L}} \cdot \cfrac{1}{1 + \cfrac{r_{\pi 3}}{R_b}} \cdot \cfrac{1}{1 + \cfrac{r_{\pi 2}}{R_C}}$$

$$\bar{B}^{(3)}_{(2,1)} = \cfrac{1}{1 + \cfrac{r_{\pi 3} + R_b \parallel (R_a + R_L)}{R_E \parallel r_{\pi 1}}}$$

(4.166a–e)

4.6 3-EET versus 1-EET. The gain of the amplifier in Problem 4.5 can be determined more easily if we do not care to express it explicitly in terms of β_1, β_2 and β_3. For example, if we designate R_E as the extra element and set it to zero, show that we can write the gain by inspection:

$$A_o = \frac{\beta_1}{r_{\pi 1}} \frac{R_C \beta_2}{R_C + r_{\pi 2}} R_L \parallel (R_a + R_b \parallel r_{\pi 2}) \tag{4.167}$$

Invoking the EET for R_E the gain of the amplifier can be written as:

$$A = A_o \frac{1 + \cfrac{R_E}{\mathscr{R}^{(E)}}}{1 + \cfrac{R_E}{R^{(E)}}} \tag{4.168}$$

Show that:

$$\mathscr{R}^{(E)} = \left(r_{e3} + \frac{R_b \parallel R_a}{1 + \beta_3} \right) \parallel \alpha_1 \beta_2 \frac{1 + R_a/R_b}{1 + r_{\pi 2}/R_C} (r_{\pi 3} + R_b \parallel R_a) \tag{4.169}$$

$$R^{(E)} = r_{e1} \parallel \left[r_{e3} + \frac{R_b \parallel (R_a + R_L)}{1 + \beta_3} \right] \parallel \left(1 + \frac{r_{\pi 3}}{R_b} \right) \left(1 + \frac{r_{\pi 2}}{R_C} \right) \left(1 + \frac{R_a + R_b \parallel r_{\pi 3}}{R_L} \right) \frac{\alpha_3 r_{e1}}{\alpha_1 \beta_3 \beta_2}$$

(4.170)

in which $r_{e1,3} = r_{\pi 1,3}/(1 + \beta_{1,3})$.

4.7 An RC network with an ideal transformer

(a) Using the results derived in Problem 4.4, show that for the circuit in Fig. 4.54a:

$$\frac{v_o(s)}{v_{in}(s)} = A_o \frac{1 + s/\omega_1}{1 + a_1 s + a_2 s^2} \tag{4.171}$$

in which:

$$A_o = \frac{R_L}{R_L + R_2 + R_1} \frac{1 - n}{1 + n(n-2)\dfrac{R_1}{R_1 + R_2 + R_L}}$$

$$= \frac{R_L(1 - n)}{R_L + R_2 + (n-1)^2 R_1}$$

$$\omega_1 = \frac{n-1}{nR_2C_2}$$

$$a_1 = \frac{C_1 R_1 \| (R_2 + R_L) + C_2 R_L \| (R_2 + R_1)\left(1 + n^2 \dfrac{R_1 \| R_2}{R_L}\right)}{1 + n(n-2)\dfrac{R_1}{R_1 + R_2 + R_L}}$$

$$a_2 = \frac{C_1 C_2 R_1 \| (R_2 + R_L) R_L \| (R_2 + R_1)}{1 + n(n-2)\dfrac{R_1}{R_1 + R_2 + R_L}}$$

$$(4.172a\text{–}d)$$

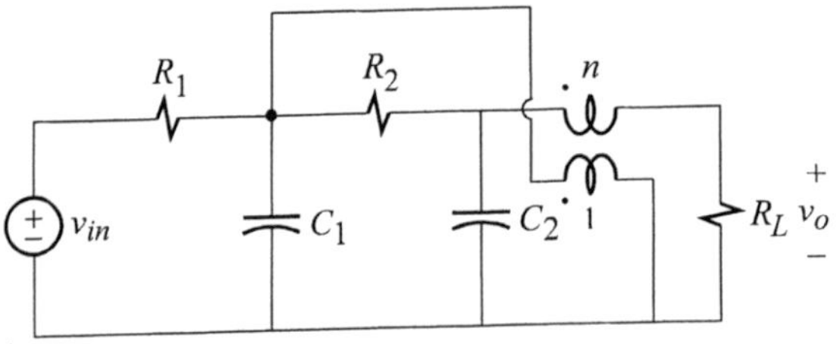

(a)
Figure 4.54

Hint: Use the 2-EET twice. First, designate C_1 and C_2 as the two extra impedance elements and take them out of the circuit as shown in Fig. 4.54b so that the transfer function, according to the 2-EET for impedance elements, can be written as:

$$A(s) = A_o \frac{1 + s[C_1 \mathcal{R}^{(1)} + C_2 \mathcal{R}^{(2)}] + s^2 C_1 C_2 \mathcal{R}^{(1)} \mathcal{R}^{(2)}_{(1)}}{1 + s[C_1 R^{(1)} + C_2 R^{(2)}] + s^2 C_1 C_2 R^{(1)} R^{(2)}_{(1)}} \qquad (4.173)$$

Next, apply the 2-EET for dependent sources to determine A_o and all the null and ordinary driving-point impedances in Eq. (4.173). Some of these calculations can be grouped together because they have many steps in common.

$A_o, R^{(1)}, R^{(2)}$: Each of these is a transfer function defined on the same circuit in Fig. 4.54b and, according to the 2-EET for dependent sources, can be expressed in terms of the turns-ratio of the transformer as:

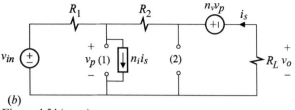

(b)

Figure 4.54 (cont.)

$$A_o = A_{o1} \frac{1 + b_1 n + b_2 n^2}{1 + \gamma_1 n + \gamma_2 n^2}$$

$$R^{(1)} = R_{o1} \frac{1 + c_1 n + c_2 n^2}{1 + \gamma_1 n + \gamma_2 n^2} \qquad (4.174a\text{--}c)$$

$$R^{(2)} = R_{o2} \frac{1 + e_1 n + e_2 n^2}{1 + \gamma_1 n + \gamma_2 n^2}$$

in which:

$$A_{o1} = \frac{R_L}{R_1 + R_2 + R_L}$$

$$R_{o1} = R_1 \,\|\, (R_2 + R_L) \qquad (4.175a\text{--}c)$$

$$R_{o2} = R_L \,\|\, (R_2 + R_1)$$

In Eq. (4.174c), e_2 is an indeterminate of the form $0 \times \infty$ which can be resolved only by connecting a resistance R_s in parallel with port (2) and letting it vanish in the product $e_2 = \lim_{R_s \to 0} \mathcal{N}_i \mathcal{N}_{v(n_i \to \infty)}$. This is shown in Fig. 4.54c.

$\mathcal{R}^{(1)}, \mathcal{R}^{(2)}, \mathcal{R}^{(2)}_{(1)}$: These are determined with $v_o = 0$ and are so simple that the 2-EET for dependent sources need not be used.

(b) Show that the maximum dc gain attainable is:

$$A_{o_{max}} = \sqrt{\frac{R_L/R_1}{1 + R_2/R_L}}$$

$$\qquad (4.176a, b)$$

$$n_{max} = 1 + \sqrt{\frac{R_L + R_2}{R_1}}$$

(c) Note that for $n = 1$, the transfer function changes from low-pass to band-pass. For the following element values, the magnitude of the transfer function is plotted in Fig. 4.54d for $n = 0, 1, n_{max}$:

$$R_1 = R_2 = 1\,\text{k}\Omega$$

$$R_L = 10\,\text{k}\Omega$$
$$C_1 = C_2 = 0.1\,\mu\text{F}$$

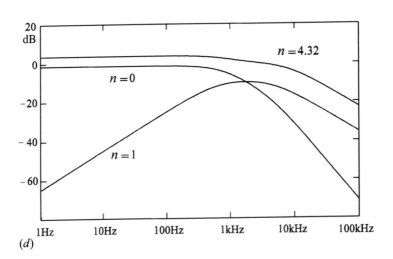

(c)

(d)

Figure 4.54 (*cont.*)

REFERENCES

1. S. Sabharwal, *The N-element Theorem: Part I and II*, Internal technical notes of students, T72 and T77, EE Department, Caltech.
2. R. D. Middlebrook, V. Vorpérian and J. Lindal, "The *N* Extra Element Theorem", *IEEE Transactions on Circuits and Systems – I: Fundamental Theory and Application*, Vol. 45, No. 9, Sept. 1998, pp. 919–935.

5 Electronic negative feedback
Low-entropy reformulation of US Patent 2,102,671

5.1 Introduction

Perhaps the best introductory remarks on electronic negative feedback are given by Harold S. Black[1] in his patent application on the *Wave Translation System*, assigned to Bell Telephone Laboratories on December 21, 1937. In this patent application not only do we learn how the inventor set out to solve all the major problems that plagued open-loop amplifiers at one fell swoop, but we also learn that negative and positive feedback were in fact employed prior to his invention to (a) build oscillators and regenerative feedback amplifiers and (b) prevent open-loop amplifiers, with parasitic positive feedback, from oscillating. Designers at the time knew quite well how to build an oscillator starting with an amplifier with a large gain A and feeding back a portion of the output in phase with the input, with a magnitude larger than the input. They also realized they could obtain very large gains from a single-stage amplifier by feeding back its output in phase with the input but with a magnitude less than the input signal – a technique known as regenerative feedback. As far as negative feedback was concerned, it was applied only in small amounts to prevent an amplifier from "singing", or oscillating, by countering the amount of positive feedback due to parasitic coupling between its input and output.

What Black discovered was that if negative feedback was applied in larger amounts than previously attempted, the closed-loop gain became independent of the gain A of the open-loop amplifier and depended only on the reciprocal of the gain of the feedback network, $1/\beta$, as long as the product of both gains was considerably larger than unity, i.e. $A\beta \gg 1$. This made the closed-loop gain independent of the distortion and variability of the gain of the open-loop amplifier. The importance of this discovery was that the feedback network was a simple passive network, often hardly more complicated than a passive voltage divider, which had none of the variability and nonlinearity of the vacuum tube amplifier. This huge discovery soon found applications in every system in which the response had to follow a control signal closely, and the field of automatic feedback control was born.

It can be speculated that the reason that the negative feedback amplifier was not discovered by anyone before Black was a psychological one. After all, vacuum tubes were the high-tech. devices of the time, which were supposed to do the amplification and not take second place to a low-tech. passive network. What led Black to his discovery were the requirements of the project he was assigned to, which was to design an amplifying system for the transcontinental telephone line. Such a system required *hundreds* of amplifiers in cascade so that it was clear to him from the outset that relying on vacuum tubes for gain was a lost cause – he literally had to find a way around the vacuum tube!

5.2 The EET for dependent sources and formulation of electronic feedback

In the introduction of his patent application Black[1] wrote ". . . the elements of a system with feedback are: an amplifying element having an input and an output; and a coupling or path for returning some of the output wave to the input of the amplifying element." In what follows, we will formulate this concept directly by applying the EET for dependent sources.

5.2.1 Gain analysis

Let A be the amplifying element in an electronic system with some kind of feedback path from output to input as shown in Fig. 5.1. The amplifying element can be any one of four dependent sources, while the input and output can be either a voltage or a current. As far as the feedback path is concerned we are not interested in its exact details: all we need to know is that there is another path from input to output other than the amplifying element A. What distinguishes this system from an arbitrary network is that the signal gain from input to output through the amplifying path A is by far greater than the signal gain through the feedback path. If this were not the case, then we would not have a system worthy of being called a feedback system and it would not deserve any special analytical treatment. Since A must be large, we consider it to be infinite and find out what is

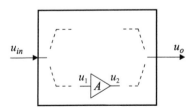

Figure 5.1

the best that feedback has to offer to the relationship between the input and output signals. Applying the EET for dependent sources, we obtain for the closed-loop transfer gain:

$$G \equiv \frac{u_o}{u_{in}} = \frac{u_o}{u_{in}}\bigg|_{A \to \infty} \frac{1 + \dfrac{1}{A\bar{\mathscr{A}}}}{1 + \dfrac{1}{A\bar{A}}} \tag{5.1}$$

The closed-loop gain with $A \to \infty$ can be thought of as the *ideal* closed-loop gain because it satisfies the primary objective of negative feedback, which is to make the closed-loop gain entirely independent of A and its variations. We shall define this as:

$$G_\infty = \frac{u_o}{u_{in}}\bigg|_{A \to \infty} \tag{5.2}$$

G_∞, $\bar{A}$ and $\bar{\mathscr{A}}$ are shown in Fig. 5.2a–c. Substituting Eq. (5.2) in (5.1) we obtain:

$$G \equiv G_\infty \frac{1 + \dfrac{1}{A\bar{\mathscr{A}}}}{1 + \dfrac{1}{A\bar{A}}} \tag{5.3}$$

Equation (5.3) yields the lowest entropy result for the flat gain of a feedback amplifier in which all the reactive elements have been removed. Hence, G in (5.3) may correspond to a low- or high-frequency asymptote, or a midband gain. To

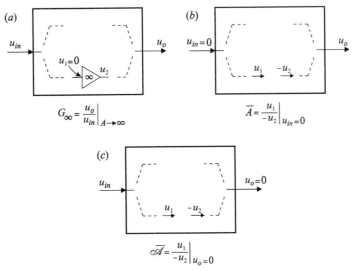

Figure 5.2

obtain the complete frequency response, we need to apply the NEET to determine the contribution of the reactive elements as explained in Chapter 4. Hence, the complete frequency response is given by:

$$G(s) = G_o \frac{N(s)}{D(s)} \tag{5.4}$$

in which G_o is determined according to Eq. (5.3). There is no useful general formulation of the effect of feedback on the frequency dependence of $G(s)$ for an arbitrary system. It is also possible to apply Eq. (5.3) to the complete response:

$$G(s) = G_\infty(s) \frac{1 + \dfrac{1}{A\mathscr{A}(s)}}{1 + \dfrac{1}{A\bar{A}(s)}} \tag{5.5}$$

Analytically, Eq. (5.5) will result in a far more complicated expression for $G(s)$ than (5.4) unless A is assumed to be infinite (as in the analysis of ideal operational amplifier circuits (see Problem 5.1)) so that $G(s) \approx G_\infty(s)$.

In the following example we shall demonstrate the application of Eqs. (5.3) and (5.4) to a single-stage FET feedback amplifier.

Example 5.1 We shall determine the gain of the FET feedback amplifier shown in Fig. 5.3a using the equivalent circuit model in Fig. 5.3b. First, we consider the circuit at low frequencies and determine the low-frequency asymptote A_o using Eq. (5.3). With the capacitor taken as an open circuit, we obtain the circuit in Fig. 5.3c

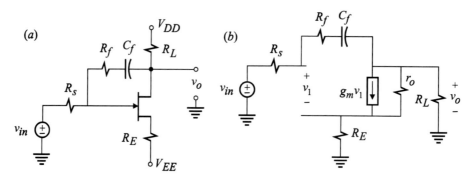

Figure 5.3

in which R_s is the feedback element. Hence, if we let $g_m \to \infty$, as shown in Fig. 5.3d, $v_1 \to 0$ and v_{in} appears directly across R_E so that the source current is given by v_{in}/R_E. Since the source and drain currents are equal, we have:

$$A_{o\infty} = -\frac{R_L}{R_E} \tag{5.6}$$

Next we determine the ordinary and null inverse gains with respect to g_m. In Fig. 5.3e, the null inverse gain is determined by nulling the output voltage. We can see,

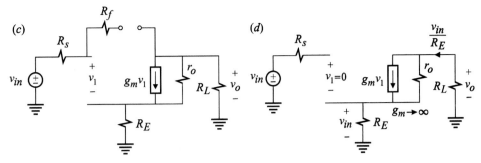

Figure 5.3 (*cont.*)

however, that the output voltage in this case cannot be nulled unless $i_T = 0$ so that the null inverse gain is infinite:

$$\mathcal{\widehat{G}}_m = \left. \frac{v_1}{i_T} \right|_{v_o = 0} \rightarrow \infty \tag{5.7}$$

In Fig. 5.3*f* the ordinary inverse gain is determined by setting $v_{in} = 0$ and replacing g_m with an independent current source pointing in the opposite direction. The voltage v_1 now appears across R_E so that the ordinary inverse gain is given by:

$$\bar{G}_m = \left. \frac{v_1}{i_T} \right|_{v_{in} = 0} = \frac{r_o R_E}{r_o + R_E + R_L} \tag{5.8}$$

Figure 5.3 (*cont.*)

Hence, according to (5.3) we have:

$$
\begin{aligned}
A_o &= A_{o\infty} \frac{1 + \dfrac{1}{g_m \mathcal{\widehat{G}}_m}}{1 + \dfrac{1}{g_m \bar{G}_m}} \\[2ex]
&= -\frac{R_L}{R_E} \frac{1}{1 + \dfrac{r_o + R_E + R_L}{g_m r_o R_E}}
\end{aligned}
\right\}
\tag{5.9a, b}
$$

The complete response is determined by reinstating C_f using the EET:

$$A(s) = A_o \frac{1 + sC_f \mathcal{R}^{(c)}}{1 + sC_f R^{(c)}} \tag{5.10}$$

The null resistance looking into port (c) with v_o nulled is given by:

$$\mathcal{R}^{(c)} = R_f + \mathcal{R}'^{(c)} \tag{5.11}$$

The null resistance $\mathcal{R}'^{(c)}$ is shown in Fig. 5.4a where we see:

$$\left. \begin{aligned} v_1 &= -(v_T + i_T R_E) \\[2mm] i_T &= g_m v_1 - \frac{i_T R_E}{r_o} \end{aligned} \right\} \tag{5.12a, b}$$

Simultaneous solution of Eqs. (5.12a) and (5.12b) yields:

$$\mathcal{R}'^{(c)} = -\frac{1}{g_m}\left(1 + g_m R_E + \frac{R_E}{r_o}\right) \tag{5.13}$$

It follows that:

$$\mathcal{R}^{(c)} = R_f - \frac{1}{g_m}\left(1 + g_m R_E + \frac{R_E}{r_o}\right) \tag{5.14}$$

The ordinary resistance looking into port (c) is given by:

$$R^{(c)} = R_f + R'^{(c)} \tag{5.15}$$

in which $R'^{(c)}$ is shown in Fig. 5.4b. We can determine $R'^{(c)}$ by nested application of the EET. Hence, if we let $g_m \to \infty$ as shown in Fig. 5.5a, the voltage across R_s, which is given by $i_T R_s$, appears across R_E and causes a current $i_e = i_T R_s / R_E$ to flow through it. The voltage across R_L is now given by $(i_T + i_e)R_L$ so that we can write:

$$v_T = i_T\left(1 + \frac{R_s}{R_E}\right)R_L + i_T R_s \tag{5.16}$$

It follows that:

$$R'^{(c)}_\infty = R_s + R_L\left(1 + \frac{R_s}{R_E}\right) = R_s\left(1 + \frac{R_L}{R_s \parallel R_E}\right) \tag{5.17}$$

According to the EET, we have:

$$R'^{(c)} = R'^{(c)}_\infty \frac{1 + \dfrac{1}{g_m \mathcal{G}'_m}}{1 + \dfrac{1}{g_m \overline{G}'_m}} \tag{5.18}$$

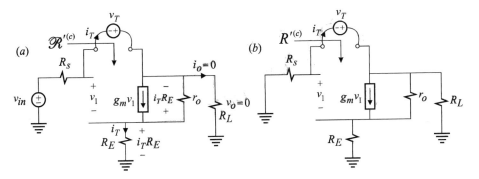

Figure 5.4

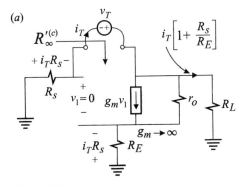

Figure 5.5

The null inverse gain $\mathscr{G}'_m$ is shown in Fig. 5.5b whence we have:

$$\mathscr{G}'_m = \frac{v_1}{i_T} = r_o \| (R_E + R_s \| R_L) \tag{5.19}$$

The ordinary inverse gain $\bar{G}'_m$ is shown in Fig. 5.5c whence we have:

$$\bar{G}'_m = \frac{v_1}{i_T} = \frac{r_o R_E}{r_o + R_E + R_L} \tag{5.20}$$

(b)   (c)

Figure 5.5 (cont.)

Substituting Eqs. (5.17), (5.19) and (5.20) in (5.18) and (5.15), we obtain:

$$R^{(c)} = R_f + R_s \left(1 + \frac{R_L}{R_s \| R_E}\right) \frac{1 + \dfrac{1}{g_m r_o \| (R_E + R_s \| R_L)}}{1 + \dfrac{r_o + R_E + R_L}{g_m r_o R_E}} \tag{5.21}$$

The complete frequency response in (5.10) is now written in pole-zero form:

$$A(s) = A_o \frac{1 + \dfrac{s}{\omega_z}}{1 + \dfrac{s}{\omega_p}} \tag{5.22}$$

The low-frequency asymptote, A_o, is given by Eq. (5.9) and ω_z and ω_p are given by:

$$\omega_z = \frac{1}{\mathscr{R}^{(c)} C_f}$$

$$\omega_p = \frac{1}{R^{(c)} C_f} \tag{5.23a, b}$$

where $\mathscr{R}^{(c)}$ and $R^{(c)}$ are given by Eqs. (5.14) and (5.21). By comparing the analytical expressions of $\mathscr{R}^{(c)}$ and $R^{(c)}$, we see that the pole occurs before the zero. Also note that $R^{(c)}$ could have been determined more easily by letting $g_m = 0$ in the EET (see Example 5.2). ☐

5.2.2   Driving-point analysis

In addition to analyzing the transfer gain of an amplifier, one is usually interested in determining the effect of feedback on the input and output impedance, or admittance, of an amplifier. Whereas the primary effect of feedback is to render the transfer gain of an amplifier insensitive to variations in A, its effect on the input and output impedance is exactly the opposite. We will show that upon application of feedback, the input and output impedances of an amplifier become dependent on A so that Eq. (5.3) is not suitable because G_∞ can be either zero or infinite. Hence, as a general rule, we shall use the following form of the EET for dependent sources to obtain the lowest entropy expression when performing driving-point analysis of a feedback amplifier:

$$G = G_0 \frac{1 + A \bar{\mathscr{A}}'}{1 + A \bar{A}'} \tag{5.24}$$

In Fig. 5.6 we show G_0, $\bar{A}'$ and $\bar{\mathscr{A}}'$ when G is a driving-point impedance. When G is a driving-point admittance, then $\bar{A}'$ and $\bar{\mathscr{A}}'$ in Figs. 5.6a and b are interchanged. In

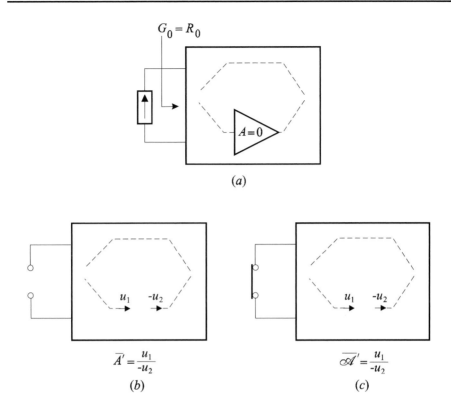

Figure 5.6

Eq. (5.24) we have used $\bar{A}'$ and $\mathscr{A}'$ in order to distinguish these from their counterparts in the expression of the transfer gain in (5.3). The complete frequency response is determined by augmenting the result in (5.24) using the NEET as in Eq. (5.4).

Example 5.2   We shall determine the output impedance of the FET amplifier in Example 5.1 to illustrate the application of Eq. (5.24). This is shown in Figs. 5.7a

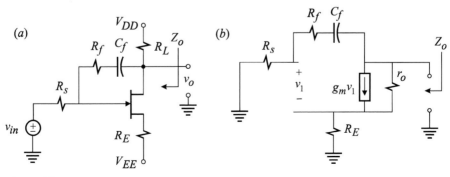

Figure 5.7

and b. First, we remove the capacitor as shown in Fig. 5.8a and determine the low-frequency asymptote of Z_o by the application of Eq. (5.24). Hence, we let $g_m = 0$ as shown in Fig. 5.8b and determine:

$$R_0 = r_o + R_E \qquad (5.25)$$

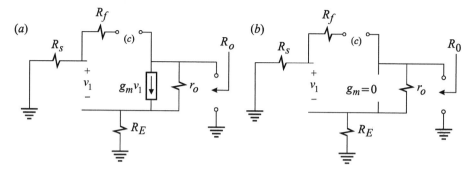

Figure 5.8

According to Eq. (5.24), we have:

$$R_o = R_0 \frac{1 + g_m \bar{\mathcal{G}}_m}{1 + g_m \bar{G}_m} \qquad (5.26)$$

in which $\bar{\mathcal{G}}_m$ and $\bar{G}_m$ are shown in Figs. 5.9a and b. In Fig. 5.9a, we can easily see that the null inverse gain is given by:

$$\bar{\mathcal{G}}_m = \frac{v_1}{i_T} = r_o \| R_E \qquad (5.27)$$

In Fig. 5.9b we can see that the ordinary inverse gain is zero:

$$\bar{G}_m = \frac{v_1}{i_T} = 0 \qquad (5.28)$$

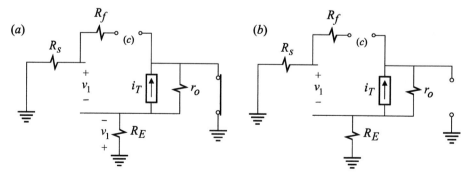

Figure 5.9

Substituting Eqs. (5.25), (5.27) and (5.28) in (5.26), we obtain:

$$R_o = (r_o + R_E)(1 + g_m r_o \| R_E) \qquad (5.29)$$

In this equation, we can see how the output resistance becomes dependent on the gain of the amplifying element g_m as a result of the feedback action caused by the resistance R_E.

Finally, we determine the frequency response of the output impedance by determining the null and ordinary resistances looking into the capacitive port as shown in Figs. 5.10a and b. According to the EET we have:

$$Z_o = R_o \frac{1 + sC\mathscr{R}^{(c)}}{1 + sCR^{(c)}} \qquad (5.30)$$

The null resistance $\mathscr{R}^{(c)}$ is shown in Fig. 5.10a and is given by:

$$\mathscr{R}^{(c)} = R_f + R_s \qquad (5.31)$$

The ordinary resistance $R^{(c)}$ is shown in Fig. 5.10b and can be deduced from the result obtained for Fig. 5.4b by letting $R_L \to \infty$ in Eq. (5.21). We shall, however, derive $R^{(c)}$ directly by applying the EET for g_m:

$$R^{(c)} = R^{(c)} \Big|_{g_m=0} \frac{1 + g_m \bar{\mathscr{G}}_m}{1 + g_m \bar{G}_m} \qquad (5.32)$$

If we let $g_m = 0$, we obtain from Fig. 5.10b:

$$R^{(c)} \Big|_{g_m=0} = R_f + R_s + r_o + R_E \qquad (5.33)$$

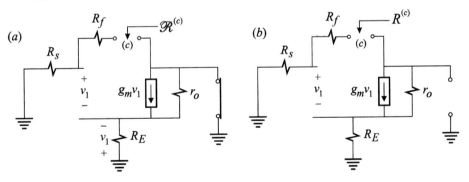

Figure 5.10

The null inverse gain $\bar{\mathscr{G}}_m$ in Eq. (5.32) is shown in Fig. 5.11a whence we have:

$$\bar{\mathscr{G}}_m = \frac{v_1}{i_T} = r_o \frac{R_s + R_E}{R_f + r_o + R_E + R_s} \qquad (5.34)$$

The ordinary inverse gain $\bar{G}_m$ in (5.32) is shown in Fig. 5.11b whence we see that:

$$\bar{G}_m = 0 \tag{5.35}$$

Substituting (5.33), (5.34) and (5.35) in Eq. (5.32), we obtain:

$$R^{(c)} = R_f + r_o + R_E + R_s + g_m r_o (R_s + R_E) \tag{5.36}$$

Hence, the output impedance in Eq. (5.30) is given by:

$$Z_o = R_o \frac{1 + \dfrac{s}{\omega_z}}{1 + \dfrac{s}{\omega_p}} \tag{5.37}$$

in which:

$$\left.\begin{array}{l} R_o = (r_o + R_E)(1 + g_m r_o \parallel R_E) \\[2mm] \omega_z = \dfrac{1}{\mathscr{R}^{(c)}C} = \dfrac{1}{(R_f + R_s)C} \\[4mm] \omega_p = \dfrac{1}{R^{(c)}C} = \dfrac{1}{[R_f + r_o + R_E + R_s + g_m r_o (R_s + R_E)]C} \end{array}\right\} \tag{5.38a–c}$$

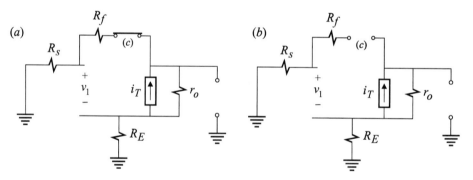

Figure 5.11

At high frequencies, when the capacitor acts as a short, both feedback paths become active. The output resistance in this case is obtained from Eq. (5.37) by letting $s \to \infty$:

$$R_{oh} = R_o \frac{\omega_p}{\omega_z} = \frac{(r_o + R_E)(1 + g_m r_o \parallel R_E)(R_f + R_s)}{R_f + r_o + R_E + R_s + g_m r_o (R_s + R_E)}$$

$$= (r_o + R_E) \parallel (R_f + R_s) \frac{1 + g_m r_o \parallel R_E}{\dfrac{R_s + R_E}{R_f + r_o + R_E + R_s}} \tag{5.39}$$

If we let $R_E = 0$ in this expression, we see that the effect of R_f (along with R_s) is to reduce the output resistance. Hence the two feedback paths have opposite effects on the output resistance. A similar analysis applies to the input impedance (see Problem 5.2). □

5.2.3 Loop gain

The product $A\bar{A}$ is defined as the loop gain:

$$T = A\bar{A} \tag{5.40}$$

In Figs. 5.12a and b we recognize this product to be the gain that the signal u_x experiences as it goes around the entire system with the input set to zero:

$$T = -\left.\frac{u_y}{u_x}\right|_{u_{in}\to 0} \tag{5.41}$$

Although there is no analytical difference between Figs. 5.12a and b, there is a practical difference. Breaking the loop and applying a signal u_x as shown in Fig. 5.12b is not physically practicable because the dc operating point of the circuit can drastically change. For example, in a transistor amplifier circuit, we simply cannot break the connection to the collector since that would disable the amplifier altogether. Injecting a signal u_z on the other hand, as shown in Fig. 5.12a, can be easily implemented in several practical ways (see Problem 5.3).

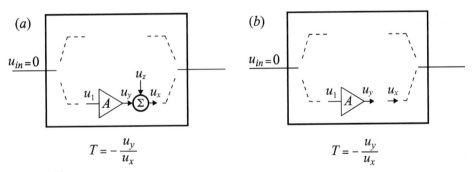

Figure 5.12

The concept of loop gain is central to any feedback system because the closed-loop transfer gain, driving-point impedance, or admittance, and stability can all be formulated in terms of the loop gain. To show this for the closed-loop transfer gain, we continue to expand Eq. (5.3) and obtain:

$$G = G_\infty \frac{A\bar{A}}{1 + A\bar{A}} + G_\infty \frac{\bar{A}}{\bar{\mathscr{A}}} \frac{1}{1 + A\bar{A}} \tag{5.42}$$

In the second term of Eq. (5.42), we immediately recognize that:

$$G_\infty \frac{\overline{A}}{\overline{\mathscr{A}}} = G_0 = \frac{u_o}{u_{in}}\bigg|_{A \to 0} \tag{5.43}$$

If G is a transfer gain, then G_0 is the amount of coupling between the input and output due to the forward gain through the feedback connection when the gain of the amplifying element is set to zero. This is shown in Fig. 5.13. If, however, G is a driving-point impedance (or admittance), then G_0 is known as the driving-point impedance (or admittance) of the open-loop amplifier modified by the loading of the feedback network but without the action of feedback. This will be explained in greater detail in Section 5.5. Substituting for T and G_0 in Eq. (5.42) we obtain:

$$G = G_\infty \frac{T}{1+T} + G_0 \frac{1}{1+T} \tag{5.44}$$

This formula is due to R. D. Middlebrook[2] and it is all that one needs to know to understand the *properties* of feedback amplifiers regarding transfer gain and

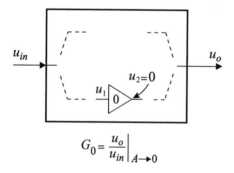

Figure 5.13

driving-point characteristics. Equation (5.44) does not yield a lower entropy expression than Eqs. (5.3) and (5.24) for the closed-loop gain and driving-point impedance, respectively, of a feedback amplifier. Its significance is that it shows how the actual closed-loop response, G, deviates from the ideal closed-loop response, G_∞, as a function of T and G_0.

Example 5.3 We shall determine the loop gain of the FET amplifier in the previous two examples using injection. Since the amplifying element is a dependent current source, we will inject an independent current source i_z in parallel with $g_m v_1$, as shown in Fig. 5.14 (see Problem 5.3), and determine the loop gain using:

$$T = -\frac{i_y}{i_x} \tag{5.45}$$

Note that in this expression of the loop gain, i_x is the excitation of T even though i_z

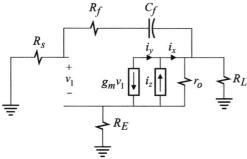

Figure 5.14

is the primary excitation. We shall designate the capacitor as the extra element and apply the EET to T in which i_y is the response and i_x is the excitation. In Fig. 5.15, we remove the capacitor and determine the low-frequency asymptote of T. In this figure, v_1 appears across R_E and is given by:

$$v_1 = i_x r_o \frac{R_E}{r_o + R_L + R_E} \tag{5.46}$$

The return current, or the response i_y, is given by:

$$i_y = -g_m v_1 \tag{5.47}$$

It follows that T_o is given by:

$$T_o = -\frac{i_y}{i_x} = \frac{g_m r_o}{1 + \dfrac{r_o + R_L}{R_E}} \tag{5.48}$$

According to the EET, the frequency response of T is given by:

$$T(s) = T_o \frac{1 + s\mathcal{R}^{(c)}C}{1 + sR^{(c)}C} \tag{5.49}$$

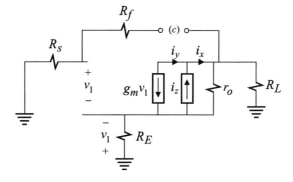

Figure 5.15

When determining $\mathscr{R}^{(c)}$ in Eq. (5.49), we null i_y as shown in Fig. 5.16a. Since $i_y = g_m v_1$, a null in i_y implies a null in v_1 so that, according to Fig. 5.16a, we have:

$$v_T = i_T \left(1 + \frac{R_s}{R_E} \right) R_L + i_T (R_s + R_f) \qquad (5.50)$$

It follows that:

$$\mathscr{R}^{(c)} = \frac{v_T}{i_T} = R_s + R_f + \left(1 + \frac{R_s}{R_E} \right) R_L \qquad (5.51)$$

When determining $R^{(c)}$ in Eq. (5.49), we set the excitation to zero, as shown in Fig. 5.16b, and obtain:

$$R^{(c)} = R_s + R_f + R_L \| (r_o + R_E) \qquad (5.52)$$

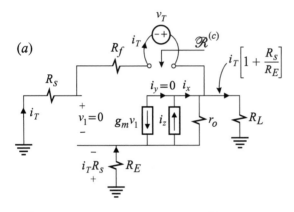

(a)

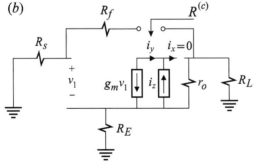

(b)

Figure 5.16

Substituting (5.52) and (5.51) in (5.49), we obtain:

$$T(s) = T_o \frac{1 + \dfrac{s}{\omega_z}}{1 + \dfrac{s}{\omega_p}} \qquad (5.53)$$

where:

$$\omega_z = \frac{1}{\mathcal{R}^{(c)}C} = \frac{1}{\left[R_s + R_f + \left(1 + \dfrac{R_s}{R_E}\right)R_L\right]C}$$

$$\omega_p = \frac{1}{R^{(c)}C} = \frac{1}{[R_s + R_f + R_L \| (r_o + R_E)]C} \qquad (5.54a\text{–}c)$$

$$T_o = \frac{g_m r_o}{1 + \dfrac{r_o + R_L}{R_E}}$$

We can see from the expressions of the pole and zero that the zero always occurs before the pole. □

5.3 Does this *circuit* have feedback or not? That is *not* the question

The main purpose of negative feedback, as mentioned earlier, is to desensitize a particular *transfer* gain, and not a "circuit", to variations in the gain of any amplifying element in that circuit. The question therefore is whether a certain response is being fed back properly or not. According to Eq. (5.44), when feedback is applied properly, the transfer gain of a certain response, G, should reduce to G_∞:

$$G = G_\infty \left(\frac{1}{1 + T^{-1}} + \frac{G_0}{G_\infty} \frac{1}{1 + T}\right)$$

$$\approx G_\infty \left(1 + \frac{G_0}{G_\infty} \frac{1}{T}\right) \qquad (5.55a\text{–}c)$$

$$\approx G_\infty$$

This is the desired outcome of negative feedback because it renders the closed-loop transfer gain independent of the gain of the amplifying element A. The first approximation in Eq. (5.55b) requires that the loop gain be much larger than unity:

$$T \gg 1 \qquad (5.56a)$$

The second approximation in (5.55c) follows automatically because the forward gain through the feedback network, G_0, must be small compared with the desired closed-loop gain G_∞, i.e:

$$\frac{G_0}{G_\infty} \ll 1 \qquad (5.56b)$$

Whereas the above argument holds for the transfer gain of a properly designed feedback amplifier, it does not hold for its input and output impedance. We know that the input or the output impedance of a strictly open-loop amplifier is entirely independent of the gain of the amplifying element A. Hence, feedback is not used to desensitize the input or output impedance of an amplifier to variations in A. In fact, application of feedback to an open-loop amplifier renders its input and output impedance dependent on the gain of the amplifying element A. This can be expressed analytically by showing $G_\infty = 0$ for the *inside* (looking into the amplifier past the source or load impedance) input and output driving-point characteristics of a feedback amplifier so that, according to Eq. (5.44), these are given by:

$$G = \frac{G_0}{1 + T'} \tag{5.57}$$

In Eq. (5.57) we have used T' to distinguish it from T used in the determination of the transfer gain. In all cases we will see that T' is derived from T simply by letting the source or the load impedance approach zero or infinity.

From a feedback point of view, Eq. (5.57) is a most undesirable outcome because it shows a closed-loop response whose variation is proportional to variations in A through the loop gain T'. This can be shown by taking partial derivatives in Eq. (5.57) and obtaining:

$$\frac{\partial G}{G} = -\frac{\partial A}{A} \tag{5.58}$$

in which we have made use of the fact that $T' \propto A$ and $T' \gg 1$. We recognize, however, that $\partial A / A$ is the sensitivity of the open-loop gain with respect to A. Hence we can say that application of feedback to an open-loop amplifier transfers the sensitivity of its open-loop gain with respect to A to its closed-loop input and output impedance. Even though this does not appear to be advantageous from a feedback control point of view, it is very useful for obtaining a very low-input impedance or conductance whose exact value may not matter once it is decreased beyond a certain point.

5.4   Gain analysis of feedback amplifiers

In this section, the four types of feedback amplifiers will be introduced and analyzed for gain using Eq. (5.3). A driving-point analysis of the same amplifiers will be given in Section 5.5. It should be made clear that the purpose of this and the following sections is not to derive any new formulas but simply to understand and demonstrate the various ways feedback can be applied to an open-loop amplifier.

Example 5.4 The feedback arrangement in the two-stage CE–CE amplifier shown in Fig. 5.17 is called *current-sampling–current-mixing* or simply *series–shunt* feedback. Before giving a detailed analysis, the basic operation of this feedback scheme is explained. Typically R_e is made much smaller than R_f so that the emitter current of Q_2 mostly flows through R_e. If we make the usual assumption that C_{E2} and all the other capacitors are shorts in the midband frequency range, then we see that the signal voltage across R_e is approximately $i_{e2}R_e$. Since the output current through R_L is very nearly equal to i_{e2}, the voltage across R_e is a copy of the output current. Hence, this type of feedback arrangement is called current sampling because it produces a signal, v_e, proportional to the output current. Next, we examine the input side at the base of Q_1 which is connected to the input through R_s and to the output through R_f. Hence, ignoring the bias resistors R_1 and R_2, the current entering the base of Q_1 is the difference between the source current and the current through the feedback path which is why this type of feedback is called current mixing. As discussed earlier, we now make the important observation that

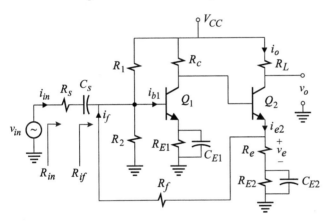

Figure 5.17

if the gain of the amplifying element in the forward path, β_1, is made very large, the current entering the base of Q_1, which is the error or difference signal, becomes nearly zero which in turn causes the signal voltage at the base of Q_1 to be nearly at ground potential. We can now quickly estimate the transfer gain of this amplifier by further observing that the current through R_f is approximately v_e/R_f and the current through R_s is v_{in}/R_s because of the virtual ground at the base of Q_1. Also, since the base current is nearly zero, the input and feedback currents are nearly identical so that we have:

$$\frac{v_e}{R_f} \approx -i_{in} \Rightarrow \frac{i_o R_e}{R_f} \approx -i_{in} \Rightarrow A_i \equiv \frac{i_o}{i_{in}} \approx -\frac{R_f}{R_e} \tag{5.59}$$

This is approximately the current gain, which is determined entirely by the feedback network. The approximate voltage gain follows from the fact that $i_{in} \approx v_{in}/R_s$ and $i_o = v_o/R_L$:

$$A_v \equiv \frac{v_o}{v_{in}} \approx -\frac{R_f}{R_e}\frac{R_L}{R_s} \tag{5.60}$$

An exact analysis will now follow using Eq. (5.3), which will yield a correction factor to the result in Eq. (5.60). The equivalent circuit diagram at midband frequencies is shown in Fig. 5.18 in which C_{E1}, C_{E2} and C_s are effectively short circuits. According to Eq. (5.3) the voltage gain is given by:

$$A_v \equiv \frac{v_o}{v_{in}} = A_{v\infty}\frac{1 + \dfrac{1}{A\tilde{\mathscr{A}}}}{1 + \dfrac{1}{A\bar{A}}} \tag{5.61}$$

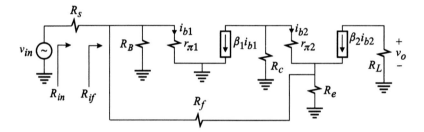

Figure 5.18

In order to determine $A_{v\infty}$, we let the amplifying element in the forward path become infinite. Although β_1 and β_2 are two such elements in the forward path, only β_1 qualifies because letting $\beta_2 \to \infty$ will not make the gain of the open-loop amplifier (the amplifier without R_f) become infinite because of the emitter feedback resistor, R_e, in the second stage. It is very important to realize that Eq. (5.61) will always yield a correct answer whether we choose β_1 or β_2. The disadvantage of choosing β_2 is that it will yield an answer which provides no useful analytical insight into the effect of feedback due to R_f.

In Fig. 5.19, we let $\beta_1 \to \infty$, which causes i_{b1}, v and hence i to be zero. It follows that the input current, given by v_{in}/R_s, flows entirely through the feedback path and impresses a voltage $-(v_{in}/R_s)R_f$ across R_e. The emitter current i_{e2} is then given by:

$$-i_{e2} = \frac{R_f}{R_e}\frac{v_{in}}{R_s} + \frac{v_{in}}{R_s} \tag{5.62}$$

Since the output voltage is given by $-\alpha_2 i_{e2} R_L$, $A_{v\infty}$ is given by:

$$A_{v\infty} = \frac{\alpha_2 R_L}{R_s}\left(1 + \frac{R_f}{R_e}\right) \tag{5.63}$$

In Fig. 5.20, we replace β_1 with an independent source pointing in the opposite direction and determine the null inverse gain $\bar{\mathscr{B}}$:

$$\bar{\mathscr{B}} = \frac{i_{b1}}{i_T} = \frac{R_c}{r_{\pi 1}}\left(1 + \frac{R_f}{R_e}\right) \tag{5.64}$$

in which we have made use of the fact that the voltage across $r_{\pi 1}$ is equal to the sum of the voltages across R_f and R_e.

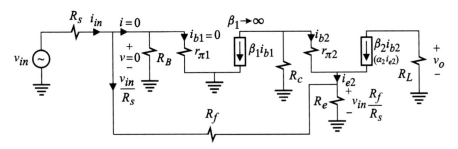

Figure 5.19

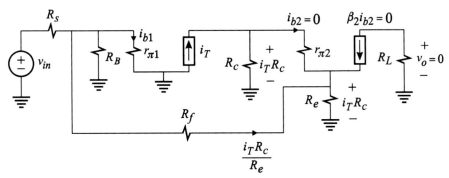

Figure 5.20

In Fig. 5.21a, we determine the inverse gain with $v_{in} = 0$. The path of the current i_T around the loop consists of several current divisions which can be determined most efficiently by taking the Norton equivalent looking back from the base of Q_1 as shown in Fig. 5.21b. It can easily be verified by inspection that the equivalent Norton source and resistance are given by:

$$i_N = i_T \frac{1 + \beta_2}{1 + \dfrac{r_{\pi 2} + (1 + \beta_2)(R_f \| R_e)}{R_c}} \cdot \frac{1}{1 + \dfrac{R_f}{R_e}}$$

$$R_N = R_B \| R_s \| \left(R_f + R_e \left\| \frac{r_{\pi 2} + R_c}{1 + \beta_2} \right. \right)$$

(5.65a, b)

(a)

(b)

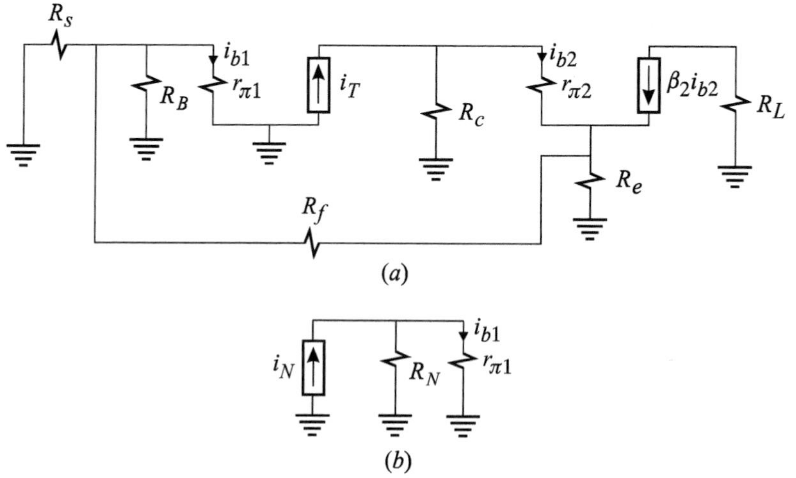

Figure 5.21

It follows that i_{b1} and, hence, $\bar{B}$ are given by:

$$i_{b1} = i_N \frac{1}{1 + \dfrac{r_{\pi 1}}{R_N}}$$

(5.66)

$$\bar{B} = \frac{i_{b1}}{i_T} = \frac{1 + \beta_2}{1 + \dfrac{r_{\pi 2} + (1 + \beta_2)(R_f \| R_e)}{R_c}} \cdot \frac{1}{1 + \dfrac{R_f}{R_e}}$$

$$\times \frac{1}{1 + \dfrac{r_{\pi 1}}{R_B \| R_s \| \left(R_f + R_e \left\| \dfrac{r_{\pi 2} + R_c}{1 + \beta_2} \right. \right)}}$$

(5.67)

With $A = \beta_1$, $\bar{A} = \bar{B}$ and $\mathscr{A} = \mathscr{B}$, the voltage gain according to Eq. (5.61) is given by:

$$A_v = \frac{\alpha_2 R_L}{R_s} \left[1 + \frac{R_f}{R_e} \right]$$

$$\times \frac{1 + \dfrac{1}{\beta_1} \dfrac{r_{\pi 1}}{R_c} \dfrac{1}{1 + R_f/R_e}}{1 + \dfrac{1 + R_f/R_e}{\beta_1(1 + \beta_2)}\left[1 + \dfrac{r_{\pi 2} + (1 + \beta_2)R_f \, \| \, R_e}{R_c}\right]\left[1 + \dfrac{r_{\pi 1}}{R_B \, \| \, R_s \, \| \, \left(R_f + R_e \left\| \dfrac{r_{\pi 2} + R_c}{1 + \beta_2}\right.\right)}\right]} \tag{5.68}$$

This is the exact expression of the voltage gain which for $\beta_1, \beta_2, R_f/R_e \gg 1$ reduces to the approximate expression derived in Eq. (5.60). Using numerical values, we can obtain useful approximations of the expressions derived above (see Problem 5.4). □

Example 5.5 The amplifier in Fig. 5.22 consists of a differential input stage, formed by Q_1 and Q_3, followed by a CE stage formed by Q_2. The output voltage is fed back through a voltage divider, R_{af} and R_{bf} into the base of Q_3. It can be seen that the voltage at the emitter of Q_3 or Q_1 is essentially a copy of the output voltage scaled by the voltage divider. This voltage is subtracted from the input voltage such that the difference between the two, less the voltage across R_s, appears across $r_{\pi 1}$ of Q_1 which is the non-inverting input to the differential amplifier stage. Hence, this type of feedback arrangement is called *voltage-sampling–voltage-mixing* or *shunt–series* feedback. Now observe that if we let β_2 become infinite, the signal current entering the base of Q_2 will become zero, which implies that the signal current of the collector of Q_1 becomes zero. This in turn implies that all the signal currents of Q_1 and Q_3 vanish simultaneously so that the signal voltages across R_s and the base-to-emitter junctions of Q_1 and Q_3 vanish too. We can now estimate

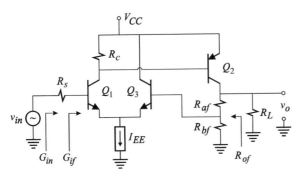

Figure 5.22

the voltage gain by recognizing that v_{in} appears directly across R_{bf} and causes a current v_{in}/R_{bf} to flow through it. Since the current into the base of Q_3 is zero, the current through R_{bf} flows through R_{af} so that the output voltage is given by $v_o = v_{in} + (v_{in}/R_{bf})R_{af}$. Hence the voltage gain for large β_2 is approximately given by:

$$A_v \approx 1 + \frac{R_{af}}{R_{bf}} \tag{5.69}$$

Note that we did not choose β_1 to become infinite because the resistance looking into the emitter of Q_3 prevents the voltage gain of Q_1 from becoming infinite.

The exact voltage gain is determined using the equivalent circuit model shown in Fig. 5.23. According to Eq. (5.3) the voltage gain is given by:

$$A_v = A_{v\infty} \frac{1 + \dfrac{1}{\beta_2 \mathscr{B}}}{1 + \dfrac{1}{\beta_2 \bar{B}}} \tag{5.70}$$

in which β_2 is chosen as the amplifying element of the open-loop amplifier. To determine $A_{v\infty}$, we let $\beta_2 \to \infty$, as shown in Fig. 5.24, in which we can see that:

$$\frac{v_{in}}{R_{bf}} = \frac{v_o}{R_{af} + R_{bf}} \Rightarrow A_{v\infty} = 1 + \frac{R_{af}}{R_{bf}} \tag{5.71}$$

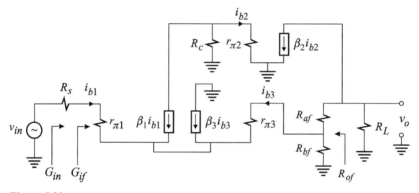

Figure 5.23

Next, $\bar{B}$ is determined as shown in Fig. 5.25. The current i_{b2} can be determined by a succession of current divisions. The resistance looking into the base of Q_3 is $r_{\pi3} + (1 + \beta_3)(r_{\pi1} + R_s)/(1 + \beta_1)$ so that i_{b3} is given by inspection:

$$i_{b3} = i_T \frac{R_L}{R_L + R_{af} + R_{bf} \left\| \left[r_{\pi3} + (1 + \beta_3)\dfrac{r_{\pi1} + R_s}{1 + \beta_1} \right] \right.}$$

$$\times \frac{R_{bf}}{R_b + r_{\pi3} + (1 + \beta_3)\dfrac{r_{\pi1} + R_s}{1 + \beta_1}} \tag{5.72}$$

The current i_{b2} is given by:

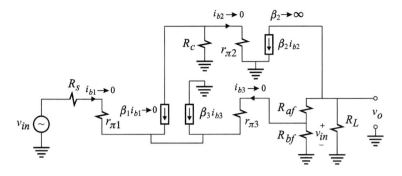

Figure 5.24

$$
\begin{rcases}
i_{b2} = -\alpha_1 i_e \dfrac{R_c}{R_c + r_{\pi 2}} \\[4mm]
\quad\;\; = \alpha_1 i_{b3}(1 + \beta_3)\dfrac{R_c}{R_c + r_{\pi 2}}
\end{rcases}
\qquad (5.73a,\,b)
$$

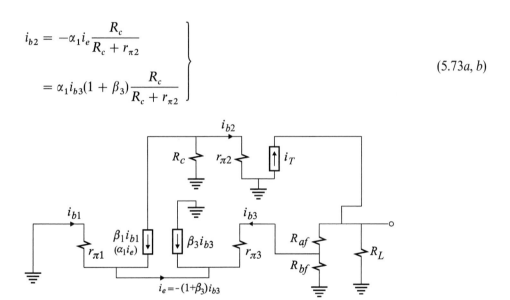

Figure 5.25

It follows that $\bar{B}$ is given by:

$$
\bar{B} = \frac{i_{b2}}{i_T} = \cfrac{\alpha_1(1 + \beta_3)}{1 + \cfrac{R_{af} + R_{bf} \left\| \left[r_{\pi 3} + (1 + \beta_3)\dfrac{r_{\pi 1} + R_s}{1 + \beta_1} \right] \right.}{R_L}}
$$

$$
\times \;\; \cfrac{1}{1 + \cfrac{r_{\pi 3} + (1 + \beta_3)\dfrac{r_{\pi 1} + R_s}{1 + \beta_1}}{R_{bf}}} \;\; \cfrac{1}{1 + \dfrac{r_{\pi 2}}{R_c}}
\qquad (5.74)
$$

If we assume that β_1, β_2 and β_3 are large and nearly equal, then Eq. (5.74) can be approximated as:

$$\bar{B} \approx \frac{\beta_3}{1 + \dfrac{R_{af} + R_{bf} \parallel (r_{\pi3} + r_{\pi1} + R_s)}{R_L}} \frac{1}{1 + \dfrac{r_{\pi3} + r_{\pi1} + R_s}{R_{bf}}} \frac{1}{1 + \dfrac{r_{\pi2}}{R_c}} \tag{5.75}$$

Next, $\bar{\mathscr{B}}$ is determined in Fig. 5.26 in which it can be seen that a null in the output voltage causes i_T to flow entirely through R_{af} and the voltage drop across R_{af} to appear directly across R_{bf}. It follows that i_{b3} is given by:

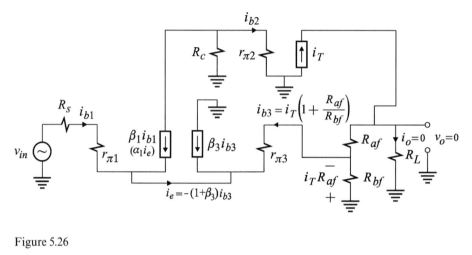

Figure 5.26

$$i_{b3} = i_T \left(1 + \frac{R_{af}}{R_{bf}} \right) \tag{5.76}$$

The relationship of i_{b2} to i_{b3} is still given by Eq. (5.73) so that substituting (5.76) in (5.73), we obtain:

$$\bar{\mathscr{B}} = \frac{i_{b2}}{i_T} = \alpha_1(1 + \beta_3) \frac{1 + \dfrac{R_{af}}{R_{bf}}}{1 + \dfrac{r_{\pi2}}{R_c}} \tag{5.77}$$

The closed-loop gain is given by Eq. (5.70) in which $A_{v\infty}$, $\bar{B}$ and $\bar{\mathscr{B}}$ are given by (5.71), (5.74) and (5.77), respectively. Using numerical values, we can obtain useful approximations for A_v (see Problem 5.5). □

Example 5.6   The load current of a two-stage CE amplifier is sensed using a $1:n$ transformer and a resistor R_{e1} as shown in Fig. 5.27 in which bias details are omitted. Since the voltage across R_{e1} is derived from the output current and is

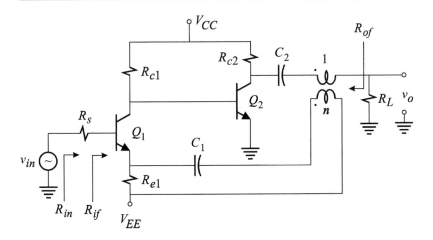

Figure 5.27

subtracted from the input voltage, this type of feedback is called *current-sampling–voltage-mixing* or simply *series–series* feedback. The small-signal equivalent circuit model is shown in Fig. 5.28 in which we have assumed an ideal $1:n$ transformer. In Fig. 5.29 we see that if we let $\beta_2 \to \infty$, the gain of the forward amplifier becomes infinite which in turn causes i_{b1} and i_{b2} to vanish causing the input voltage to appear directly across R_{e1} so that we have:

$$\frac{i_L}{n} = \frac{v_{in}}{R_{e1}} \tag{5.78}$$

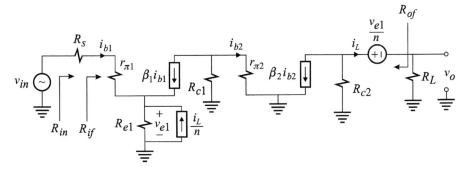

Figure 5.28

Hence, the output load current follows the input signal as determined by the feedback circuit (n and R_e) regardless of the load R_L and other variations in the amplifier. The ideal voltage gain follows from Eq. (5.78) and is given by:

$$A_{v\infty} = \frac{n}{R_{e1}} R_L = G_{t\infty} R_L \tag{5.79}$$

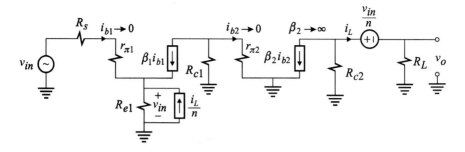

Figure 5.29

in which $G_{t\infty}$ is the ideal transconductance, i_L/v_{in}. Next, we determine the exact transconductance by applying the feedback formula:

$$G_t = G_{t\infty} \frac{1 + \dfrac{1}{\beta_2 \mathscr{B}}}{1 + \dfrac{1}{\beta_2 \bar{B}}} \tag{5.80}$$

The determination of $\bar{B}$ is shown in Fig. 5.30 in which i_T goes through a succession of current divisions which can be determined by inspection:

$$\bar{B} = \frac{R_{c2}}{R_{c2} + R_L + \dfrac{1}{n^2} R_{e1} \left\| \dfrac{r_{\pi 1} + R_s}{1 + \beta_1} \right.} \cdot \frac{1}{n} \cdot \frac{\alpha_1}{1 + \dfrac{R_s + r_{\pi 1}}{R_{e1}(1 + \beta_1)}} \cdot \frac{1}{1 + \dfrac{r_{\pi 2}}{R_{c1}}} \tag{5.81}$$

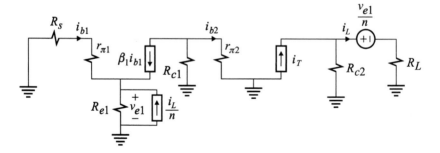

Figure 5.30

The determination of $\mathscr{B}$ is shown in Fig. 5.31 in which we see that a null in the output current causes i_T to flow entirely through R_{c2} resulting in a voltage drop of $i_T R_{c2}$ across R_{c2} which is equal to v_{e1}/n. Hence:

$$\frac{v_{e1}}{n} = i_T R_{c2} \Rightarrow v_{e1} = n i_T R_{c2} \tag{5.82}$$

The collector current of Q_1 is given by:

$$i_{c1} = \alpha_1 \frac{v_{e1}}{R_{e1}} = i_T \frac{n\alpha_1 R_{c2}}{R_{e1}} \tag{5.83}$$

It follows that the null inverse gain is given by:

$$\bar{\mathscr{B}} = \frac{i_{b2}}{i_T} = -\alpha_1 n \frac{R_{c2}}{R_{e1}} \frac{1}{1 + \dfrac{r_{\pi2}}{R_{c1}}} \tag{5.84}$$

Substituting (5.81) and (5.84) in Eq. (5.80), we obtain:

$$G_t = \frac{n}{R_{e1}} \cdot \frac{1 - \dfrac{1}{\beta_2 n \alpha_1} \dfrac{R_{e1}}{R_{c2}} \left(1 + \dfrac{r_{\pi2}}{R_{c1}}\right)}{1 + \dfrac{n}{\alpha_1 \beta_2} \left(1 + \dfrac{R_L}{R_{c2}}\right)\left(1 + \dfrac{R_s + r_{\pi1}}{R_{e1}(1 + \beta_1)}\right)\left(1 + \dfrac{r_{\pi2}}{R_{c1}}\right)} \tag{5.85}$$

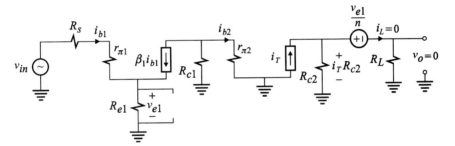

Figure 5.31

Typically $R_{e1}/R_{c2} < 1$ and $\alpha \approx 1$ so that Eq. (5.85) can be approximated as:

$$G_t \approx \frac{n}{R_{e1}} \cdot \frac{1}{1 + \dfrac{n}{\beta_2} \left(1 + \dfrac{R_L}{R_{c2}}\right)\left(1 + \dfrac{R_s + r_{\pi1}}{R_{e1}\beta_1}\right)\left(1 + \dfrac{r_{\pi2}}{R_{c1}}\right)} \tag{5.86}$$

Other approximations are possible depending on the particular set of component values (see Problem 5.6). ☐

Example 5.7 The amplifier shown in Fig. 5.32 is a current-loaded cascode amplifier with a feedback connection from the output voltage to the input current at the base of Q_1. The details of the bias circuit are omitted. Since the signal voltage at the base of Q_1 is small in comparison with the output voltage, the fedback current is proportional to the output voltage. We can also see that the current entering the base of Q_1 which drives the amplifier stage is equal to the difference between the input source current and the fedback current. Hence this type of feedback arrangement is called *voltage-sensing–current-mixing* or simply *shunt–shunt* feedback. We can easily estimate the voltage gain by realizing that the signal voltage at the base of Q_1 is essentially at ground potential in comparison with v_{in} or v_o so that the input current is approximately $i_{in} \approx v_{in}/R_s$ while the fedback current is approximately $i_f \approx v_o/R_f$. Assuming the amplifier stage has a large open-loop gain, the current entering the base of Q_1 can be ignored so that $i_f \approx -i_{in}$ which leads to:

$$\frac{v_o}{v_{in}} \approx -\frac{R_f}{R_s} \qquad (5.87)$$

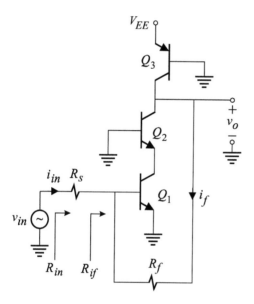

Figure 5.32

This, of course, is $A_{v\infty}$, which is derived formally in Fig. 5.33 by letting $\beta_1 \to \infty$ (since letting $\beta_2 \to \infty$ does not make the gain of the open-loop amplifier become infinite) which causes $i_{b1}, v_{be1} \to 0$. It follows that:

$$\frac{v_{in}}{R_s} = -\frac{v_o}{R_f} \Rightarrow A_{v\infty} = -\frac{R_f}{R_s} \qquad (5.88)$$

The closed-loop voltage gain is given by:

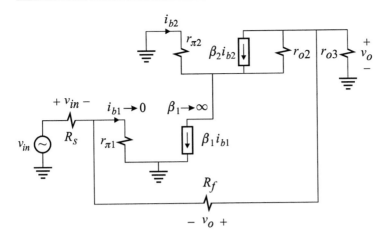

Figure 5.33

$$A_v = A_{v\infty} \frac{1 + \dfrac{1}{\beta_1 \mathscr{B}}}{1 + \dfrac{1}{\beta_1 \overline{B}}} \tag{5.89}$$

in which $\mathscr{B}$ and $\overline{B}$ are determined in Figs. 5.34 and 5.35, respectively. In Fig. 5.34, $\beta_1 i_{b1}$ is replaced with an independent current source pointing in the opposite direction and the output voltage is nulled. A null in the output voltage causes the voltage across $r_{\pi 2}$ to appear directly across r_{o2} so that the current i_T can be expressed in terms of i_{b2}:

$$i_T = -\left(i_{b2} + \beta_2 i_{b2} + \frac{i_{b2} r_{\pi 2}}{r_{o2}} \right) \Rightarrow i_{b2} = -\frac{i_T}{1 + \beta_2 + \dfrac{r_{\pi 2}}{r_{o2}}} \tag{5.90}$$

A null in the output voltage also causes the output current to be zero so that the fedback current is given by:

$$i_f = i_T + i_{b2} \tag{5.91}$$

which when substituted in (5.90) yields:

$$i_f = i_T \frac{\beta_2 + \dfrac{r_{\pi 2}}{r_{o2}}}{1 + \beta_2 + \dfrac{r_{\pi 2}}{r_{o2}}} \tag{5.92}$$

Because of the null in the output voltage, the voltage across R_f is equal and opposite to v_{be1} so that we have:

$$i_{b1}r_{\pi 1} = -i_f R_f \tag{5.93}$$

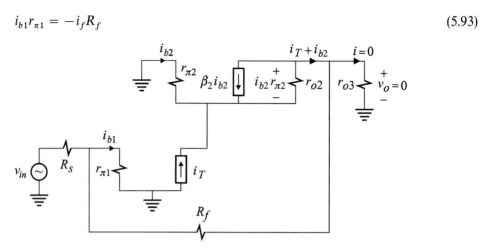

Figure 5.34

Hence, the null inverse gain is given by:

$$\hat{\mathcal{B}} = \frac{i_{b1}}{i_T} = -\frac{R_f}{r_{\pi 1}} \frac{\beta_2 + \dfrac{r_{\pi 2}}{r_{o2}}}{1 + \beta_2 + \dfrac{r_{\pi 2}}{r_{o2}}} \tag{5.94}$$

The inverse gain with $v_{in} = 0$ is shown in Fig. 5.35 in which i_T goes through a succession of current divisions which can be determined relatively easily by applying the EET to β_2:

$$\bar{B} = \frac{i_{b1}}{i_T} = \bar{B}|_{\beta_2 \to \infty} \frac{1 + \dfrac{1}{\beta_2 \bar{\mathcal{B}}_2}}{1 + \dfrac{1}{\beta_2 \bar{B}_2}} \tag{5.95}$$

which (see Problem 5.7) yields:

$$\bar{B} = \frac{1}{1 + \dfrac{R_f + R_s \| r_{\pi 1}}{r_{o3}}} \frac{1}{1 + \dfrac{r_{\pi 1}}{R_s}} \frac{1 + \dfrac{r_{\pi 2}}{\beta_2 r_{o2}}}{1 + \dfrac{1}{\beta_2}\left[1 + \dfrac{r_{\pi 2} + r_{o3} \| (R_f + R_s \| r_{\pi 2})}{r_{o2}}\right]} \tag{5.96}$$

Since $r_\pi \ll r_o$, Eqs. (5.94) and (5.96) can be approximated:

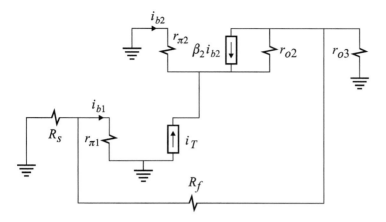

Figure 5.35

$$\bar{B} \approx \frac{1}{1 + \dfrac{R_f}{r_{o3}}} \frac{1}{1 + \dfrac{r_{\pi1}}{R_s}} \tag{5.97}$$

$$\bar{\mathscr{B}} \approx -\frac{R_f}{r_{\pi1}} \alpha_2 \tag{5.98}$$

so that the closed-loop voltage gain is given by:

$$A_v \approx -\frac{R_f}{R_s} \frac{1 - \dfrac{1}{\alpha_2 \beta_1} \dfrac{r_{\pi1}}{R_f}}{1 + \dfrac{1}{\beta_1}\left(1 + \dfrac{r_{\pi1}}{R_s}\right)\left(1 + \dfrac{R_f}{r_{o3}}\right)} \tag{5.99}$$

Other approximations may apply depending on the particular values of the components used (see Problem 5.7).   ☐

5.5   Driving-point analysis of feedback amplifiers

The *inside* input and output impedances of the four types of feedback amplifiers will be discussed in this section. Particular emphasis will be placed on the loop gain, T, because, unlike the closed-loop gain, the input and output impedances of a feedback amplifier are sensitive to the open-loop gain, A, and hence to the loop gain T. Note that we study either an impedance or an admittance function depending upon the type of mixing and sampling.

5.5.1   Input *impedance* for current mixing

Consider an open-loop amplifier with input impedance Z_i in which a current signal, i_f, proportional to the output, u_o, is fed back to the input side as shown in Fig. 5.36a. The closed-loop response, which can be either a voltage gain or a transconductance, can be expressed in terms of the loop gain:

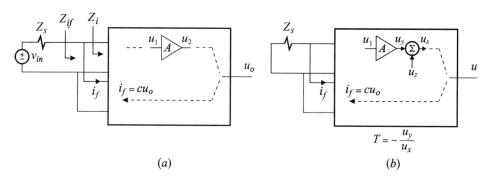

(a)   (b)

Figure 5.36

$$G \equiv \frac{u_o}{v_{in}} = G_\infty \frac{T}{1 + T} + G_0 \frac{1}{1 + T} \qquad (5.100)$$

In this equation the loop gain, T, is determined by setting $v_{in} = 0$ as shown in Fig. 5.36b. The impedance seen by the source v_{in} can be written as the sum of the source impedance, Z_s, and the *inside* input impedance Z_{if}, as shown in Fig. 5.36a. Our objective is to study Z_{if} as a function of the loop gain since Z_s is fixed. To do so, we connect a test current source i_t to the input of the amplifier as shown in Fig. 5.37a and study the response v_t by applying the feedback formula in Eq. (5.44) to obtain:

$$Z_{if} = Z_{if\infty} \frac{T'}{1 + T'} + Z_{if0} \frac{1}{1 + T'} \qquad (5.101)$$

in which the loop gain T' is obtained by setting $i_t = 0$ as shown in Fig. 5.37b. A comparison of Figs. 5.36b and 5.37b shows that the two loop gains T and T' are related by:

$$T' = T|_{z_s \to \infty} \qquad (5.102)$$

It is also clear from Figs. 5.36b and 5.37b that T is smaller than T' simply because the fedback signal in Fig. 5.36b splits between Z_s and the input of the amplifier, whereas in Fig. 5.37b it goes entirely into the amplifier. Hence:

$$T' > T \qquad (5.103)$$

The impedance Z_{if_∞} is shown in Fig. 5.38b and is determined by letting $A \to \infty$ so

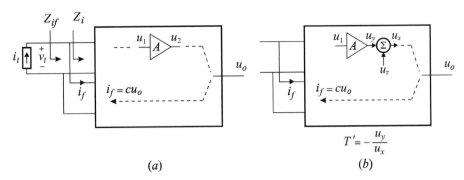

Figure 5.37

that the input to the amplifying element, u_1, and hence the input to the amplifier, $v_i = v_t$, both become zero. It follows that:

$$Z_{if\infty} = \left.\frac{v_t}{i_t}\right|_{A\to\infty} = 0 \tag{5.104}$$

The impedance Z_{if_0} is shown in Fig. 5.38b and is determined by letting $A = 0$:

$$Z_{if_0} = \left.\frac{v_t}{i_t}\right|_{A\to 0} \tag{5.105}$$

In Fig. 5.38b we see that Z_{if_0} is the impedance looking into the amplifier *in the presence of the loading of the feedback network but without the action of feedback* simply because A and, hence, T are both zero. Since the feedback network appears in shunt with Z_i, we can see from Fig. 5.38b that:

$$Z_{if_0} = Z_i \,\|\, \{\text{loading of feedback network}\} < Z_i \tag{5.106}$$

When Eq. (5.104) is substituted in Eq. (5.101), the inside input impedance reduces to:

$$Z_{if} = \frac{Z_{if_0}}{1 + T'} \tag{5.107}$$

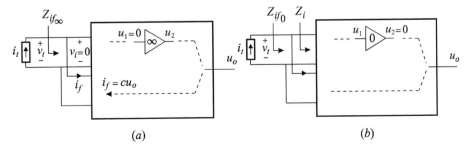

Figure 5.38

Making use of the inequalities in Eqs. (5.103) and (5.106), we deduce:

$$Z_{if} = \frac{Z_{ifo}}{1 + T'} < \frac{Z_i}{1 + T'} < \frac{Z_i}{1 + T} \tag{5.108}$$

What this result says is that if you add current feedback to an open-loop amplifier with source impedance Z_s and input impedance Z_i, then the inside input imped-ance of the resulting feedback amplifier is smaller than Z_i *at least* by a factor of $1 + T$ where T is the loop gain associated with the transfer gain.

Example 5.8 The input impedance of the two-stage, series–shunt feedback amplifier, discussed in Example 5.4 and shown in Fig. 5.17, can be written as:

$$R_{in} = R_s + R_{if} \tag{5.109}$$

Since R_s is fixed, we study R_{if} by connecting a test current source as shown in Fig. 5.39 and applying the feedback formula:

$$R_{if} = R_{if\infty}\frac{T'}{1 + T'} + R_{ifo}\frac{1}{1 + T'} \tag{5.110}$$

in which T' can be determined either as shown in Fig. 5.40 or by simply letting $R_s \to \infty$ in the expression of the loop gain derived in Example 5.4:

$$T' = T|_{R_s \to \infty}$$

$$= \beta_1 \bar{B}|_{R_s \to \infty}$$

$$= \frac{\beta_1(1 + \beta_2)}{1 + \dfrac{r_{\pi 2} + (1 + \beta_2)(R_f \| R_e)}{R_c}} \frac{1}{1 + \dfrac{R_f}{R_e}} \tag{5.111}$$

$$\times \frac{1}{1 + \dfrac{r_{\pi 1}}{R_B \| \left(R_f + R_e \left\| \dfrac{r_{\pi 2} + R_c}{1 + \beta_2} \right. \right)}}$$

in which $\bar{B}$ was determined in Eq. (5.67).

To determine $R_{if\infty}$, we let $\beta_1 \to \infty$ in Fig. 5.39 and observe that this causes $i_{b1} \to 0$ so that $v_T = i_{b1}r_{\pi 1} \to 0$. Hence we have:

$$R_{if\infty} = 0 \tag{5.112}$$

Equation (5.110) now reduces to:

$$R_{if} = \frac{R_{ifo}}{1 + T'} \tag{5.113}$$

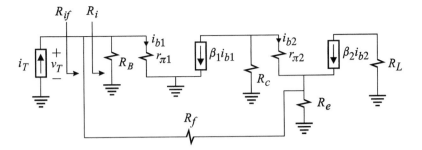

Figure 5.39

Next we determine R_{if_o}, which is the input resistance with the loading of the feedback but without the action of feedback. To do so, we set $\beta_1 = 0$ as shown in Fig. 5.41 whence we have by inspection:

$$R_{if_o} = R_B \| r_{\pi 1} \| \left(R_f + R_e \left\| \frac{r_{\pi 2} + R_c}{1 + \beta_2} \right. \right) \tag{5.114}$$

We can see in this expression that $R_B \| r_{\pi 1}$ is indeed the input impedance, R_i, of the amplifier without feedback and that $R_f + R_e \| [(r_{\pi 2} + R_c)/(1 + \beta_2)]$ is the contribution of the loading of the feedback connection.

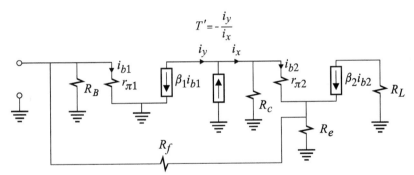

Figure 5.40

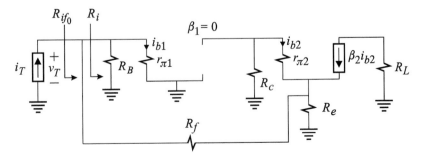

Figure 5.41

Substituting Eqs. (5.111) and (5.114) in (5.113) we obtain the complete expression for the input resistance. Using numerical values, these expressions can be further simplified. Also note that the input impedance can be determined directly by applying the EET to β_1 and using Eq. (5.24) without referring to the loop gain (see Problem 5.8).   □

5.5.2   Output *impedance* for voltage sensing

In this section we shall determine the effect of feedback on the output impedance of feedback amplifiers whose output voltage is sampled as shown Figure 5.42a. The open-loop amplifier, prior to the application of feedback, is assumed to have an output impedance of Z_o. The transfer function, which can be either a voltage gain or a transresistance, can be expressed according to the feedback formula:

$$G \equiv \frac{v_o}{u_{in}} = G_\infty \frac{T}{1+T} + G_0 \frac{1}{1+T} \tag{5.115}$$

in which the loop gain, T, is obtained by setting $u_{in} = 0$ as shown in Fig. 5.42b.

The inside output impedance is determined by setting the input to zero and connecting a test current source at the output as shown in Fig. 5.43a. Applying the feedback formula in Eq. (5.44) to the output impedance function we obtain:

$$Z_{of} = Z_{of\infty} \frac{T'}{1+T'} + Z_{ofo} \frac{1}{1+T'} \tag{5.116}$$

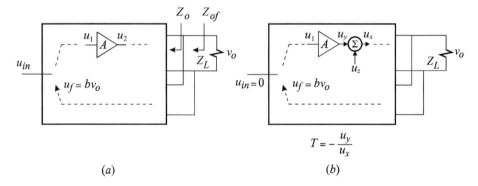

(a)   (b)

Figure 5.42

in which T' is the loop gain obtained by setting $i_t = 0$ as shown in Fig. 5.43b. According to Figs. 5.42b and 5.43b the two loop gains, T and T', are related by:

$$T' = T|_{Z_L \to \infty} \tag{5.117}$$

We can also see by comparing Figs. 5.42b and 5.43b that:

$$T' > T \tag{5.118}$$

because in Fig. 5.43*b* the signal enters the feedback path without the shunting effect of the load impedance in Fig. 5.42*b*.

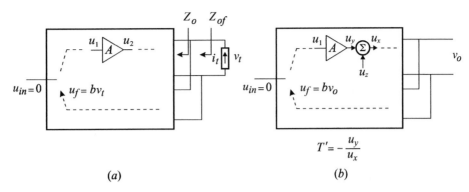

$$T' = -\frac{u_y}{u_x}$$

(*a*)   (*b*)

Figure 5.43

The impedance $Z_{of\infty}$ is shown in Fig. 5.44*a* whence we see that as $A \rightarrow \infty$ the output voltage, or v_t, approaches zero so that we have:

$$Z_{of\infty} = \left.\frac{v_t}{i_t}\right|_{A \rightarrow \infty} = 0 \tag{5.119}$$

In order to see how v_t vanishes as $A \rightarrow \infty$, notice that the input u_1 to the amplifying element A is a linear combination of the input signal, u_{in}, and the feedback signal, u_f, so that:

$$u_1 = au_{in} + bu_f \tag{5.120}$$

Since u_f is proportional to the output voltage, i.e. $u_f = cv_o$, Eq. (5.120) becomes:

$$u_1 = au_{in} + bcv_o \tag{5.121}$$

Since $u_{in} = 0$ and $u_1 \rightarrow 0$ as $A \rightarrow \infty$, we have according to Eq. (5.121) $v_t = v_o \rightarrow 0$.

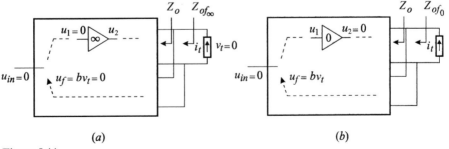

(*a*)   (*b*)

Figure 5.44

Hence it follows that $Z_{of\infty} = 0$ and the closed-loop output impedance in Eq. (5.116) reduces to:

$$Z_{of} = \frac{Z_{ofo}}{1 + T'} \tag{5.122}$$

The impedance Z_{ofo} is determined by setting $A = 0$ as shown in Fig. 5.44b. We can see that Z_{ofo} is the output impedance *in the presence of the loading of the feedback network but without the action of feedback* since $A, T = 0$. In fact, Z_{ofo} is given by the parallel combination of Z_o and the impedance loading of the feedback network so that we have:

$$Z_{ofo} = Z_o \,\|\, \{\text{Impedance loading of feedback network}\} < Z_o \tag{5.123}$$

Making use of the inequalities in Eqs. (5.118) and (5.123), we deduce:

$$Z_{of} = \frac{Z_{ofo}}{1 + T'} < \frac{Z_{ofo}}{1 + T} < \frac{Z_o}{1 + T} \tag{5.124}$$

What this result says is that the inside output impedance of a voltage-sampled feedback amplifier, with loop gain T and feeding a load Z_L, is less than the inside output impedance of the open-loop amplifier at least by a factor of $1 + T$. This result is expected because sampling the output voltage causes it to follow the input signal according to $v_o \approx u_{in} G_\infty$ making the output appear as a voltage source. In other words, this feedback scheme automatically compensates for any voltage drop in the output impedance of the amplifier, Z_o, as the load current changes.

Example 5.9   The inside output impedance of the amplifier discussed in Example 5.5 is determined by connecting a test current source at its output and setting $v_{in} = 0$ as shown in Fig. 5.45. According to the feedback formula:

$$R_{of} = R_{of\infty} \frac{T'}{1 + T'} + R_{ofo} \frac{1}{1 + T'} \tag{5.125}$$

In this equation, we shall show that $R_{of\infty} = 0$ by observing in Fig. 5.45 that letting $\beta_2 \to \infty$ causes $i_{b2} \to 0$ which in turn causes the following sequence of events:

$$i_{b1} \to 0 \Rightarrow i_{b3} \to 0 \Rightarrow v_{bf} = \to 0 \Rightarrow v_{af} = \to 0 \Rightarrow v_T \to 0 \tag{5.126}$$

It follows that $R_{of\infty} = v_T / i_T = 0$ so that we have:

$$R_{of} = \frac{R_{ofo}}{1 + T'} \tag{5.127}$$

The loop gain T' is shown in Fig. 5.46 and can be deduced from the loop gain T determined earlier for the voltage gain in Example 5.5 simply by letting $R_L \to \infty$:

$$T' = T|_{R_L \to \infty} = \beta_2 \bar{B}|_{R_L \to \infty} \tag{5.128}$$

Substituting for $\bar{B}$ given in Eq. (5.74) and letting $R_L \to \infty$, we obtain for T':

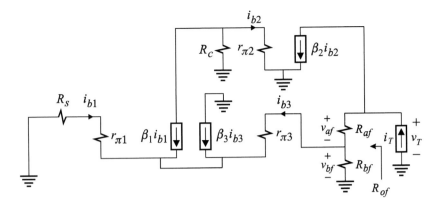

Figure 5.45

$$T' = \frac{\beta_2 \alpha_1 (1 + \beta_3)}{1 + \dfrac{r_{\pi3} + \dfrac{1 + \beta_3}{1 + \beta_1}(r_{\pi1} + R_s)}{R_{bf}}} \cdot \frac{1}{1 + \dfrac{r_{\pi2}}{R_c}} \tag{5.129}$$

Next we determine R_{ofo}, which is the inside output impedance with the loading of the feedback network but without the action of feedback. To do so, we set $\beta_2 = 0$ as shown in Fig. 5.47 and obtain by inspection:

$$R_{ofo} = R_{af} + R_{bf} \left\| \left[r_{\pi3} + (1 + \beta_3) \frac{r_{\pi1} + R_s}{1 + \beta_1} \right] \right. \tag{5.130}$$

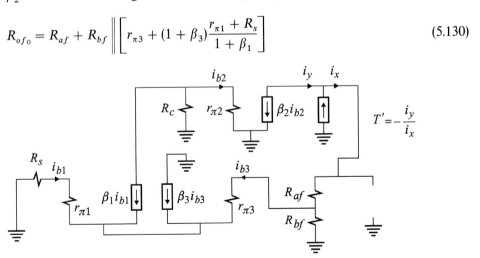

Figure 5.46

Note the inside output impedance, R_o, of the open-loop amplifier is infinite because it is the impedance looking into the collector of Q_3. Hence R_{ofo} is entirely due to the loading of the feedback network. Substituting Eqs. (5.129) and (5.130) in

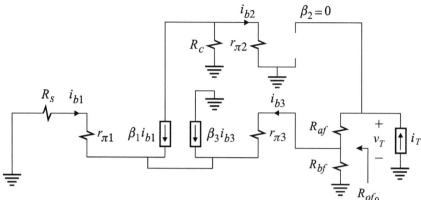

Figure 5.47

(5.127) we obtain the complete expression for the inside output resistance. Using numerical values, useful approximations for the expressions derived above can be obtained. Also note that the output impedance can be determined directly by applying the EET to β_2 and using Eq. (5.24) without referring to the loop gain (see Problem 5.9). □

5.5.3   Input *admittance* for voltage mixing

A general diagram of voltage mixing is shown in Fig. 5.48a in which a voltage signal, v_f, proportional to the output of an amplifier, is fed back to the input side and combined with the input voltage in a loop. The transfer gain of the amplifier in Fig. 5.48a can be either a voltage gain or a transconductance which according to the feedback formula can be expressed in terms of the loop gain:

$$G \equiv \frac{u_o}{v_{in}} = G_\infty \frac{T}{1 + T} + G_o \frac{1}{1 + T} \qquad (5.131)$$

The loop gain T is determined by setting $v_{in} = 0$ as shown in Fig. 5.48b.

The impedance seen by the source in Fig. 5.48a can be written as the sum of the source impedance and the *inside* input impedance:

$$Z_{in} = Z_s + Z_{if} \qquad (5.132)$$

Since Z_s is fixed, we shall study Z_{if}. As we shall see, it is more appropriate in this case to work with admittance functions rather than impedance functions. Hence, to determine Y_{if}, we connect a voltage source, v_t, as shown in Fig. 5.49a and apply the feedback formula in Eq. (5.44) to the input admittance function:

$$Y_{if} = Y_{if\infty} \frac{T'}{1 + T'} + Y_{if o} \frac{1}{1 + T'} \qquad (5.133)$$

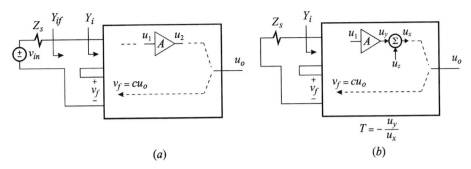

Figure 5.48

in which T' is determined as shown in Fig. 5.49b and is related to the loop gain, T, of the transfer gain shown in Fig. 5.48b by:

$$T' = T|_{Z_s \to 0} \tag{5.134}$$

In Fig. 5.48b, v_f goes through a voltage division between Z_s and Z_i before it enters the amplifier, whereas in Fig. 5.49b v_f goes straight into the amplifier. If follows that the loop gain in Fig. 5.49b must be larger than the one in Fig. 5.48b:

$$T' > T \tag{5.135}$$

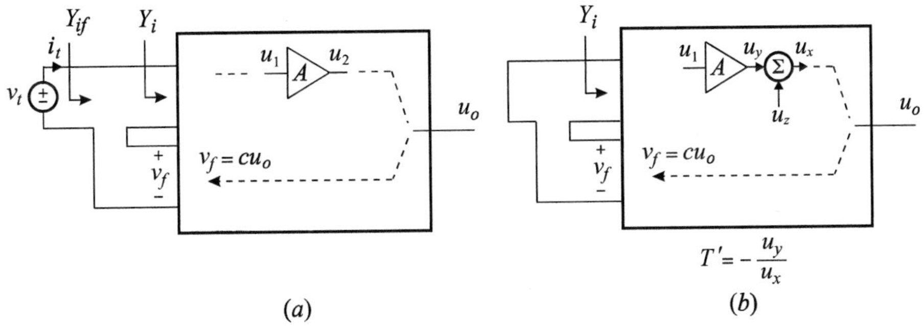

Figure 5.49

The admittance $Y_{if\infty}$ is determined by letting $A \to \infty$ as shown in Fig. 5.50a in which the input voltage to the amplifier, v_i, approaches zero because u_1 approaches zero. It follows that the current entering the amplifier, $i_i = Y_i v_i$, approaches zero so that:

$$Y_{if\infty} = \frac{i_i}{v_t}\bigg|_{A \to \infty} \to 0 \tag{5.136}$$

Hence, Eq. (5.133) reduces to:

$$Y_{if} = \frac{Y_{ifo}}{1 + T'} \tag{5.137}$$

Now we see that had we considered Z_{if} instead of Y_{if}, we would have obtained $Z_{if\infty} \to \infty$ and $T' = 0$ (because we would have had to open the input side in Fig. 5.49b rather than short it) which would have caused the first term in the feedback formula to be indeterminate.

The admittance Y_{if_0} is determined by letting $A = 0$ as shown in Fig. 5.50b whence we see that:

$$Y_{if_0} = Y_i \,\|\, \{\text{admittance loading of feedback network}\} \tag{5.138}$$

It follows that:

$$Y_{if_0} < Y_i \Rightarrow Z_{if_0} > Z_i \tag{5.139}$$

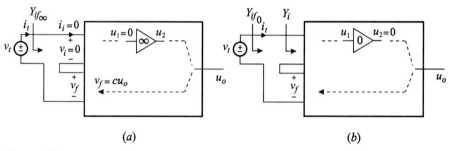

(a)   (b)

Figure 5.50

The impedance Z_{if_0} is the input impedance *with the loading of the feedback network but without the action of feedback* since A and, hence, T are both zero.

Now we can compare the inside input admittance of the feedback amplifier with the inside input admittance of the amplifier before connecting the feedback network by making use of Eqs. (5.135), (5.137) and (5.138):

$$Y_{if} = \frac{Y_{if_0}}{1 + T'} < \frac{Y_{if_0}}{1 + T} < \frac{Y_i}{1 + T} \tag{5.140}$$

The last inequality can be written in its more popular form:

$$Z_{if} > Z_i(1 + T) \tag{5.141}$$

which states that the inside input impedance of an amplifier with voltage mixing is at least $1 + T$ times larger than the inside input impedance of the open-loop amplifier where T is the loop gain associated with the transfer gain. This increase in the input impedance is expected because negative feedback reduces the input voltage, v_i, to the amplifier stage by v_f so that the amplifier draws less current from the source. In the limit, as $A \to \infty, v_i \to 0$ and the current drawn by the amplifier from the source vanishes and the input impedance becomes infinite.

Example 5.10 To determine the inside input conductance of the amplifier in

Example 5.5, we connect a test voltage source at the input terminals as shown in Fig. 5.51. According to the feedback formula we have:

$$G_{if} = G_{if\infty}\frac{T'}{1+T'} + G_{if0}\frac{1}{1+T'} \tag{5.142}$$

in which T' is determined as shown in Fig. 5.52. As in the previous examples, we can deduce T' from the loop gain for the voltage transfer function, $T = \beta_2\bar{B}$:

$$T' = T|_{R_s\to 0} = \beta_2\bar{B}|_{R_s\to 0} \tag{5.143}$$

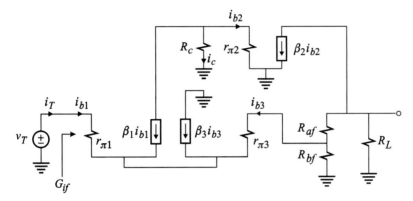

Figure 5.51

in which $\bar{B}$ is given by Eq. (5.74) in Example 5.5. Hence we have:

$$T' = \frac{\alpha_1(1+\beta_3)\beta_2}{1+\dfrac{R_{af}+R_{bf}\left\|\left(r_{\pi3}+\dfrac{1+\beta_3}{1+\beta_1}r_{\pi1}\right)\right.}{R_L}}$$

$$\times \frac{1}{1+\dfrac{r_{\pi3}+\dfrac{1+\beta_3}{1+\beta_1}r_{\pi1}}{R_{bf}}}\cdot\frac{1}{1+\dfrac{r_{\pi2}}{R_c}} \tag{5.144}$$

In Fig. 5.51, if we let $\beta_2\to\infty$ we can see that $G_{if\infty}\to 0$ because of the following sequence of events:

$$\beta_2\to\infty\Rightarrow i_{b2}=0\Rightarrow i_c=0\Rightarrow\beta_1 i_{b1}=0\Rightarrow i_{b1}=i_T=0 \tag{5.145}$$

Equation (5.142) now reduces to:

$$G_{if}=G_{if0}\frac{1}{1+T'} \tag{5.146}$$

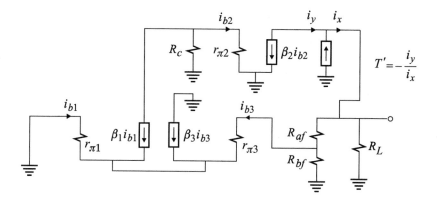

Figure 5.52

It should be clear that had we chosen to study Z_{if} in Eq. (5.142) instead of G_{if}, then $Z_{if\infty} \to \infty$ and $T' \to 0$ (because in this case T' would have been derived from T by letting $R_s \to \infty$) and the first term in Eq. (5.141) would have been indeterminate.

Equation (5.145) is written in its more popular form:

$$R_{if} = R_{if_0}(1 + T') \tag{5.147}$$

In Fig. 5.53 we let $\beta_2 = 0$ so that R_{if_0} can be determined by inspection:

$$R_{if_0} = r_{\pi 1} + \frac{r_{\pi 3} + R_{fb} \| (R_{fa} + R_L)}{1 + \beta_3}(1 + \beta_1) \tag{5.148}$$

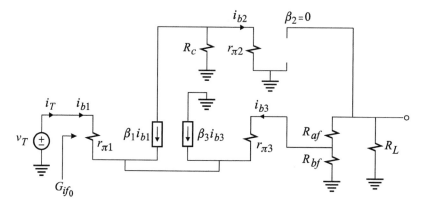

Figure 5.53

Substituting Eqs. (5.144) and (5.148) in (5.147), we obtain the complete expression of R_{if} which is seen to increase proportional to the open-loop gain. There is a rather simple interpretation for the increase in the input impedance: by increasing the loop gain the voltage at the base of Q_3 becomes more and more identical to the voltage at the base of Q_1 so that the differential voltage across $r_{\pi 1}$ and $r_{\pi 2}$ approaches zero and the input current reduces to zero.

If we assume that all βs are large, then R_{if_o} and T' can be simplified:

$$T' \approx \frac{\alpha_1 \beta_3 \beta_2}{1 + \dfrac{R_{af} + R_{bf} \parallel (r_{\pi 3} + r_{\pi 1})}{R_L}} \frac{1}{1 + \dfrac{r_{\pi 3} + r_{\pi 1}}{R_{bf}}} \frac{1}{1 + \dfrac{r_{\pi 2}}{R_c}} \tag{5.149}$$

$$R_{if_o} \approx r_{\pi 1} + r_{\pi 3} + R_{fb} \parallel (R_{fa} + R_L) \tag{5.150}$$

Note that the input resistance, R_{in}, seen by the source can be determined directly by applying the EET to β_2 (see Problem 5.10). ☐

5.5.4 Output *admittance* for current sensing

Figure 5.54a shows a general diagram for an amplifier in which a signal u_f proportional to the output current is fed back to the input side. Since the output current is the response, a transfer function of this amplifier can be either a current gain or a transconductance which according to the feedback formula in Eq. (5.44) is given by:

$$G \equiv \frac{i_o}{u_{in}} = G_\infty \frac{T}{1 + T} + G_o \frac{1}{1 + T} \tag{5.151}$$

The loop gain, T, in Eq. (5.151) is determined by setting the input to zero and injecting a signal u_z as shown in Fig. 5.54b. Before studying the output admittance,

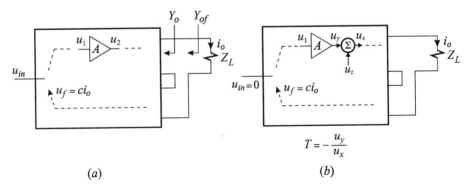

$$T = -\frac{u_y}{u_x}$$

(a) $\qquad\qquad\qquad\qquad$ (b)

Figure 5.54

we establish the fact that u_1, which is the input signal to the amplifying element A, is a linear combination of the output current and the input voltage:

$$u_1 = au_{in} + bu_f \tag{5.152}$$

Since u_f is proportional to the output current, $u_f = ci_o$, Eq. (5.152) becomes:

$$u_1 = au_{in} + bci_o \tag{5.153}$$

The *inside* output admittance is determined by setting the input to zero, $u_{in} = 0$, and connecting a test voltage source at the output as shown in Fig. 5.55a. Applying the feedback formula in Eq. (5.44) to the output admittance function, we obtain:

$$Y_{of} = Y_{of_\infty} \frac{T'}{1 + T'} + Y_{oo} \frac{1}{1 + T'} \tag{5.154}$$

in which T' is the loop gain obtained by setting $v_t = 0$ as shown in Fig. 5.55b because v_t is the excitation of Y_{of}. Immediately we can see that T' is related to T in Fig. 5.54b by:

$$T' = T|_{Z_L \to 0} \geq T \tag{5.155}$$

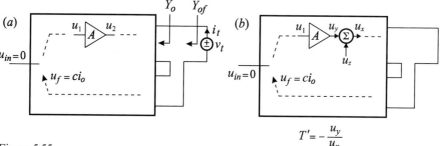

Figure 5.55

The reason why $T' > T$ is simply because the output of the amplifying element A in Fig. 5.55b encounters a smaller impedance than in Fig. 5.54b and hence can push more current into the feedback path than in Fig. 5.54b. It follows that the signal gain around the loop is larger in Fig. 5.55b than in Fig. 5.54b. Another way to see this is to apply the EET to T, in which u_x is the excitation, u_y is the response and Z_L is the extra element (see Problem 5.11). Note that when the amplifying element A is a dependent current generator directly feeding the load Z_L, then $T' = T$.

In Fig. 5.56a, Y_{of_∞} is determined by letting $A \to \infty$, which causes $u_1 = 0$. Since $u_{in} = 0$, we have according to Eq. (5.153):

$$i_t = 0 \tag{5.156}$$

It follows that:

$$Y_{of_\infty} = \frac{i_t}{v_t}\bigg|_{A \to \infty} = 0 \tag{5.157}$$

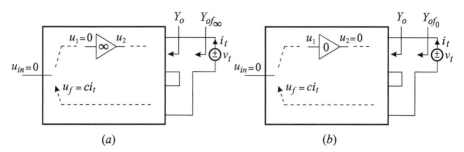

Figure 5.56

Hence the feedback formula once again reduces to:

$$Y_{of} = \frac{Y_{ofo}}{1 + T'} \tag{5.158}$$

The output admittance Y_{ofo} is determined by letting $A = 0$ as shown in Fig. 5.56b in which we see that Y_{ofo} is the *output admittance in the presence of the loading of the feedback network but without the action of feedback*. In fact we have:

$$Y_{ofo} = Y_o \parallel \{\text{admittance loading of feedback network}\} \tag{5.159}$$

so that we have:

$$Y_{ofo} < Y_o \tag{5.160}$$

Applying the results in Eqs. (5.155) and (5.160) to Eq. (5.158), we conclude that:

$$Y_{of} = \frac{Y_{ofo}}{1 + T'} < \frac{Y_{ofo}}{1 + T} < \frac{Y_o}{1 + T} \tag{5.161}$$

It is not surprising that sampling the output current should lower the output admittance because this causes the output current to follow the input signal through $i_o \approx u_{in}G_\infty$ which in turn makes the output appear as a current source. In this case, feedback compensates for any current shunted by open-loop output admittance, Y_o.

Example 5.11 To determine the inside output conductance seen by the load R_L of the amplifier in Example 5.6, we connect a test voltage source at the output as shown in Fig. 5.57 and apply the feedback formula to obtain:

$$G_{of} = \frac{i_T}{v_T} = G_{of_\infty} \frac{T'}{1 + T'} + G_{ofo} \frac{1}{1 + T'} \tag{5.162}$$

The loop gain T' can either be determined as shown in Fig. 5.58 or deduced from the results derived for T in Example 5.6:

$$T' = T|_{R_L \to 0} = \beta_2 \bar{B}|_{R_L \to 0} \tag{5.163}$$

Substituting for $\bar{B}$, given by Eq. (5.81), in Eq. (5.163) and taking the limit we obtain:

$$T' = \frac{R_{c2}}{R_{c2} + \frac{1}{n^2}R_{e1}} \cdot \frac{1}{n} \left| \frac{r_{\pi1} + R_s}{1 + \beta_1} \right| \cdot \frac{\alpha_1\beta_2}{1 + \frac{R_s + r_{\pi1}}{R_{e1}(1 + \beta_1)}} \cdot \frac{1}{1 + \frac{r_{\pi2}}{R_{c1}}}$$ (5.164)

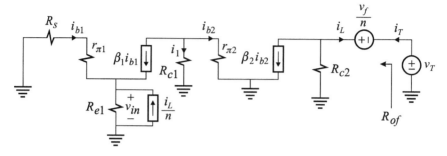

Figure 5.57

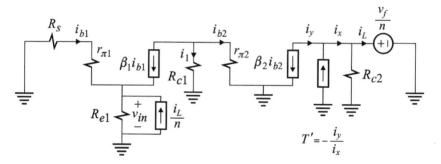

Figure 5.58

To determine G_{of_∞}, we let $\beta_2 \to \infty$, which causes the following sequence of events as shown in Fig. 5.59:

$$i_{b2} = 0 \Rightarrow i_1 = 0 \Rightarrow \beta_1 i_{b1} = 0 \Rightarrow i_{b1} = 0 \Rightarrow v_f = 0 \Rightarrow \frac{i_L}{n} = 0$$ (5.165)

It follows that $G_{of_\infty} = 0$ and the inside output conductance reduces to:

$$G_{of}\frac{1}{1 + T'}$$ (5.166)

This is written in its more popular form:

$$R_{of} = R_{of_o}(1 + T')$$ (5.167)

To determine G_{of_o}, we let $\beta_2 = 0$ as shown in Fig. 5.60 whence we see that the output impedance is given by the series combination of R_{c2} and the impedance seen looking into the emitter of Q_1 reflected through the transformer:

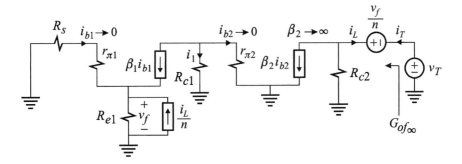

Figure 5.59

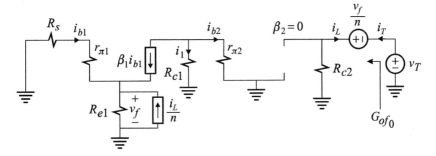

Figure 5.60

$$R_{ofo} = R_{c2} + \frac{1}{n^2}\left(R_{e1} \left\| \frac{r_{\pi1} + R_s}{1 + \beta_1} \right)\right. \tag{5.168}$$

Substituting Eqs. (5.164) and (5.168) in Eq. (5.167), we obtain the complete expression of the output impedance. Before the application of feedback, the impedance seen by the load is R_{c2}, which is nearly identical to R_{ofo}. The output impedance, after feedback, is seen to be increased by a factor of $1 + T'$ as given by Eq. (5.167). The output impedance discussed in this example can also be derived without any reference to loop gain simply by applying the EET to β_2 (see Problem 5.12). □

5.6   Loop gain: a more detailed look

The definition of the loop gain $T = A\bar{A}$ given earlier corresponds to injecting a signal, u_{inj}, *immediately after the amplifying element A* and determining the ratio $T = -u_y/y_x$ as shown earlier in Fig. 5.4a. For example, when the amplifying element is a dependent voltage source Au_g, a voltage source is inserted in series with Au_g as shown in Fig. 5.61a and the loop gain is given by the ratio of the return voltage to the forward voltage:

$$T = -\frac{v_y}{v_x}\Bigg|_{u_{in}=0} = A\bar{A}$$ (5.169a)

When the amplifying element is a dependent current source, Au_ε, a current source is injected in parallel with Au_ε as shown in Fig. 5.61b and the loop gain is given by the ratio of return current to the forward current:

$$T = -\frac{i_y}{i_x}\Bigg|_{u_{in}=0} = A\bar{A}$$ (5.169b)

The determination of the loop gain by the ratio of the return signal to the forward signal can be generalized to injecting a signal at any point in the circuit and not necessarily immediately after the amplifying element. For an arbitrary point of injection the expression of the loop gain obtained is no longer equal to $A\bar{A}$ and is different for different points of injection. Hence, if the feedback formula is to yield the same closed-loop gain, G, for different loop gains, then G_∞ and G_0 must be different at different points of injection. The proof of the feedback formula for an arbitrary point of injection is left as an exercise (see Problem 5.13). When the signal is injected immediately after the amplifying element as shown in Figs. 5.61a

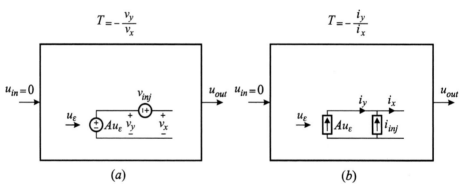

(a) (b)

Figure 5.61

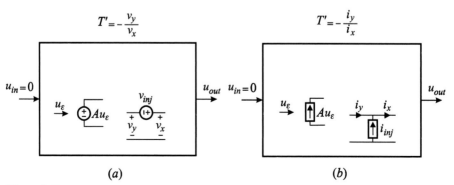

(a) (b)

Figure 5.62

and b, the corresponding G_∞ and G_0 are determined by letting $A \to \infty$ and $A = 0$, respectively, as explained earlier. For an arbitrary point of injection, G_∞ and G_0 are determined using null double injection as shown in Figs. 5.63a and b (voltage injection) and Figs. 5.64a and b (current injection). In Fig. 5.63a, v_{inj} and u_{in} are used to null v_y so that G_∞ is determined according to:

$$G_\infty = \left. \frac{u_{out}}{u_{in}} \right|_{v_y = 0} \tag{5.170}$$

In Fig. 5.63b v_{inj} and u_{in} are used to null v_x so that G_0 is determined according to:

$$G_0 = \left. \frac{u_{out}}{u_{in}} \right|_{v_x = 0} \tag{5.171}$$

The same is repeated in Figs. 5.64a and b for current injection.

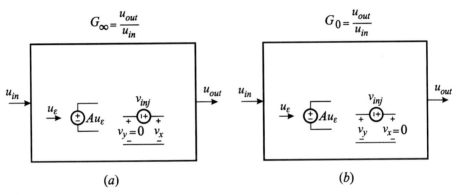

Figure 5.63

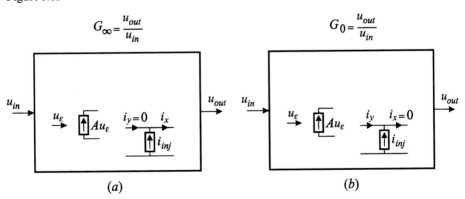

Figure 5.64

The generalization of the loop gain as described above does not necessarily lead to better analytical answers: it simply generalizes the feedback formula in Eq. (5.44). The following simple example compares the expressions of the gain obtained by the use of the EET and the feedback formula using two different points of injection.

Example 5.12   The voltage gain of the single-stage common-emitter amplifier with output voltage feedback shown in Fig. 5.65 is obtained by using the EET for dependent sources:

$$
\left.
\begin{aligned}
A_v &= A_{v\infty} \frac{1 + \dfrac{1}{\beta \bar{\bar{\mathscr{B}}}}}{1 + \dfrac{1}{\beta \bar{B}}} \\[2em]
&= -\frac{R_f}{R_s} \frac{1 - \dfrac{1}{\beta}\dfrac{r_\pi}{R_f}}{1 + \dfrac{1}{\beta}\left(1 + \dfrac{r_\pi}{R_s}\right)\left(1 + \dfrac{R_f + R_s \| r_\pi}{R_L}\right)}
\end{aligned}
\right\}
\qquad (5.172a, b)
$$

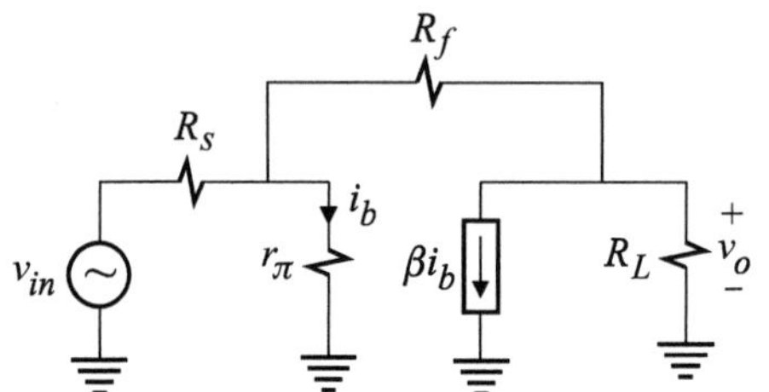

Figure 5.65

Next, we use the feedback formula and inject a current signal directly in parallel with the β generator as shown in Fig. 5.66. In Fig. 5.66a the loop gain obtained is the same as $\beta \bar{B}$ in Eq. (5.172b):

$$
T = \frac{\beta}{\left(1 + \dfrac{r_\pi}{R_s}\right)\left(1 + \dfrac{R_f + R_s \| r_\pi}{R_L}\right)}
\qquad (5.173)
$$

In Fig. 5.66b $A_{v\infty}$ is obtained by nulling i_y:

$$
A_{v\infty} = -\frac{R_f}{R_s}
\qquad (5.174)
$$

which is the same expression as the one in Eq. (5.172b). In Fig. 5.66c A_{v0} is obtained by nulling i_x:

$$
A_{v0} = \frac{1}{1 + \dfrac{R_s}{r_\pi \| (R_f + R_L)}} \cdot \frac{1}{1 + \dfrac{R_f}{R_L}}
\qquad (5.175)
$$

Substituting the above in the feedback formula:

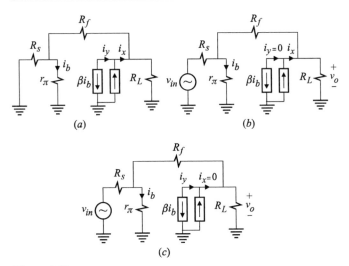

Figure 5.66

$$A_v = A_{v\infty} \frac{1}{1 + T^{-1}} + A_{v0} \frac{1}{1 + T} \qquad (5.176)$$

we obtain another expression for the voltage gain which is equivalent to the expression in Eq. (5.172) but somewhat more complicated.

Finally, we inject a signal at the base of the transistor as shown in Fig. 5.67a and obtain for the loop gain:

$$T' = -\frac{i_y}{i_x} = \frac{r_\pi}{R_s} + \frac{r_\pi + \beta R_L}{R_f + R_L} \qquad (5.177)$$

For $A'_{v\infty}$ we obtain from Fig. 5.67b:

$$A'_{v\infty} = -\frac{R_f}{R_s} \frac{1 - \dfrac{r_\pi}{\beta R_f}}{1 + \dfrac{r_\pi}{\beta R_s}\left(1 + \dfrac{R_f + R_s}{R_L}\right)} \qquad (5.178)$$

For A'_{v0} we obtain from Fig. 5.67c:

$$A'_{v0} = 0 \qquad (5.179)$$

Substituting Eqs. (5.177), (5.178) and (5.179) in the feedback formula we obtain a third equivalent expression for the closed-loop voltage gain. The only problem with this last expression is that $A'_{v\infty}$ does not represent an ideal closed-loop gain as does $A_{v\infty}$ and cannot be as easily interpreted. □

It is clear that at an arbitrary point of injection, either a current signal or a

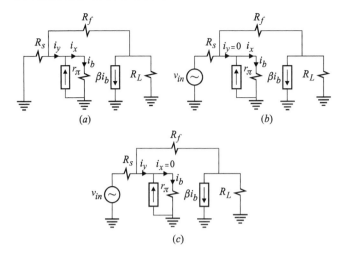

Figure 5.67

voltage signal can be used and two different loop gains can be obtained: a current loop gain, T_i, and a voltage loop gain, T_v. In general there is no useful relationship between T_i and T_v, but in an important special case a useful relationship between the two can be obtained (see Problem 5.14).

5.7   Stability

A system is unstable when any of its poles, simple or complex, is located in the right-half plane (RHP). For a complex pole-pair in the RHP, the time-domain response is oscillatory whose amplitude is limited by the nonlinearity of the circuit. For a simple pole in the RHP, the time-domain response is a growing exponential whose growth is also bounded by the nonlinearity of the circuit. In this section, a brief theory of the stability of linear circuits and the role of the loop gain in assessing stability will be discussed.

The stability of a system can be determined from the number of net encircle-ments of the polar plot of the loop gain $T(j\omega)$, also known as the Nyquist plot, of the $(-1, j0)$ point. If the net number of encirclements less the number of RHP poles of the open-loop plant is zero, then the system is stable, otherwise it is unstable. A Bode plot of the magnitude and phase of the loop gain is also used to study the stability of linear systems whose open-loop zeros and poles are in the left-half plane (LHP). Such systems, also known as *minimum phase* systems, be-come unstable when the phase of the loop gain exceeds $-180°$ at the frequency at which the magnitude of the loop gain equals unity. Except in decoupled circuits, *any* loop gain will yield exactly the same condition of instability or the same condition and frequency of oscillation.

Example 5.13   This is an example of two decoupled circuits in which the loop gain yields no information about the stability of the output voltage response. Recall that feedback is applied to a particular response and not to every possible response in the entire circuit. The loop gain of the circuit in Fig. 5.68 is given by:

$$T(s) = \beta\bar{B} = \frac{T_o}{1 + s/\omega_1} \tag{5.180}$$

where:

$$\left.\begin{array}{c} T_o = \dfrac{\beta}{1 + \dfrac{R_s + r_\pi}{R_E}} \\[2em] \omega_1 = \dfrac{1}{R_E \parallel (r_\pi + R_s)C_E} \end{array}\right\} \tag{5.181a, b}$$

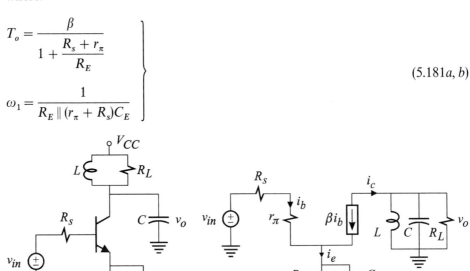

Figure 5.68

A Nyquist plot of the loop gain is shown in Fig. 5.69, which shows no possible sign of instability no matter how large or small β or T_o may be even if R_L in the output circuit may be infinite or negative which would cause the output voltage to become unstable. There is actually nothing wrong with this loop gain because it is the emitter current and not the output voltage which is being fed back. In other words, the output voltage is outside the feedback loop in which the amplifying element βi_b is acting. Hence, the emitter current can exhibit a very stable response, as predicted by the loop gain, while the output voltage may be oscillatory. A block diagram representation of decoupled systems in general is shown in Fig. 5.70 in which the loop gain T_1 determines the behaviour of the feedback system G_1 and H_1 and *not* G_2. Hence, even if u_1 exhibits a perfectly stable response, u_2 may exhibit a marginally stable or unstable response depending on the poles of G_2.   □

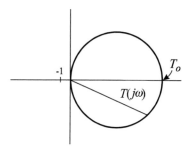

Figure 5.69

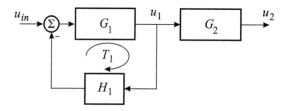

Figure 5.70

Example 5.14   The circuit in the previous example can be modified in such a way as to feed back the output voltage positively in order to make an oscillator. This is shown in Fig. 5.71a and the circuit is known as the Colpitts oscillator. Using the equivalent circuit model shown in Fig. 5.71b, we can determine the loop gain (see Problem 5.15) to be:

$$T(s) = g_m \bar{G} = T_o \frac{1 + \dfrac{s}{\omega_1}}{1 + a_1 s + a_2 s^2 + a_3 s^3} \tag{5.182}$$

where:

$$T_o = g_m R_E \| r_\pi$$

$$\omega_1 = \frac{R_L}{L}$$

$$a_1 = R_E \| r_\pi (C_c + C_E)$$

$$a_2 = LC_c \frac{R_E \| r_\pi}{R_L \| R_E \| r_\pi} + LC_E \frac{R_E \| r_\pi}{R_L}$$

$$a_3 = LC_c C_E R_E \| r_\pi$$

$$\left. \right\} \tag{5.183a–e}$$

If we set $T(j\omega) = -1$, we can find the condition and frequency at which this circuit will oscillate. Before doing so however, we shall give a simple physical argument for the operation of the circuit. If at the desired frequency of oscillation, the reactance of C_E is much smaller than R_E and the resistance looking into the emitter, then the voltage at the emitter junction is essentially given by the capacitive voltage divider:

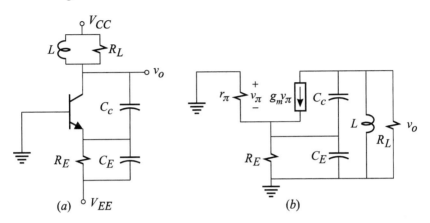

Figure 5.71

$$v_E \approx v_o \frac{C_c}{C_c + C_E} = \gamma v_o \tag{5.184}$$

Now, the collector current is given by:

$$i_c = -g_m v_\pi = -\gamma g_m v_o \tag{5.185}$$

If this current replenishes the losses in the resonant tank mainly due to R_L then the circuit can sustain oscillation. Hence, equating i_c to v_o/R_L should yield the condition for the amount of gain needed to sustain oscillation:

$$\gamma g_m v_o = v_o/R_L \Rightarrow g_m = \frac{1}{\gamma R_L} \tag{5.186}$$

For typical loads and current bias levels, γ turns out to be much less than unity so that for a typical design of this circuit we have:

$$\gamma \ll 1 \Rightarrow C_E \gg C_c \tag{5.187}$$

It follows that the series combination of C_E and C_c is essentially given by C_c so that the frequency of oscillation of the resonant tank is given by:

$$\omega_o \approx \frac{1}{\sqrt{LC_c}} \tag{5.188}$$

Now, using the concept of loop gain, we derive the same result by requiring that $T(j\omega_o) = -1$ in Eq. (5.182):

$$T_o \frac{1 + \dfrac{j\omega_o}{\omega_1}}{1 - a_2\omega_o^2 + j(a_1\omega_o - a_3\omega_o^3)} = -1 \qquad (5.189)$$

Multiplying out the terms in Eq. (5.189) we obtain:

$$1 + T_o - a_2\omega_o^2 + j\omega_o\left(\frac{T_o}{\omega_1} + a_1 - a_3\omega_o^2\right) = 0 \qquad (5.190)$$

First, we set the imaginary part in Eq. (5.190) to zero and obtain the frequency of oscillation:

$$\omega_o = \sqrt{\frac{1}{LC_c \parallel C_E} + \frac{g_m}{C_c C_E R_L}} \qquad (5.191)$$

Using the fact that the Q of the tank circuit is fairly respectable, Eq. (5.191) can be approximated (see Problem 5.16) as:

$$\omega_o \approx \sqrt{\frac{1}{LC_c \parallel C_E}} \qquad (5.192)$$

Second, we set the real part in Eq. (5.190) to zero and obtain:

$$1 + T_o = a_2\omega_o^2 \qquad (5.193)$$

Substituting for T_o, a_2 and ω_o (see Problem 5.16), we obtain the minimum required value for g_m to sustain oscillation:

$$G_m = \frac{1}{R_L \left\|\left(\dfrac{R_E}{\gamma^2}\right)\right.} \frac{1}{\gamma\left(1 - \dfrac{\gamma}{\alpha}\right)} \qquad (5.194)$$

in which $\alpha = \beta/(1 + \beta)$. Since $\gamma \ll 1$ and $\alpha \approx 1$, the condition on g_m to sustain oscillation in Eq. (5.194) simplifies to the condition obtained earlier in Eq. (5.186). □

The stability of an electronic circuit can also be determined using impedance functions. Hence, the condition and frequency of oscillation can be obtained by requiring that the real and imaginary parts of the impedance looking into any port across a capacitor or an inductor to be infinite. Alternatively, the same conditions can be obtained by requiring the impedance looking into a loop containing an inductor or a capacitor to be zero. These two conditions are shown in Figs. 5.72a and b.

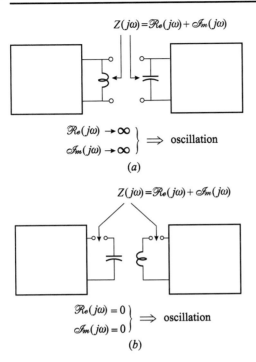

$$Z(j\omega) = \mathcal{R}e(j\omega) + \mathcal{I}m(j\omega)$$

$$\left.\begin{array}{l} \mathcal{R}e(j\omega) \rightarrow \infty \\ \mathcal{I}m(j\omega) \rightarrow \infty \end{array}\right\} \Rightarrow \text{oscillation}$$

(a)

$$Z(j\omega) = \mathcal{R}e(j\omega) + \mathcal{I}m(j\omega)$$

$$\left.\begin{array}{l} \mathcal{R}e(j\omega) = 0 \\ \mathcal{I}m(j\omega) = 0 \end{array}\right\} \Rightarrow \text{oscillation}$$

(b)

Figure 5.72

Example 5.15 The same condition and the frequency of oscillation in the previous example can also be deduced if we examine the impedance looking into any capacitive or inductive port of the circuit and requiring its real and imaginary parts be infinite. Since the inductor is connected across the output port, we determine the output impedance:

$$Z_o(s) = R_L \parallel sL \parallel Z'_o(s) \tag{5.195}$$

Instead of determining $Z_o(s)$, we can determine $Z'_o(s)$ and require that its real and imaginary parts be equal and opposite to those of $R_L \parallel j\omega L$. Using the 2-EET, we can show (see Problem 5.17) that $Z'_o(s)$ is given by:

$$Z'_o(s) = R_E \parallel r_\pi \frac{1 + \dfrac{C_E}{C_c} + \dfrac{1}{sC_c r_\pi \parallel R_E \parallel g_m^{-1}}}{1 + sC_E R_E \parallel r_\pi} \tag{5.196}$$

Making use of γ in Eq. (5.184) and $g_m r_\pi = \beta = \alpha/(1 - \alpha)$, we can rewrite $Z'_o(s)$ as:

$$Z'_o(s) = R_E \parallel r_\pi \dfrac{\dfrac{1}{\gamma} + \dfrac{1}{sC_cR_E \parallel (\alpha g_m^{-1})}}{1 + sC_ER_E \parallel r_\pi}$$

$$= \dfrac{1}{\gamma} \dfrac{R_E \parallel r_\pi}{1 + sC_ER_E \parallel r_\pi} + \dfrac{\dfrac{R_E \parallel r_\pi}{sC_cR_E \parallel (\alpha g_m^{-1})}}{1 + sC_ER_E \parallel r_\pi}$$

$$(5.197a, b)$$

The real and imaginary parts of $Z'_o(s)$ in the vicinity of the resonant frequency are determined as follows. The first term in the expression of $Z'_o(s)$ is rationalized to yield:

$$\dfrac{1}{\gamma} \left[\dfrac{R_E \parallel r_\pi}{1 + (\omega C_E R_E \parallel r_\pi)^2} - \dfrac{j\omega C_E(R_E \parallel r_\pi)^2}{1 + (\omega C_E R_E \parallel r_\pi)^2} \right] \qquad (5.198a)$$

In the vicinity of the resonant frequency $\omega C_E R_E \parallel r_\pi \gg 1$ so that Eq. (5.198a) can be approximated:

$$\dfrac{1}{\gamma \omega^2 C_E^2 R_E \parallel r_\pi} + \dfrac{1}{j\omega\gamma C_E} \qquad (5.198b)$$

The second term in the expression of $Z'_o(s)$ can be approximated:

$$\dfrac{\dfrac{R_E \parallel r_\pi}{j\omega C_c R_E \parallel (\alpha g_m^{-1})}}{1 + j\omega C_E R_E \parallel r_\pi} \approx -\dfrac{1}{\omega^2 C_c C_E R_E \parallel (\alpha g_m^{-1})} \qquad (5.198c)$$

Combining Eqs. (5.198b) and (5.198c) we obtain for the real and imaginary parts of $Z'_o(s)$:

$$\dfrac{1}{(\omega\gamma C_E)^2} \left[\dfrac{\gamma}{R_E \parallel r_\pi} - \dfrac{\gamma^2 C_E}{C_c R_E \parallel (\alpha g_m^{-1})} \right] + \dfrac{1}{j\omega\gamma C_E} \qquad (5.199)$$

The effective capacitance in Eq. (5.199) is seen to be $\gamma C_E = C_c \parallel C_E$. The real and imaginary parts of $R \parallel j\omega L$ in the vicinity of resonance can be approximated using a high-Q assumption:

$$R \parallel j\omega L = \dfrac{j\omega L}{1 + j\dfrac{\omega L}{R}} = \dfrac{j\omega L \left(1 - j\dfrac{\omega L}{R}\right)}{1 + \left(\dfrac{\omega L}{R}\right)^2} \approx j\omega L + \dfrac{(\omega L)^2}{R} \qquad (5.200)$$

Requiring that the imaginary parts in Eqs. (5.199) and (5.200) be equal and opposite we obtain the frequency of oscillation:

$$-j\omega L = \frac{1}{j\omega\gamma C_E} \Rightarrow \omega_o = \frac{1}{\sqrt{LC_c \| C_E}} \tag{5.201}$$

Requiring that the real parts in Eqs. (5.199) and (5.200) be equal and opposite we obtain the condition for oscillation:

$$\frac{1}{(\omega\gamma C_E)^2}\left[\frac{\gamma}{R_E \| r_\pi} - \frac{\gamma^2 C_E}{C_c R_E \| (\alpha g_m^{-1})}\right] = -\frac{(\omega L)^2}{R_L}$$

$$= -\frac{1}{R_L(\omega\gamma C_E)^2} \tag{5.202}$$

whence we can solve the condition on g_m for oscillation:

$$g_m = \frac{1}{R_L \left\|\left(\dfrac{R_E}{\gamma^2}\right)} \frac{1}{\gamma\left(1 - \dfrac{\gamma}{\alpha}\right)} \tag{5.203}$$

The high-Q approximations discussed above for tapped resonant circuits are so routine in tuned amplifier and oscillator circuit analysis that approximate equivalent circuits are derived to render the analysis straightforward and avoid repetitious approximations. These circuits and techniques will be discussed in Chapter 7. □

Example 5.16   As an illustration of the Nyquist encirclement condition for stability, consider an open-loop plant, which has a RHP pole, with negative feedback applied to it as shown in Fig. 5.73. Although the open-loop plant is unstable, it can be easily shown that the closed-loop system is stable if $\beta A_o > 1$. The closed-loop response in Fig. 5.73 is given by:

$$\left.\begin{aligned}\frac{u_o}{u_{in}} &= \frac{A}{1 + A\beta} \\[2ex] &= \frac{A_o}{A_o\beta - 1}\ \frac{1}{1 + \dfrac{s}{\omega_1(A_o\beta - 1)}}\end{aligned}\right\} \tag{5.204a, b}$$

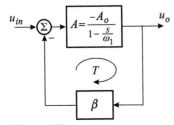

Figure 5.73

Hence, this feedback system is stable if the closed-loop pole in Eq. (5.205b) is in the LHP or simply if $\beta A_o > 1$. The same conclusion can be reached from the Nyquist plot of the loop gain which is shown in Fig. 5.74 and is given by:

$$T(j\omega) = A\beta = -\frac{A_o\beta}{1 - \frac{s}{\omega_1}} \tag{5.205}$$

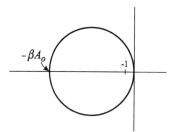

Figure 5.74

We can see in Fig. 5.74 that if $\beta A_o > 1$, then the number of net encirclements of the $(-1, j0)$ point is one. Hence, the system is stable because the number of encirclements less the number of RHP poles of the open-loop system is zero.

An excellent example of a circuit with open-loop RHP poles operating in a stable closed-loop configuration is two regulating dc-to-dc converters connected in cascade (see Chapter 8). □

5.8   Phase and gain margins

Phase and gain margins are rather simple and intuitive marginal stability criteria for simple minimum-phase systems. Figure 5.75 shows two different loop gains which have the same magnitude when the phase of each reaches $-180°$ at a particular frequency. The factor by which either gain has to be increased to encircle the $(-1, j0)$ point and make the system unstable is called *gain margin*. Hence, the gain margin for both loop gains in Fig. 5.75 is given by:

$$G_M = 20\log\left[\frac{1}{|T_1(j\omega_{\phi1})|}\right] = 20\log\left[\frac{1}{|T_2(j\omega_{\phi2})|}\right] \tag{5.206}$$

We can see, however, that the system with loop gain T_2 is more susceptible to an instability simply because it comes closer to the critical point. Hence, we define another marginal stability figure, called the *phase margin*, which is the amount of extra phase required to encircle the critical point at the frequency at which the loop gain becomes unity. Hence, the phase margins for T_1 and T_2 in Fig. 5.75 are given by:

$$P_{M1} = 180 + \angle T_1(j\omega_{c1})$$
$$P_{M2} = 180 + \angle T_2(j\omega_{c2})$$

$$(5.207a, b)$$

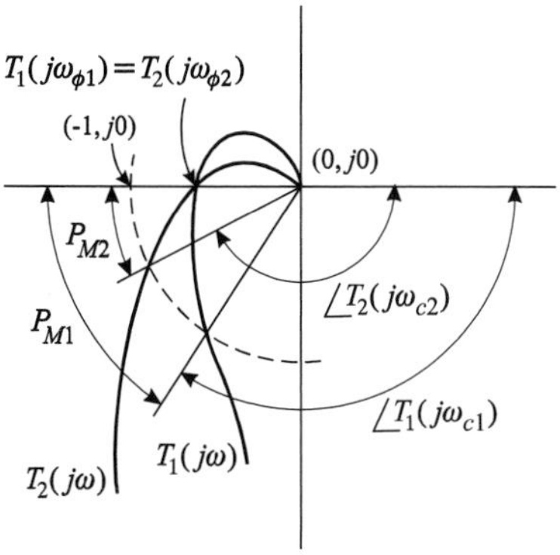

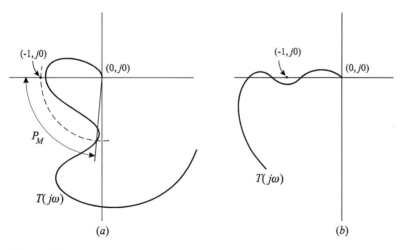

Figure 5.75

If the magnitude of the loop gain crosses the unity gain point and has a shape such as the one shown in Fig. 5.76a, then the relevant stability margin is the gain margin and not the phase margin (see Problem 5.16). In conditionally stable systems, such as the one shown in Fig. 5.76b, phase margin has no meaning. Although somewhat cumbersome, two gain margins may be defined for such loop gains if necessary.

Figure 5.76

If a response is fed back with a loop gain which has a small stability margin, then that response tends to have an underdamped oscillatory transient. Hence, we can think of stability margin as an effective damping constant of the dominant response. In fact, for a second-order system with no zeroes, there is an *exact* relationship between the damping factor and the phase margin (which can be found in most introductory control theory books).

Example 5.17   In micro-electromechanical sensors, signals are commonly developed across micro-capacitive transducers in the form of equivalent current generators whose strength is proportional to the velocity of the movable plate, or plates, and the applied bias voltage. The most suitable way of biasing a capacitive micro-sensor and amplifying the current signal is to use a transimpedance amplifier, shown in Fig. 5.77a. The small-signal equivalent circuit is shown in Fig. 5.77b in which the input capacitance of the operational amplifier is included in C_t. Since the current signal is of the order of femto or pico amperes, the feedback resistor is typically of the order of 100 MΩ. Such an amplifier can easily oscillate, or have a sharp resonant peak in its transimpedance function, because the combined phase shift from the open-loop gain of the amplifier and the input circuit, $r_{in}C_t$, can reach $-180°$ as the loop gain approaches unity.

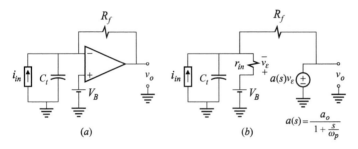

Figure 5.77

The loop gain is determined using the equivalent circuit model shown in Fig. 5.77b whence we have:

$$T(s) = \frac{T_o}{\left(1 + \frac{s}{\omega_p}\right)\left(1 + \frac{s}{\omega_f}\right)} \tag{5.208}$$

where

$$\left.\begin{aligned} T_o &= \frac{a_o}{1 + \frac{R_f}{r_{in}}} \\[2em] \omega_f &= \frac{1}{C_t r_{in} \| R_f} \end{aligned}\right\} \tag{5.209a, b}$$

For a typical application using an electrometer-grade amplifier, such as the

AD549, we have the following numerical values:

$$\left.\begin{array}{l} a_o = 10^6 \\ \omega_p = 1\,\text{Hz} \\ r_{in} = 10^{13}\,\Omega \\ R_f = 10^8\,\Omega \\ C_t = 5\,\text{pF} \end{array}\right\} \qquad (5.210a\text{–}e)$$

Using these numerical values, we determine:

$$\left.\begin{array}{l} T_o \approx a_o = 10^6 \\[2mm] \omega_f \approx \dfrac{1}{C_t R_f} = 2\pi(318)\,\text{rad/s} \end{array}\right\} \qquad (5.211a, b)$$

An asymptotic magnitude and phase plot of the loop gain is shown in Fig. 5.78. The phase margin is given by:

$$\phi_M = 180° - \tan^{-1}\frac{f_x}{f_p} - \tan^{-1}\frac{f_x}{f_f}$$

$$= 180° - 90° - 88.98° = 1.02°$$

Such a small phase margin at the crossover frequency simply means that the response, v_o, has a very sharp resonant peak at the crossover frequency ω_x. The crossover frequency can be easily solved to an excellent approximation:

$$|T(j\omega)| \approx \frac{T_o \omega_p \omega_f}{\omega_x^2} = 1 \Rightarrow \omega_x = \sqrt{a_o \omega_p \omega_f} = 2\pi(17\,832)\,\text{rad/s} \qquad (5.212)$$

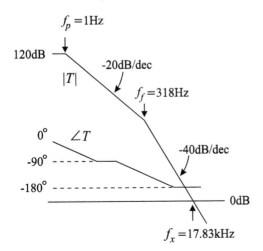

Figure 5.78

To show the correspondence between the phase margin and the peaking in the closed-loop response, we apply the feedback formula:

$$\frac{v_o}{i_{in}} = G_\infty \frac{1}{1 + T^{-1}} + G_0 \frac{1}{1 + T} \tag{5.213}$$

in which we can easily see that:

$$\left.\begin{aligned} G_\infty &= -R_f \\ G_0 &= 0 \end{aligned}\right\} \tag{5.214a, b}$$

Performing the necessary substitutions, we obtain:

$$\frac{v_o}{i_{in}} = \frac{R_f}{1 + T_o^{-1}} \frac{1}{1 + \dfrac{s}{1 + T_o}\dfrac{\omega_f + \omega_p}{\omega_f \omega_p} + \dfrac{s^2}{(1 + T_o)\omega_f \omega_p}} \tag{5.215}$$

Equation (5.215) can be written as:

$$\frac{v_o}{i_{in}} = \frac{R_{fo}}{1 + \dfrac{s}{\omega_o Q} + \dfrac{s^2}{\omega_o^2}} \tag{5.216}$$

in which:

$$\left.\begin{aligned} R_{fo} &= \frac{R_f}{1 + T_o^{-1}} \\ \omega_o &= \sqrt{(1 + T_o)\omega_f \omega_p} \\ Q &= \frac{\omega_o}{\omega_f + \omega_p} \end{aligned}\right\} \tag{5.217a–c}$$

Since $T_o \gg 1$ we have:

$$\left.\begin{aligned} \omega_o &\approx \omega_x \\ R_{fo} &\approx R_f \end{aligned}\right\} \tag{5.218a, b}$$

Substituting the values for ω_o and ω_f above, we find:

$$Q \approx \frac{\omega_o}{\omega_f} = 56 \tag{5.219}$$

Hence we see that a small phase margin at the crossover frequency simply means that there is an undamped resonant peak at the crossover frequency in the closed-loop response. Since the operational amplifier has another pole at a higher frequency, the phase margin can easily become negative causing the amplifier to oscillate at the crossover frequency.   □

Example 5.18   A simple remedy for the problem encountered in the previous example is shown in Fig. 5.79 in which a small capacitance is added in parallel with R_f to provide some phase lead in the feedback path. The loop gain in this case is given by:

$$T(s) = T_o \frac{\left(1 + \dfrac{s}{\omega_z}\right)}{\left(1 + \dfrac{s}{\omega_p}\right)\left(1 + \dfrac{s}{\omega'_f}\right)} \tag{5.220}$$

in which:

$$\left. \begin{aligned} \omega_z &= \frac{1}{R_f C_f} \\[2ex] \omega'_f &= \frac{1}{R_f \| r_{in}(C_f + C_t)} \approx \frac{1}{R_f(C_f + C_t)} \end{aligned} \right\} \tag{5.221a, b}$$

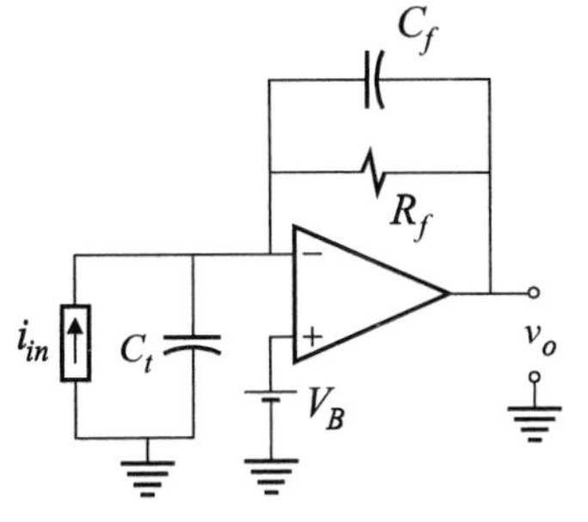

Figure 5.79

If we wish to design for a 45° phase margin, then we place ω_z at the crossover frequency (which is the same as ω_o) and obtain for C_f:

$$\omega_z = \omega_x = 2\pi(17\,832) \Rightarrow C_f = 0.089\,\text{pF} \tag{5.222}$$

Since $C_f \ll C_t$, the pole at ω'_f remains essentially at the same place as ω_f and the loop gain, given by Eq. (5.220), looks as shown in Fig. 5.80. It should be noted that, although microwave chip capacitors at 0.1 pF are readily available, the value of C_f determined above is an extremely small value and can be easily masked by the amount of stray capacitance on the circuit board. Hence great care and awareness of stray capacitances must be exercised when laying out and building such circuits.

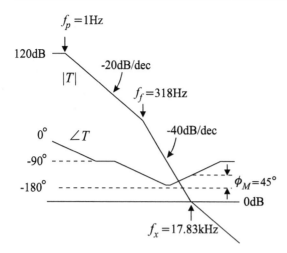

Figure 5.80

The closed-loop response is now given by:

$$\frac{v_o}{i_{in}} = G_\infty(s)\frac{1}{1 + T^{-1}(s)} \tag{5.223}$$

in which:

$$G_\infty(s) = -\frac{R_f}{1 + sC_fR_f} = -\frac{R_f}{1 + \dfrac{s}{\omega_z}} \tag{5.224}$$

Substituting Eq. (5.220) in (5.223), we obtain, similar to Eq. (5.216):

$$\frac{v_o}{i_{in}} = -\frac{R_f}{1 + T_o^{-1}}\frac{1}{1 + s\left(\dfrac{1}{\omega_z} + \dfrac{1}{\omega_oQ}\right) + \dfrac{s^2}{\omega_o^2}} \tag{5.225}$$

in which Q and ω_o are as defined as in Example 5.17. Hence, since $\omega_z \approx \omega_o$ and T_o and Q are both much larger than unity, Eq. (5.225) reduces to:

$$\frac{v_o}{i_{in}} = -\frac{R_f}{1 + \dfrac{s}{\omega_o} + \dfrac{s^2}{\omega_o^2}} \tag{5.226}$$

Hence, in comparison to Eq. (5.216) the Q has been reduced from 56 to unity simply by adding C_f to increase the phase margin to 45°. This choice of C_f results in the maximum possible bandwidth without peaking in the frequency response (ideally Q should be 0.707). $\square$

5.9   Review

The primary objective of designing a feedback amplifier is to achieve a transfer gain which is insensitive to the inherent variabilities found in all amplifying devices such as transistors and vacuum tubes. This is accomplished by designing an amplifier stage with a large gain, A (often consisting of several transistor stages), whose output is the desired response and whose input is the difference between the applied excitation and a fraction, β, of the response obtained by a feedback network which has very little variability. If the amplifier stage is designed with a very large gain, then, despite its variability, the input difference signal to the amplifier nearly vanishes and, hence, the response follows the excitation magnified by a factor of $1/\beta$. Ideally, if $A \rightarrow \infty$ the transfer gain reduces to $1/\beta$ and becomes entirely independent of A. It is natural then to formulate electronic feedback using the EET for dependent sources in which A is taken as the dependent source. This formulation is given in Eq. (5.3) in which G_∞ is the ideal closed-loop transfer gain obtained by letting $A \rightarrow \infty$. The actual closed-loop transfer gain in Eq. (5.3) differs from G_∞ by the bilinear factor in A.

The formulation of electronic feedback using the EET for dependent sources can be manipulated further to yield another formulation in which the closed-loop transfer gain is expressed as a linear combination of the ideal closed-loop gain, G_∞, and a nonideal gain, G_0, which represents the amount of coupling between the excitation and the response through the feedback network when the gain of the amplifier stage is set to zero. This formulation is given in Eq. (5.44) in which we can see the explicit dependence of the closed-loop gain on the loop gain T through the coefficients of G_∞ and G_0.

The concept of loop gain is central to the study of stability of any feedback system. A polar plot of the loop gain, known as a Nyquist plot, reveals the stability of a system by virtue of its encirclement of the $(-1, j0)$ point. A Nyquist plot also gives an indication of relative stability by virtue of the closeness of the polar plot to the $(-1, j0)$ point.

Both feedback formulations can be applied to study the input and output driving-point characteristics of feedback amplifiers. It is shown that whereas feedback renders the transfer gain insensitive to variations in the gain of the amplifier stage, A, it renders the input and output impedance functions directly sensitive to variations in A.

Problems

5.1 Ideal closed-loop gain. If the gain of the operational amplifier in Fig. 5.81a is infinite, the voltage-gain transfer function is given by the ideal closed-loop gain $G_\infty(s)$:

$$\frac{v_o(s)}{v_{in}(s)} = -\frac{Z_2(s)}{Z_1(s)} = G_\infty(s) \tag{5.227}$$

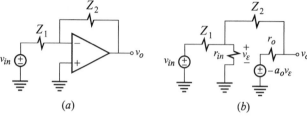

(a)   (b)

Figure 5.81

(a) If the operational amplifier has a finite gain a_o, an input resistance r_{in} and an output resistance r_o, as shown in Fig. 5.81b, then show that application of the EET to the gain a_o directly yields the following high-entropy result (as described in Eq. (5.5)):

$$\frac{v_o(s)}{v_{in}(s)} = -\frac{Z_2(s)}{Z_1(s)} \frac{1 - \dfrac{r_o}{a_o Z_2(s)}}{1 + \dfrac{1}{a_o}\left[1 + \dfrac{r_o + Z_2(s)}{Z_1(s) \| r_{in}}\right]} \tag{5.228}$$

(b) As a specific example, determine the transfer function of the circuit in Fig. 5.81c using the procedure in Eq. (5.4). First, remove the capacitor and apply the EET to a_o and determine the low-frequency gain of the response. Second, reinstate the capacitor using the EET for the capacitor. Following this procedure, show that:

$$\frac{v_o(s)}{v_{in}(s)} = G_\infty \frac{1 + \dfrac{1}{a_o \bar{a}} \dfrac{1 + sC\bar{\mathcal{R}}^{(c)}}{1 + sC\bar{R}^{(c)}}}{1 + \dfrac{1}{a_o \bar{a}}} \tag{5.229}$$

in which:

$$G_\infty = -\frac{R_2}{R_1}$$

$$\tilde{\alpha} = -\frac{R_2}{r_o}$$

$$\tilde{a} = \frac{R_1 \| r_{in}}{r_o + R_2 + R \| r_{in}}$$

$$\bar{\mathcal{R}}^{(c)} = R_2 \left\| \frac{-r_o}{a_o} \right.$$

$$\bar{R}^{(c)} = R_2 \| [r_o + (1 + a_o)R_1 \| r_{in}]$$

(5.230a–e)

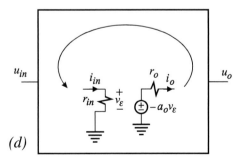

(c)

Figure 5.81 (cont.)

(c) Show that, in a general feedback configuration, all it takes for a nonideal operational amplifier to behave as an ideal one is $a_o \to \infty$ without further requiring that $r_{in} \to \infty$ and $r_o = 0$. To do so, show that the transfer function of an arbitrary feedback circuit shown in Fig. 5.81d is independent of r_{in} and r_o.

(d)

Figure 5.81 (cont.)

Hint: (a) Show that $i_{in} = 0$. (b) Show that i_o is independent of r_o. To do so, apply the EET to r_o:

$$\frac{i_o}{u_{in}} = \frac{i_o}{u_{in}}\bigg|_{r_o=0} \frac{1 + \dfrac{r_o}{\mathscr{R}^{(o)}}}{1 + \dfrac{r_o}{R^{(o)}}} \tag{5.231}$$

Show that $\mathscr{R}^{(o)}, R^{(o)} \to \infty$. To determine $R^{(o)}$, apply the EET for a_o.

5.2 Input impedance of FET amplifier with shunt–shunt feedback. Show that the input impedance of the FET amplifier in Example 5.3 is given by:

$$Z_{in}(s) = R_h \left(1 + \frac{\omega_z}{s}\right) \tag{5.232}$$

where:

$$R_h = R_s + [R_f + R_L \| (r_o + R_E)] \frac{1 + \dfrac{g_m r_o}{1 + \dfrac{r_o + R_L \| R_f}{R_E}}}{1 + g_m r_o \| (R_L + R_f)} \tag{5.233}$$

$$\omega_z = \frac{1}{C_f \left\{R_f + [R_s + R_L \| (r_o + R_E)] \dfrac{1 + g_m r_o \| (R_E + R_L \| R_s)}{1 + g_m \dfrac{r_o R_E}{r_o + R_E + R_L}}\right\}} \tag{5.234}$$

5.3 Loop gain measurement by current injection. A signal generator in series with a resistance and a blocking capacitor is used to inject a current signal into the drain of the FET amplifier discussed in Example 5.3 as shown in Fig. 5.82a. The response obtained using OrCAD/Pspice simulation is shown in Fig. 5.82b. Using the circuit values shown and the small-signal data of the FET, verify Eqs. (5.53) and (5.54) derived for the loop gain.
Dc operating point and small-signal parameters of J1:

NAME	J_J1
MODEL	J2N4867
ID	7.26E − 04
VGS	− 7.26E − 02
VDS	5.41E + 00
GM	1.35E − 03
GDS	4.90E − 06
CGS	1.17E − 11
CGD	4.08E − 12

5.4 Two-stage CE–CE amplifier with current sampling and current mixing. A simulation of the amplifier in Example 5.4 using OrCAD/Pspice is shown in Fig. 5.83a. The small-signal parameters of each transistor evaluated at its dc operating point are given in the table below and the frequency response of the voltage gain

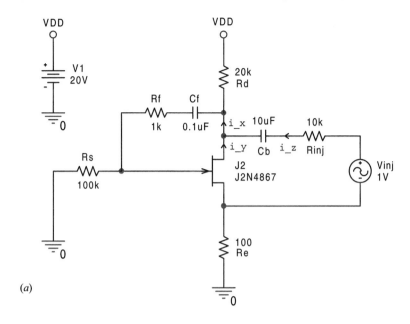

(a)

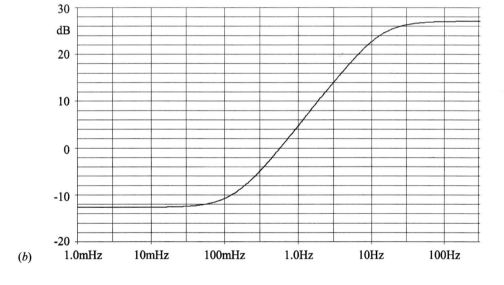

(b)   1.0mHz   10mHz   100mHz   1.0Hz   10Hz   100Hz

Figure 5.82

transfer function is shown in Fig. 5.83b. Using the given numerical values, verify that the flat gain is $A_v = 39.7$dB and obtain various approximations to A_v in Eq. (5.68).

Dc operating point and small-signal parameters of Q_1 and Q_2:

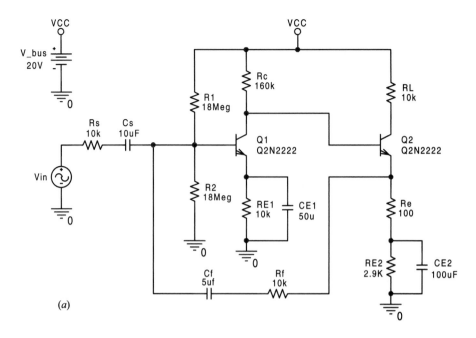

(a)

Figure 5.83

NAME	Q_Q2	Q_Q1
MODEL	Q2N2222	Q2N2222
IB	3.48E − 06	9.26E − 07
IC	5.55E − 04	1.07E − 04
VBE	6.27E − 01	5.88E − 01
VBC	− 1.21E + 01	− 6.35E − 01
VCE	1.28 + 01	1.22E + 00
BETADC	1.60E + 02	1.16E + 02
GM	2.14E − 02	4.14E − 03
RPI	8.34E + 03	3.21E + 04
RX	1.00E + 01	1.00E + 01
RO	1.55E + 05	6.97E + 05
CBE	4.46E − 11	3.64E − 11
CBC	2.77E − 12	5.92E − 12
CJS	0.00E + 00	0.00E + 00
BETAAC	1.79E + 02	1.33E + 02
CBX/CBX2	0.00E + 00	0.00E + 00
FT/FT2	7.20E + 07	1.56E + 07

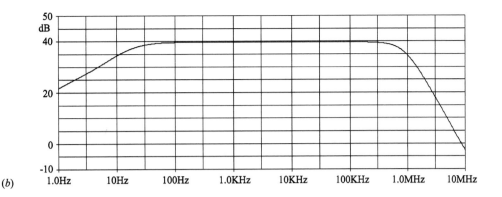

(b)

Figure 5.83 (cont.)

5.5 Differential input amplifier with voltage sampling and voltage mixing.

A simulation of the amplifier in Example 5.5 using OrCAD/Pspice is shown in Fig. 5.84a. The small-signal parameters of each transistor evaluated at its dc operating point are given in the table below and the frequency response of the voltage gain transfer function are shown in Fig. 5.84b. Using the numerical values given below, verify that the flat gain is $A_v = 33.5\,\text{dB}$ and obtain various approximations to A_v in Eq. (5.70).

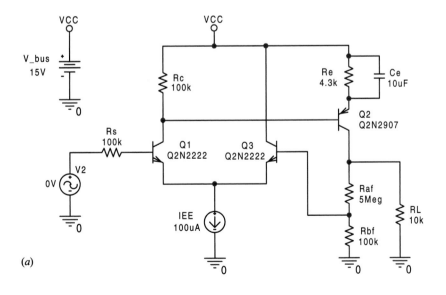

(a)

Figure 5.84

Dc operating point and small-signal parameters of Q_1, Q_2 and Q_3:

NAME	Q_Q2	Q_Q3	Q_Q1
MODEL	Q2N2907	Q2N2222	Q2N2222
IB	$-1.49E-06$	$5.91E-07$	$2.17E-07$
IC	$-3.39E-04$	$7.61E-05$	$2.31E-05$
VBE	$-6.96E-01$	$5.74E-01$	$5.44E-01$
VBC	$9.45E+00$	$-1.50E+01$	$-1.29+01$
VCE	$-1.01E+01$	$1.56E+01$	$1.34E+01$
BETADC	$2.28E+02$	$1.29E+02$	$1.06E+02$
GM	$1.31E-02$	$2.94E-03$	$8.93E-04$
RPI	$1.81E+04$	$5.07E+04$	$1.40E+05$
RX	$1.00E+01$	$1.00E+01$	$1.00E+01$
RO	$3.69E+05$	$1.17E+06$	$3.76E+06$
CBE	$4.01E-11$	$3.55E-11$	$3.38E-11$
CBC	$3.62E-12$	$2.58E-12$	$2.71E-12$
CJS	$0.00E+00$	$0.00E+00$	$0.00E+00$
BETAAC	$2.38E+02$	$1.49E+02$	$1.25E+02$
CBX/CBX2	$0.00E+00$	$0.00E+00$	$0.00E+00$
FT/FT2	$4.77E+07$	$1.23E+07$	$3.78E+06$

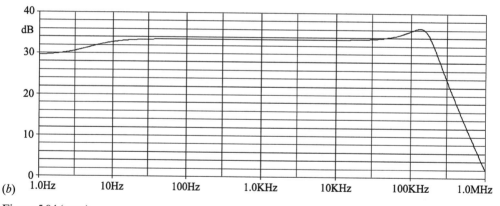

(b)

Figure 5.84 (*cont.*)

5.6 Current-sense feedback amplifier. A simulation of the amplifier in Example 5.6 using OrCAD/Pspice is shown in Fig. 5.85a. The small-signal parameters of each transistor evaluated at its dc operating point are given in the table below and the frequency response of the transconductance are shown in Fig. 5.85b.

(a) Using the given numerical values, determine the value of the midband gain G_t shown in Fig. 5.85b and given by Eq. (5.85).

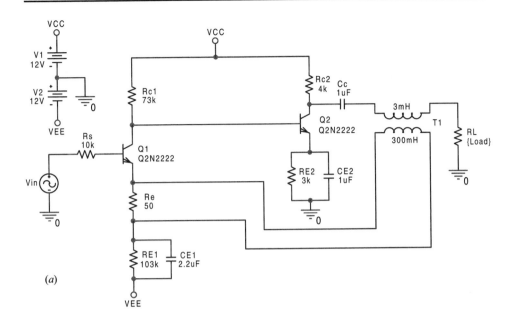

Figure 5.85

Dc operating point and small-signal parameters of Q_1 and Q_2:

NAME	Q_Q1	Q_Q2
MODEL	Q2N2222	Q2N2222
IB	9.15E$-$07	6.16E$-$06
IC	1.10E$-$04	9.58E$-$04
VBE	5.87E$-$01	6.43E$-$01
VBC	$-$3.54E$+$00	$-$4.63E$+$00
VCE	4.13E$+$00	5.28E$+$00
BETADC	1.20E$+$02	1.55E$+$02
GM	4.24E$-$03	3.69E$-$02
RPI	3.25E$+$04	4.67E$+$03
RX	1.00E$+$01	1.00E$+$01
RO	7.07E$+$05	8.21E$+$04
CBE	3.64E$-$11	5.15E$-$11
CBC	4.03E$-$12	3.73E$-$12
CJS	0.00E$+$00	0.00E$+$00
BETAAC	1.38E$+$02	1.72E$+$02
CBX/CBX2	0.00E$+$00	0.00E$+$00
FT/FT2	1.67E$+$07	1.06E$+$08

(b) Apply the NEET up to second-order terms to determine and evaluate the

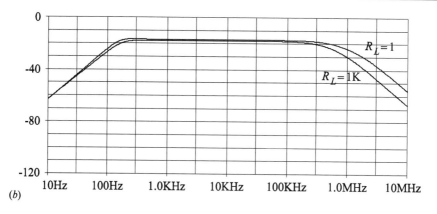

(b)

Figure 5.85 (*cont.*)

low-frequency roll-off in Fig. 5.85*b*. Use inverted notation for the complex frequency *s*:

$$G_t(s) \sim \frac{1}{1 + \dfrac{a_1}{s} + \dfrac{a_2}{s^2}}$$ (5.235)

5.7 Cascode amplifier. A simulation of the amplifier in Example 5.7 using Or-CAD/Pspice is shown in Fig. 5.86*a*. The quiescent operating point of each transistor and its small-signal parameters are given in the table below and the frequency response of the voltage gain transfer function is shown in Fig. 5.86*b*.

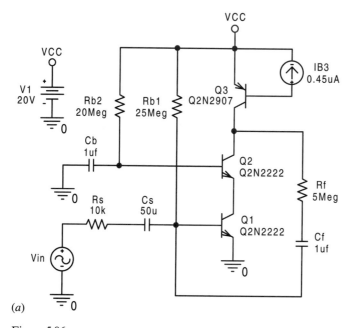

(a)

Figure 5.86

Dc operating point and small-signal parameters of Q_1 and Q_2:

NAME	Q_Q1	Q_Q2	Q_Q3
MODEL	Q2N2222	Q2N2222	Q2N2907
IB	7.77E−07	7.23E−07	−4.50E−07
IC	9.19E−05	9.12E−05	−9.12E−05
VBE	5.83E−01	5.80E−01	−6.63E−01
VBC	−4.39E+00	−1.05E+01	3.26E+00
VCE	4.97E+00	1.11E+01	−3.92E+00
BETADC	1.18E+02	1.26E+02	2.03E+02
GM	3.55E−03	3.52E−03	3.53E−02
RPI	3.84E+04	4.13E+04	6.16E+04
RX	1.00E+01	1.00E+01	1.00E+01
RO	8.53E+05	9.27E+05	1.30E+06
CBE	3.60E−11	3.59E−11	3.36E−11
CBC	3.79E−12	2.89E−12	5.99E−12
BETAAC	1.36E+02	1.46E+02	2.17E+02
FT/FT2	1.42E+07	1.44E+07	1.42E+07

(a) Using the given numerical values determine the value of the flat gain A_v shown in Fig. 5.86b and given by Eq. (5.89).

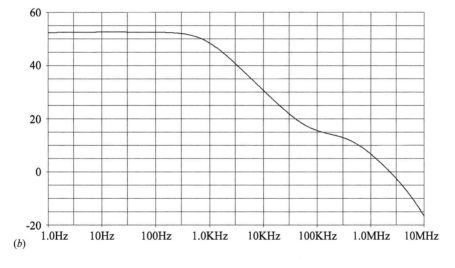

(b)

Figure 5.86 (cont.)

(b) Apply the NEET and the numerical values of the capacitances of Q1 and Q2 in the table above to obtain an expression of the dominant pole in Fig. 5.86b.

5.8 Input impedance of CE–CE amplifier with current mixing. The frequency response of the inside input impedance Z_{if} of the amplifier in Example 5.8 is shown in Fig. 5.87.

(a) Using the numerical values in Problem 5.4 and the results derived in Example 5.8, verify the midband value of R_{if} shown in Fig. 5.87.

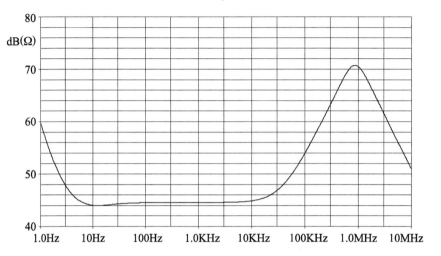

Figure 5.87

(b) Determine the outside input resistance seen by the source by applying the EET to β_1 and show that it is given by:

$$R_{in} = R_s \frac{1 + \dfrac{1 + R_f/R_e}{\beta_1(1 + \beta_2)}\left[1 + \dfrac{r_{\pi 2} + (1 + \beta_2)R_f \parallel R_e}{R_c}\right]\left[1 + \dfrac{r_{\pi 1}}{R_B \parallel R_s \parallel \left(R_f + R_e \left\| \dfrac{r_{\pi 2} + R_c}{1 + \beta_2}\right.\right)}\right]}{1 + \dfrac{1 + R_f/R_e}{\beta_1(1 + \beta_2)}\left[1 + \dfrac{r_{\pi 2} + (1 + \beta_2)R_f \parallel R_e}{R_c}\right]\left[1 + \dfrac{r_{\pi 1}}{R_B \parallel \left(R_f + R_e \left\| \dfrac{r_{\pi 2} + R_c}{1 + \beta_2}\right.\right)}\right]}$$

(5.236)

5.9 Output impedance for voltage sensing. The frequency response of the inside output impedance Z_{of} of the amplifier in Example 5.9 is shown in Fig. 5.88.

(a) Using the numerical values in Problem 5.5 and the results derived in Example 5.9 verify the midband value of R_{of} shown in Fig. 5.88.

(b) Determine the outside output resistance R_o looking into the load R_L directly by applying the EET to β_2 as explained in Eq. (5.24).

5.10 Input admittance with voltage mixing. The frequency response of the inside

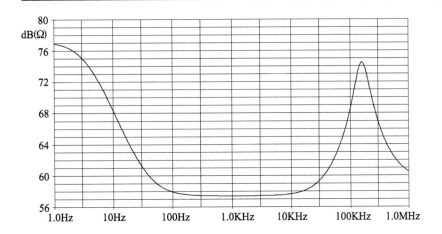

Figure 5.88

input admittance Y_{if} of the amplifier in Example 5.10 is shown in Fig. 5.89.

(a) Using the numerical values in Problem 5.5 and the results derived in Example 5.10, verify the midband value of Y_{if} in Fig. 5.89.

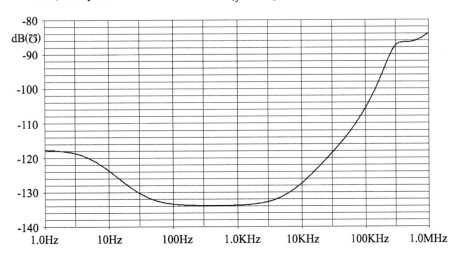

Figure 5.89

(b) Determine the outside input resistance, R_{in}, seen by the source by applying the EET directly to β_2 as explained in Eq. (5.24).

5.11 T and T' as functions of Z_L. The two loop gains, T and T', in Figs. 5.54b and 5.55b are related by:

$$T = T' \frac{1 + \dfrac{Z_L}{\mathscr{L}^{(L)}}}{1 + \dfrac{Z_L}{Z^{(L)}}} \tag{5.237}$$

in which (L) is the port across which the load impedance Z_L is connected. Using this fact, show that $T' > T$.

5.12 Output conductance of current-sense feedback amplifier. The frequency response of the output conductance Z_{of} of the amplifier in Example 5.11 is shown in Fig. 5.90.

(a) Using the numerical values in Problem 5.6 and the results derived in Example 5.11 verify the midband value of $G_{of} = -81.7\,\text{dB}\,(\Omega^{-1})$ shown in Fig. 5.90.

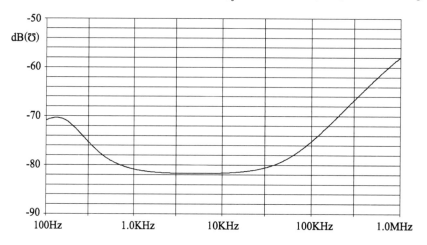

Figure 5.90

(b) Determine R_{of} directly by applying the EET to β_2 as explained in Eq. (5.24).

5.13 Proof of feedback formula for an arbitrary point of injection. Consider an arbitrary LTI system in which a signal u_z is injected at an arbitrary point as shown in Fig. 5.91. Prove the feedback formula in Eq. (5.44) using superposition to obtain:

$$\left. \begin{aligned} u_o &= a u_{in} + b u_z \\ u_y &= c u_{in} + d u_z \\ u_x &= e u_{in} + f u_z \end{aligned} \right\} \tag{5.238a-c}$$

We also have the constraint:

$$u_x = u_z + u_y \tag{5.239}$$

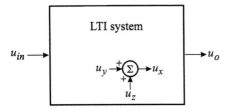

Figure 5.91

Hint: Express each of the following in terms of the coefficients in Eq. (5.238):

$$
\left.
\begin{aligned}
T &= \left.\frac{-u_y}{u_x}\right|_{u_{in}=0} \\[2ex]
G &= \left.\frac{u_o}{u_{in}}\right|_{u_z=0} \\[2ex]
G_\infty &= \left.\frac{u_o}{u_{in}}\right|_{u_y=0} \\[2ex]
G_o &= \left.\frac{u_o}{u_{in}}\right|_{u_x=0}
\end{aligned}
\right\}
\tag{5.240}
$$

Also, recognize that when $u_z = 0$, $u_y = u_x$.

5.14 Relationship between T, T_i and T_v in loop gain measurement of feedback systems. To measure the loop gain in an experimental circuit,[3] one cannot inject a signal immediately after a dependent source (ideal point) because of the inevitable internal impedance associated with all physical generators. Hence, either a voltage or a current source is injected immediately after the output impedance of the gain element (nonideal point) and a voltage or a current loop gain is determined as shown in Fig. 5.92. The impedance looking immediately to the right of z_o is represented by Z and the loop gain is given by:

$$
T = a_o \bar{a}_o = -a_o \beta \frac{Z}{Z + z_o}
\tag{5.241}
$$

(*a*) Show that the measurements T_v and T_i are related to T by:

$$
\left.
\begin{aligned}
T_v &= T\left(1 + \frac{z_o}{Z}\right) + \frac{z_o}{Z} \\[2ex]
T_i &= T\left(1 + \frac{Z}{z_o}\right) + \frac{Z}{z_o}
\end{aligned}
\right\}
\tag{5.242a, b}
$$

It follows that:

$$T_i = \frac{1 + T_v}{1 + \dfrac{z_o}{Z}} \tag{5.243}$$

Note that, in order for T_v to approximate T closely, we require that $z_o/Z \ll 1$ and $z_o/Z \ll T$ in the measured frequency range. Similarly, if T_i is to approximate T closely, we require that $Z/z_o \ll 1$ and $Z/z_o \ll T$ in the measured frequency range.

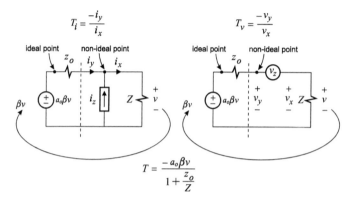

$$T_i = \frac{-i_y}{i_x} \qquad\qquad T_v = \frac{-v_y}{v_x}$$

$$T = \frac{-a_o\beta v}{1 + \dfrac{z_o}{Z}}$$

Figure 5.92

(b)   Show that the desired loop gain T can be obtained from the two measurements T_i and T_v according to:[3]

$$T = \frac{T_i T_v - 1}{2 + T_i + T_v} \tag{5.244}$$

5.15 Loop gain of the Colpitts oscillator. To determine the loop gain, replace the dependent current source with an independent current generator pointing in the opposite direction and apply the 3-EET to the reference circuit shown in Fig. 5.93.

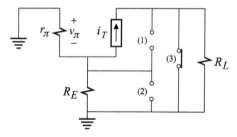

Figure 5.93

5.16 Approximate frequency of the Colpitts oscillator.

(a) Factor Eq. (5.191):

$$\omega_o = \sqrt{\frac{1}{LC_c \parallel C_E}\left[1 + \frac{Lg_m}{(C_c + C_E)R_L}\right]}$$ (5.245)

Use Eq. (5.186) to show that:

$$1 + \frac{Lg_m}{(C_c + C_E)R_L} = 1 + \frac{1}{Q^2} \approx 1$$ (5.246)

(b) A Colpitts oscillator is simulated using OrCAD/Pspice as shown in Fig. 5.94a. The output voltage is shown in Fig. 5.94b. Verify the condition and the frequency of oscillation. The amplitude of oscillation is limited by the non-linearity of the transconductance of the transistor.[4]

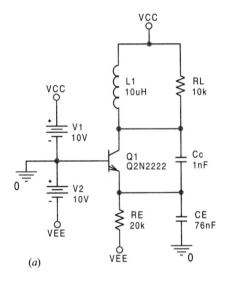

(a)

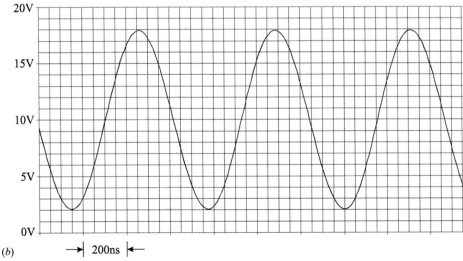

(b)　→| 200ns |←

Figure 5.94

(c) Wouldn't it be nice if you could measure the loop gain of the Colpitts oscillator and observe the fact that its gain margin is less than zero? But how can you do this if the circuit is oscillating? Well, the technique[3] discussed in Problem 5.14 can be applied in a way that the internal resistance of the injecting signal stabilizes the circuit while the loop gain is measured outside the internal resistance as shown in Fig. 5.94c.

Determine the loop gain using the simplest model of the transistor and compare it against the measurement shown in Fig. 5.94d. Also, make a Nyquist plot and observe the encirclement of the $(-1, j0)$ point.

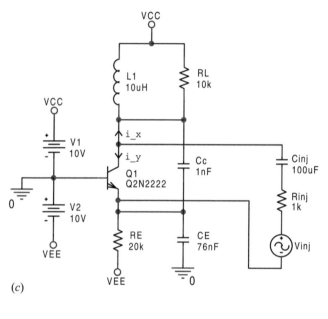

(c)

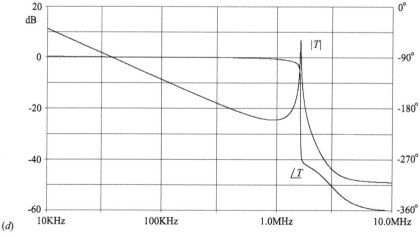

(d)

Figure 5.94 (cont.)

5.17 The 2-EET for the determination of $Z_o'(s)$. Using the reference circuit shown in Fig. 5.95 derive Eq. (5.196).

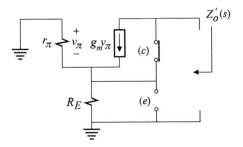

Figure 5.95

REFERENCES

1. H. S. Black, *Wave Translation System*, United States Patent No. 2,102,671. December 21, 1937.
2. R. D. Middlebrook, "Design-oriented analysis of feedback amplifiers", *Proc. of National Electronics Conference*, Vol. XX, Oct. 1964, pp. 1–4.
3. R. D. Middlebrook, "Measurement of loop gain in feedback systems", *Int. J. of Electronics*, Vol. 38, No. 4, April 1975, pp. 485–512.
4. Kenneth K. Clark and Donald T. Hess, *Communication Circuits: Analysis and Design*, Reading, Addison-Wesley, 1978.

6 High-frequency and microwave circuits
Doing it all in your head

6.1 Introduction

In this chapter, we shall apply the NEET to determine the frequency response of high-order networks using *minimum* algebra which in most cases can be worked out by inspection. Three representative examples of high-frequency circuits are worked out in detail. As is typical of the method of NEET, all the algebra is done on a set of very simple circuits derived from the original circuit using the rules of the NEET.

6.2 Cascode MOS amplifier

The current-loaded cascode MOS amplifier shown in Fig. 6.1 is a basic building block in many operational amplifier designs. We shall determine its high-frequency response using the small-signal model of the MOS transistor shown in Fig. 6.2. The equivalent circuit model of the amplifier is shown in Fig. 6.3 in which the small-signal elements of M_1 and M_2 are combined:

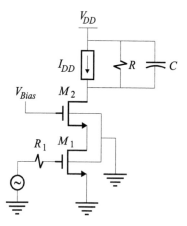

Figure 6.1

$$\left.\begin{aligned} C_1 &= C_{gs1} + C_{gb1} \\ C_2 &= C_{db1} + C_{sb2} + C_{gs2} \\ C_3 &= C_{gd1} \\ C_4 &= C + C_{gd2} + C_{db2} \\ g_{m2} &= g_{m2} + g_{mb2} \\ g_{m1} &= g_{m1} \end{aligned}\right\} \qquad (6.1a\text{–}f)$$

C_{gd}
G
D
+
C_{gs}
v_{gs}
−
$g_m v_{gs}$
$g_{mb} v_{bs}$
r_o
C_{db}
S
−
v_{bs}
C_{sb}
+
C_{gb}
B

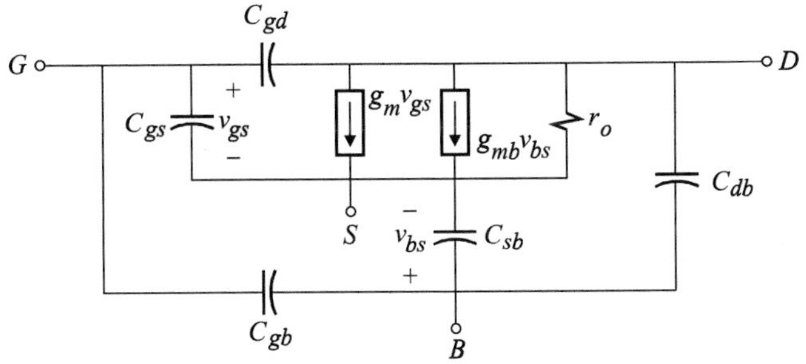

Figure 6.2

$g_{m2} v_2$
R_s
C_3
+
r_{o2}
+
v_{in}
C_1
v_1
$g_{m1} v_1$
r_{o1}
C_2
v_2
C_4
R
v_o
−
−
−

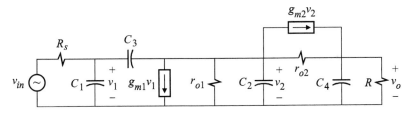

Figure 6.3

The voltage gain transfer function of the circuit in Fig. 6.3 is given by:

$$\frac{v_o(s)}{v_{in}(s)} = A_o \frac{N(s)}{D(s)} \qquad (6.2)$$

in which A_o, $N(s)$ and $D(s)$ will be determined separately.

The low-frequency asymptote A_o

To determine A_o, we open all the capacitors in Fig. 6.3 so that $v_1 = v_{in}$ and $g_{m1}v_{in}$ feeds the rest of the circuit. Now we see that if g_{m2} were infinite, then $v_2 \to 0$ (virtual ground) and all the current of $g_{m1}v_{in}$ would flow into the output resistor R and we would have:

$$v_o = -g_{m1}v_{in}R \Rightarrow A_o|_{g_{m2}\to\infty} = -g_{m1}R \qquad (6.3)$$

Hence, the gain A_o according to the EET is given by:

$$A = A_o|_{g_{m2} \to \infty} \frac{1 + \dfrac{1}{g_{m2}\bar{\bar{\mathscr{G}}}}}{1 + \dfrac{1}{g_{m2}\bar{G}}} \tag{6.4}$$

in which $\bar{\bar{\mathscr{G}}}$ is the inverse gain with respect to g_{m2} determined with the output voltage (the response) nulled and $\bar{G}$ is the inverse gain with respect to g_{m2} with v_{in} (the excitation) set to zero.

To determine $\bar{G}$, we set $v_{in} = 0$, replace $g_{m2}v_2$ with an independent current source i_T pointing in the opposite direction, take the Thevenin equivalent of i_T and r_{o2}, and finally determine v_2 by simple voltage division:

$$v_2 = i_T r_{o2} \frac{r_{o1}}{r_{o1} + r_{o2} + R} \tag{6.5}$$

It follows that:

$$\bar{G} \equiv \frac{v_2}{i_T}\bigg|_{v_{in}=0} = r_{o2} \frac{r_{o1}}{r_{o1} + r_{o2} + R} \tag{6.6}$$

To determine $\bar{\bar{\mathscr{G}}}$, we replace $g_{m2}v_2$ with an independent current source i_T pointing in the opposite direction and null the output voltage, which causes the current through R to be nulled and i_T to flow entirely through r_{o2}. Now the voltage across r_{o2} is $i_T r_{o2}$ and it appears directly across r_{o1} because v_o has been nulled. Since the voltage across r_{o1} is equal to v_2, it follows that:

$$\bar{\bar{\mathscr{G}}} \equiv \frac{v_2}{i_T}\bigg|_{v_o=0} = r_{o2} \tag{6.7}$$

Substituting these results in the expression of A in Eq. (6.4), we obtain:

$$A_o = -g_{m1}R \frac{1 + \dfrac{1}{g_{m2}r_{o2}}}{1 + \dfrac{1}{g_{m2}r_{o2}}\left(1 + \dfrac{R + r_{o2}}{r_{o1}}\right)} \tag{6.8}$$

The numerator $N(s)$

The numerator is determined by examining the transform circuit for a null in the response, i.e. $v_o(s) = 0$. This is shown in Fig. 6.4 in which we can easily see that:

$$g_{m2}v_2(s)r_{o2} = -v_2(s) \Rightarrow v_2(s) = 0 \tag{6.9}$$

Hence a null in $v_o(s)$ requires a null in $v_2(s)$ which in turn requires a null in $i_2(s)$

because $i_2(s) = v_2(s)/[r_{o1} \| (1/sC_2)]$. It follows that the current through C_3 must be equal to $g_{m1}v_1(s)$ and the voltage across C_1 must be equal to $v_1(s)$. Hence we have:

$$v_1(s)sC_3 = g_{m1}v_1(s) \Rightarrow 1 - \frac{sC_3}{g_{m1}} = 0 \Rightarrow N(s) = 1 - \frac{sC_3}{g_{m1}} \tag{6.10}$$

This is the usual RHP zero associated with the parasitic feedback capacitor in a common-source or common-emitter amplifier stage.

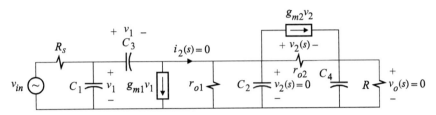

Figure 6.4

The denominator $D(s)$

The denominator is determined by setting the excitation to zero and applying the 4-EET to the reference circuit shown in Fig. 6.5 in which all the capacitors are taken as open circuits. Since there are only three linearly independent capacitors, we have a third-order system and $D(s)$ is given by:

$$D(s) = 1 + a_1s + a_2s^2 + a_3s^3 \tag{6.11}$$

The coefficients a_i are determined next.[†]

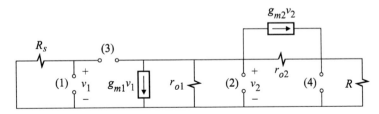

Figure 6.5

The coefficient a_1: This is given by:

$$a_1 = R^{(1)}C_1 + R^{(2)}C_2 + R^{(3)}C_3 + R^{(4)}C_4 \tag{6.12}$$

$R^{(1)}$: The driving-point port resistance $R^{(1)}$ can be seen by inspection to be given by:

[†] *Note:* Since we are going to refer to Fig. 6.5 in all the following calculations, it would be convenient to make a copy of it for handy reference.

$$R^{(1)} = R_s \tag{6.13}$$

$R^{(2)}$: When looking into port (2) with ports (3) and (1) open, the dependent current source $g_{m1}v_1$ is inactive so that the driving-point resistance $R^{(2)}$ consists of the parallel combination of r_{o1} with the resistance to the right of port (2). The latter can be determined using the EET for g_{m2} as follows. If we let $g_{m2} \rightarrow 0$, then looking to the right of port (2) we have $r_{o2} + R$. According to the EET, we replace $g_{m2}v_2$ with an independent current source pointing in the opposite direction and determine v_2 once with port (2) open and once with port (2) short. When port (2) is short, $v_2 = 0$ and the inverse gain is zero. When port (2) is open, $v_2 = i_T r_{o2}$ and the inverse gain is $v_2/i_T = r_{o2}$. Hence, the resistance looking to the right of port (2) is:

$$(r_{o2} + R) \frac{1 + g_{m2}.0}{1 + g_{m2}r_{o2}} \tag{6.14}$$

It follows that $R^{(2)}$ is given by:

$$R^{(2)} = r_{o1} \left\| \frac{r_{o2} + R}{1 + g_{m2}r_{o2}} \right. \tag{6.15}$$

$R^{(3)}$: When determining $R^{(3)}$, we can first set R_s to zero so that v_1 and $g_{m1}v_1$ both vanish to yield:

$$R^{(3)}\big|_{R_s=0} = R^{(2)} = r_{o1} \left\| \frac{r_{o2} + R}{1 + g_{m2}r_{o2}} \right. \tag{6.16}$$

Now, according to the EET, R_s is reinstated in $R^{(3)}$:

$$R^{(3)} = R^{(3)}\big|_{Rs=0} \frac{1 + R_s/\mathscr{R}^{(s)}}{1 + R_s/R^{(s)}} \tag{6.17}$$

in which $R^{(s)}$ and $\mathscr{R}^{(s)}$ are the resistances looking in from the port across which R_s is connected with port (3) once open and once short, respectively. When port (3) is opened, $R^{(s)} \rightarrow \infty$ and the denominator in Eq. (6.17) becomes unity. When port (3) is short, v_1 appears directly across the dependent source $g_{m1}v_1$ so that $g_{m1}v_1$ acts like a conductance g_{m1} in parallel with the rest of the circuit to its right which has an input resistance $R^{(2)}$. Hence, $\mathscr{R}^{(s)}$ is given by:

$$\mathscr{R}^{(s)} = \frac{1}{g_{m1}} \left\| R^{(2)} \Rightarrow \frac{1}{\mathscr{R}^{(s)}} = g_{m1} + \frac{1}{R^{(2)}} \right. \tag{6.18}$$

Substituting these results in $R^{(3)}$, we obtain:

$$R^{(3)} = R^{(2)} \left[1 + R_s \left(g_{m1} + \frac{1}{R^{(2)}} \right) \right]$$

$$= R_s + (1 + g_{m1}R_s)R^{(2)} \tag{6.19a, b}$$

Substituting for $R^{(2)}$:

$$R^{(3)} = R_s + (1 + g_{m1}R_s)\left(r_{o1} \left\| \frac{r_{o2} + R}{1 + g_{m2}r_{o2}} \right.\right) \tag{6.20}$$

$R^{(4)}$: Looking into port (4), we see R in parallel with the resistance looking in from the right of port (4) which, when $g_{m2} = 0$, is equal to $r_{o1} + r_{o2}$. To reinstate g_{m2}, we replace $g_{m2}v_2$ with an independent current source i_T pointing in the opposite direction and determine v_2 once with port (4) open and once with port (4) short. (Remember that R is open because it has already been accounted for.) Now, when port (4) is open, $v_2 = 0$; and when port (4) is short, $v_2 = i_T r_{o1} \| r_{o2}$. Hence, $\bar{G} = 0$ and $\mathscr{G} = r_{o1} \| r_{o2}$ and according to the EET we have:

$$R^{(4)} = R \left\| \left[(r_{o1} + r_{o2}) \frac{1 + g_{m2}\mathscr{G}}{1 + g_{m2}\bar{G}} \right] \right.$$

$$= R \| [(r_{o1} + r_{o2})(1 + g_{m2}r_{o1} \| r_{o2})] \tag{6.21a, b}$$

It is worthwhile to perform the following simplification:

$$(r_{o1} + r_{o2})(1 + g_{m2}r_{o1} \| r_{o2}) = r_{o1} + r_{o2} + g_{m2}r_{o1}r_{o2}$$

$$= r_{o2} + r_{o1}(1 + g_{m2}r_{o2}) \tag{6.22}$$

We rewrite $R^{(4)}$ as:

$$R^{(4)} = R \| [r_{o2} + r_{o1}(1 + g_{m2}r_{o2})] \tag{6.23}$$

The coefficient a_1 is obtained by substituting for $R^{(i)}$ in Eq. (6.12):

$$a_1 = C_1 R_s + C_2 r_{o1} \left\| \frac{r_{o2} + R}{1 + g_{m2}r_{o2}} \right.$$

$$+ C_3 \left[R_s + (1 + g_{m1}R_s)\left(r_{o1} \left\| \frac{r_{o2} + R}{1 + g_{m2}r_{o2}} \right.\right) \right] \tag{6.24}$$

$$+ C_4 R \| [r_{o2} + r_{o1}(1 + g_{m2}r_{o2})]$$

The coefficient a_2: This is given by:

$$a_2 = C_1 R^{(1)} C_2 R_{(1)}^{(2)} + C_1 R^{(1)} C_3 R_{(1)}^{(3)} + C_1 R^{(1)} C_4 R_{(1)}^{(4)}$$

$$+ C_2 R^{(2)} C_3 R_{(2)}^{(3)} + C_2 R^{(2)} C_4 R_{(2)}^{(4)} + C_3 R^{(3)} C_4 R_{(3)}^{(4)} \tag{6.25}$$

$R_{(1)}^{(2)}$: This is the same as $R^{(2)}$ since the condition of port (1) does not affect the resistance looking into port (2). In both cases $v_1 = 0$ and the dependent source $g_{m1}v_1$ is inactive. Hence:

$$R_{(1)}^{(2)} = R^{(2)} \tag{6.26}$$

$R_{(1)}^{(3)}$: Upon shorting port (1), we see that ports (3) and (2) become coincident so that:

$$R_{(1)}^{(3)} = R^{(2)} \tag{6.27}$$

$R_{(1)}^{(4)}$: This is the same as $R^{(4)}$ because the condition of port (1) in this case does not affect the resistance looking into port (4). Hence:

$$R_{(1)}^{(4)} = R^{(4)} \tag{6.28}$$

$R_{(2)}^{(3)}$: When port (2) is shorted, pretty much everything to the right of port (3) is wiped out and all we see looking into port (3) is R_s:

$$R_{(2)}^{(3)} = R_s \tag{6.29}$$

$R_{(2)}^{(4)}$: When port (2) is shorted, $v_2 = 0$ and $g_{m2}v_2$ vanishes. Hence, looking into port (4) we see:

$$R_{(2)}^{(4)} = r_{o2} \parallel R \tag{6.30}$$

$R_{(3)}^{(4)}$: When port (3) is shorted, v_1 is impressed across $g_{m1}v_1$ which in turn acts as a conductance g_{m1} in parallel with r_{o1} and R_s. Hence, looking into port (4) in this case, we see the same arrangement as in $R^{(4)}$ except for the fact that r_{o1} is replaced with $r_{o1} \parallel R_s \parallel g_{m1}^{-1}$. Hence we have:

$$R_{(3)}^{(4)} = R^{(4)} \big|_{r_{o1} \to r_{o1} \parallel R_s \parallel g_{m1}^{-1}} = R \parallel [r_{o2} + (r_{o1} \parallel R_s \parallel g_{m1}^{-1})(1 + g_{m2}r_{o2})] \tag{6.31}$$

Substituting for $R_{(i)}^{(j)}$ in the expression for a_2, we obtain:

$$
\begin{aligned}
a_2 = C_1 R_s &\left\{ (C_2 + C_3)r_{o1} \left\| \frac{r_{o2} + R}{1 + g_{m2}r_{o2}} \right. \right. \\
&\left. + C_4 R \parallel [r_{o2} + r_{o1}(1 + g_{m2}r_{o2})] \right\} \\
&+ C_2 r_{o1} \left\| \frac{r_{o2} + R}{1 + g_{m2}r_{o2}} (C_3 R_s + C_4 r_{o2} \parallel R) \right. \\
&+ C_3 C_4 R \parallel [r_{o2} + (r_{o1} \parallel R_s \parallel g_{m1}^{-1})(1 + g_{m2}r_{o2})] \\
&\times \left[R_s + (1 + g_{m1}R_s) \left(r_{o1} \left\| \frac{r_{o2} + R}{1 + g_{m2}r_{o2}} \right. \right) \right]
\end{aligned}
\tag{6.32}
$$

The coefficient a_3: This is given by:

$$
\begin{aligned}
a_3 = &\, C_1 R^{(1)} C_2 R_{(1)}^{(2)} C_3 R_{(1,2)}^{(3)} + C_1 R^{(1)} C_2 R_{(1)}^{(2)} C_4 R_{(1,2)}^{(4)} \\
&+ C_1 R^{(1)} C_3 R_{(1)}^{(3)} C_4 R_{(1,3)}^{(4)} + C_2 R^{(2)} C_3 R_{(2)}^{(3)} C_4 R_{(2,3)}^{(4)}
\end{aligned}
\tag{6.33}
$$

$R_{(1,2)}^{(3)}$: Clearly, this is zero:

$$R_{(1,2)}^{(3)} = 0 \tag{6.34}$$

$R_{(1,2)}^{(4)}$, $R_{(1,3)}^{(4)}$, $R_{(2,3)}^{(4)}$: All of these are easily seen to be equal to r_{o2}:

$$R_{(1,2)}^{(4)} = R_{(1,3)}^{(4)} = R_{(2,3)}^{(4)} = r_{o2} \tag{6.35}$$

Substituting these results in the expression of a_3, we obtain:

$$a_3 = [C_1(C_2 + C_3) + C_2 C_3]C_4 r_{o2} R_s \left(r_{o1} \left\| \frac{r_{o2} + R}{1 + g_{m2} r_{o2}} \right. \right) \tag{6.36}$$

This completes the determination of the transfer function. In the following example, numerical values are used to obtain some useful analytical approximations.

Example 6.1   For a certain NMOS biased at $I_D = 100\,\mu A$, $V_{DS} = 5\,V$ and $V_{SB} = 2\,V$ the following element values are determined along with R_s, R and C:

$C_{gs} = 0.08\,pF$	$g_m = 98\,\mu\Omega^{-1}$	$R_s = 100\,\Omega$
$C_{gb} = 0.05\,pF$	$g_{mb} = 15\,\mu\Omega^{-1}$	$R = 25\,M\Omega$
$C_{gd} = 0.01\,pF$	$r_o = 500\,k\Omega$	$C = 0.05\,pF$
$C_{sb} = 0.05\,pF$		
$C_{db} = 0.03\,pF$		

The values in the equivalent circuit model of the amplifier in Fig. 6.5 as given in Eqs. (6.1a–f) are computed to be:

$C_1 = 0.13\,pF$	$g_{m1} = 98\,\mu\Omega^{-1}$
$C_2 = 0.16\,pF$	$g_{m2} = 113.2\,\mu\Omega^{-1}$
$C_3 = 0.01\,pF$	
$C_4 = 0.09\,pF$	

The transfer function is given by:

$$A(s) = A_o \frac{1 - s/\omega_z}{1 + a_1 s + a_2 s^2 + a_3 s^3} \tag{6.37}$$

in which

$$\left. \begin{aligned} A_o &= 1.299 \times 10^3 \ (62.275\,dB) \\ \omega_z &= 2\pi(1.56 \times 10^9)\,rad/s \\ a_1 &= 1.254 \times 10^{-6}\,s \\ a_2 &= 1.779 \times 10^{-15}\,s^2 \\ a_3 &= 3.074 \times 10^{-27}\,s^3 \end{aligned} \right\} \tag{6.38a–e}$$

A Bode plot of the magnitude response $|A(j\omega)|$ is shown in Fig. 6.6 in which the bandwidth is seen to be 126 kHz, which is almost entirely dictated by the dominant pole.

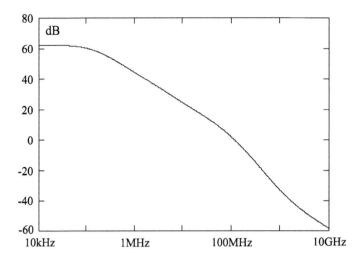

Figure 6.6

Since the roots of the denominator are very well separated, they can be factored to an excellent approximation (see Problem 6.1):

$$A(s) \approx A_o \frac{1 - \dfrac{s}{\omega_z}}{(1 + a_1 s)\left(1 + \dfrac{a_2}{a_1}s\right)\left(1 + \dfrac{a_3}{a_2}s\right)} \tag{6.39}$$

A magnitude plot of $|A(j\omega)|$ in Eq. (6.39) essentially lies on top of the exact plot in Fig. 6.6 so that the dominant pole to an excellent approximation is given by:

$$f_{pl} = \frac{1}{2\pi} \frac{1}{a_1} = 126.9 \text{ kHz} \tag{6.40}$$

which is seen to be in excellent agreement with Fig. 6.6. If desired, the expression of a_1 can be approximated further using the numerical values given by:

$$a_1 \approx C_2 r_{o1} \left\| \frac{r_{o2} + R}{1 + g_{m2} r_{o2}} + C_4 R \right\| [r_{o2} + r_{o1}(1 + g_{m2} r_{o2})] \tag{6.41}$$

Hence a simple expression for the bandwidth (BW), or the dominant pole, is given by:

$$BW = f_{p1} \approx \frac{1}{2\pi} \frac{1}{C_2 r_{o1} \left\| \dfrac{r_{o2} + R}{1 + g_{m2} r_{o2}} + C_4 R \right\| [r_{o2} + r_{o1}(1 + g_{m2} r_{o2})]} \tag{6.42}$$

The value of f_{p1} obtained from the approximate expression in Eq. (6.42) is 127.2 kHz, which is less than 0.1% from the exact value of 127.1 kHz. The following approximate analytical expression for the second pole can also be obtained:

$$f_{p2} \approx \frac{a_1}{a_2}$$

$$\approx \frac{1}{1 + \dfrac{C_2}{C_4} \, r_{o1} \left\| \dfrac{r_{o2}}{1 + g_{m2} r_{o2}} \right.} \frac{1}{(C_1 + C_3)R_s + (C_2 + C_3)r_{o1} \left\| \dfrac{r_{o2}}{1 + g_{m2} r_{o2}} \right.} \tag{6.43}$$

The value of f_{p2} obtained from this approximation is 710.7 MHz, which is less than 1% away from the exact value of 704.9 MHz. It is possible to obtain an analytical expression for f_{p3} but since it falls outside the range of validity of the model (92 GHz), the expression will not be of much value. ☐

6.3 Fifth-order Chebyshev low-pass filter

Lumped-parameter circuit models are often used to model and synthesize, quite satisfactorily, distributed microwave structures that act as various types of filters. An example of such a circuit is the low-pass ladder network shown in Fig. 6.7. The elements of this circuit can be chosen in such a way to yield a Chebyshev response with a specified pass-band ripple. Since neither the pass elements of this ladder network have any poles nor its shunt elements have any zeros, the transfer function has no zeros and is given by:

$$A(s) = \frac{v_o(s)}{v_{in}(s)} = \frac{1}{1 + R_1/R_5} \frac{1}{1 + \displaystyle\sum_{n=1}^{5} a_n s^n} \tag{6.44}$$

The coefficient a_1: This is given by:

$$a_1 = C_1 R^{(1)} + C_3 R^{(3)} + C_5 R^{(5)} + \frac{L_2}{R^{(2)}} + \frac{L_4}{R^{(4)}} \tag{6.45}$$

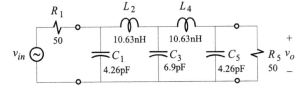

Figure 6.7

in which $R^{(i)}$ are determined using the reference circuit in Fig. 6.8.

$R^{(1)}$, $R^{(3)}$, $R^{(5)}$: It is clear from Fig. 6.8 that all of these are equal and are given by:

$$R^{(1)} = R^{(3)} = R^{(5)} = R_1 \parallel R_5 \tag{6.46}$$

$R^{(2)}$, $R^{(4)}$: We can see from Fig. 6.8 that these two are equal and are given by:

$$R^{(2)} = R^{(4)} = R_1 + R_5 \tag{6.47}$$

Substituting, we obtain:

$$a_1 = (C_1 + C_3 + C_5)R_1 \parallel R_5 + \frac{L_2 + L_4}{R_1 + R_5} \tag{6.48}$$

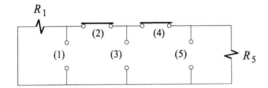

Figure 6.8

The coefficient a_2: This is given by:

$$a_2 = C_1 R^{(1)} \frac{L_2}{R_{(1)}^{(2)}} + C_1 R^{(1)} C_3 R_{(1)}^{(3)} + C_1 R^{(1)} \frac{L_4}{R_{(1)}^{(4)}}$$

$$+ C_1 R^{(1)} C_5 R_{(1)}^{(5)} + \frac{L_2}{R^{(2)}} C_3 R_{(2)}^{(3)} + \frac{L_2}{R^{(2)}} \frac{L_4}{R_{(2)}^{(4)}}$$

$$+ \frac{L_2}{R^{(2)}} C_5 R_{(2)}^{(5)} + C_3 R^{(3)} \frac{L_4}{R_{(3)}^{(4)}} + C_2 R^{(3)} C_5 R_{(3)}^{(5)} \tag{6.49}$$

$$+ \frac{L_4}{R^{(4)}} C_5 R_{(4)}^{(5)}$$

$R_{(2)}^{(4)}$: This is seen to be infinite.

$R_{(1)}^{(3)}$, $R_{(1)}^{(5)}$, $R_{(3)}^{(5)}$: These are all easily seen to be zero.

$R_{(1)}^{(2)}$, $R_{(1)}^{(4)}$, $R_{(2)}^{(3)}$, $R_{(2)}^{(5)}$, $R_{(3)}^{(4)}$, $R_{(4)}^{(5)}$: These are all equal to R_5.

Substituting these results in Eq. (6.49), we obtain:

$$a_2 = [C_1(L_2 + L_4) + C_3 L_4] \frac{1}{1 + \dfrac{R_5}{R_1}}$$

$$+ [L_2(C_3 + C_5) + L_4 C_5] \frac{1}{1 + \dfrac{R_1}{R_5}} \tag{6.50}$$

The coefficient a_3: This is given by the sum of the product of three time constants formed by taking three ports at a time which can be written in the following compact way:

$$a_3 = (1,2).C_3R^{(3)}_{(1,2)} + (1,2).\frac{L_4}{R^{(4)}_{(1,2)}} + (1,2).C_5R^{(5)}_{(1,2)}$$

$$+ (1,3).\frac{L_4}{R^{(4)}_{(1,3)}} + (1,3).C_5R^{(5)}_{(1,3)} + (1,4).C_5R^{(5)}_{(1,4)}$$

$$+ (2,3).\frac{L_4}{R^{(4)}_{(2,3)}} + (2,3).C_5R^{(5)}_{(2,3)} + (2,4).C_5R^{(5)}_{(2,4)} \qquad (6.51)$$

$$+ (3,4).C_5R^{(5)}_{(3,4)}$$

Here, we have used the notation (i,j) to denote $\tau^{(i)}\tau^{(j)}_{(i)}$ in which τ represents the proper time constant formed at each port. Note that we have retained $(1,3)$ and $(2,4)$ in the expression of a_3 even though these were determined to be zero in our previous calculation of a_2 because we would like to make sure that we do not miss any indeterminacy. For example if $R^{(4)}_{(1,3)} = 0$, then we will have an indeterminacy of the type $0/0$ in the coefficient of $(1,3)$ in Eqn. (6.51). We shall find out in this example that there are no indeterminate terms.

The following are verified easily from Fig. 6.8:

$$R^{(3)}_{(1,2)} = R^{(5)}_{(1,2)} = R^{(4)}_{(1,3)} = R^{(4)}_{(2,3)} = R^{(5)}_{(1,4)} = R^{(5)}_{(2,4)} = R^{(5)}_{(3,4)} = R_5 \qquad (6.52)$$

$$R^{(4)}_{(1,2)} \to \infty \qquad (6.53)$$

$$R^{(5)}_{(1,3)} = R^{(5)}_{(2,3)} = 0 \qquad (6.54)$$

Substituting these results in Eq. (6.51), we obtain:

$$a_3 = [C_1L_2(C_3 + C_5) + C_5L_4(C_1 + C_3)]R_1 \| R_5 + \frac{L_2L_4C_3}{R_1 + R_5} \qquad (6.55)$$

The coefficient a_4: This is given by:

$$a_4 = (1,2,3).\frac{L_4}{R^{(4)}_{(1,2,3)}} + (1,2,3).C_5R^{(5)}_{(1,2,3)}$$

$$+ (1,2,4).C_5R^{(5)}_{(1,2,4)} + (1,3,4).C_5R^{(5)}_{(1,3,4)} \qquad (6.56)$$

$$+ (2,3,4).C_5R^{(5)}_{(2,3,4)}$$

The following are verified easily from Fig. 6.8:

$$R^{(4)}_{(1,2,3)} = R^{(5)}_{(1,2,4)} = R^{(5)}_{(2,3,4)} = R^{(5)}_{(2,3,4)} = R_5 \tag{6.57}$$

$$R^{(4)}_{(1,2,3)} = 0 \tag{6.58}$$

We also have from previous calculations for a_2 and a_3 that $(1,2,4)$ and $(1,3,4)$ are both zero so that a_4 in Eq. (6.56) reduces to:

$$a_4 = C_3 L_2 L_4 \left(\frac{C_1}{1 + R_5/R_1} + \frac{C_5}{1 + R_1/R_5} \right) \tag{6.59}$$

The coefficient a_5: Finally we have for a_5:

$$a_5 = (1,2,3,4) . C_5 R^{(5)}_{(1,2,3,4)} \tag{6.60}$$

Since $R^{(5)}_{(1,2,3,4)} = R_5$ we have:

$$a_5 = C_1 C_3 C_5 L_2 L_4 R_1 \| R_5 \tag{6.61}$$

This completes the determination of the voltage gain transfer function or equivalently the S_{21} parameter of the filter. In the following example, numerical values are given to demonstrate a Chebyshev response.

Example 6.2   For the element values[2] shown in Fig. 6.7, the following numerical values for the coefficients a_i are obtained:

$$a_1 = 0.598 \times 10^{-9}\, s \qquad a_3 = 3.307 \times 10^{-29}\, s^3 \qquad a_5 = 3.537 \times 10^{-49}\, s^5$$
$$a_2 = 1.639 \times 10^{-19}\, s^2 \qquad a_4 = 3.321 \times 10^{-39}\, s^4$$

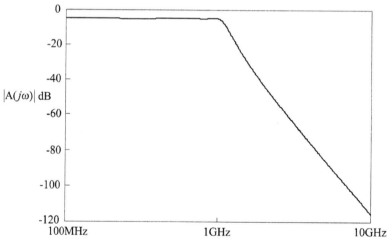

Figure 6.9

A magnitude plot is shown in Fig. 6.9. An expanded view of the 0.2-dB variation in the passband is shown in Fig. 6.10.   □

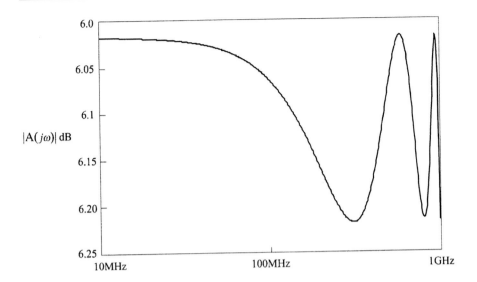

Figure 6.10

6.4   MESFET amplifier

The equivalent circuit model of a single-stage MESFET amplifier[2] is shown in Fig. 6.11. We shall determine the voltage gain or equivalently the S_{21} parameter of the MESFET. Although there are seven reactive elements, the circuit is only fifth-order because only two of the three capacitors (C_2, C_3, C_4) in a loop are linearly independent and only two of the three inductors (L_1, L_5, L_6) connected to the ground node are linearly independent. There are numerous calculations all of which will be performed by visual inspection of the circuit.

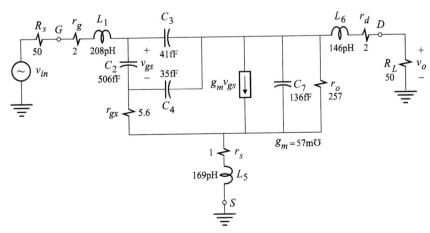

Figure 6.11

The transfer function is given by:

$$\frac{v_o(s)}{v_{in}(s)} = A_o \frac{N(s)}{D(s)} \tag{6.62}$$

The low-frequency asymptote A_o

This is determined by replacing all the capacitors with open circuits and the inductors with short circuits as shown in Fig. 6.12. Now, if we let $g_m \rightarrow \infty$, then $v_{gs} \rightarrow 0$ and v_{in} appears across r_s and causes a current v_{in}/r_s to flow through it. Since r_s carries the output load current, the output voltage is given by:

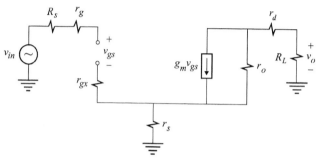

Figure 6.12

$$v_o = -R_L(v_{in}/r_s) \tag{6.63a}$$

It follows that:

$$A_o|_{g_m \rightarrow \infty} = -\frac{R_L}{r_s} \tag{6.63b}$$

To reinstate g_m, $g_m v_{gs}$ is replaced with an independent current source i_T pointing in the opposite direction. The gain from i_T to v_{gs} with $v_{in} = 0$ is the same as the gain to the voltage across r_s which can be seen (after taking a Thevenin equivalent of i_T and r_o) to be:

$$\bar{G} = \frac{v_{gs}}{i_T} = r_o \frac{r_s}{r_s + r_d + r_o + R_L} \tag{6.64}$$

The same gain with v_o nulled is infinite because the only way v_o can be nulled in this case is by letting $i_T = 0$. Hence, $\mathcal{G} \rightarrow \infty$ and we have:

$$A_o = A_o|_{g_m \rightarrow \infty} \frac{1 + 1/(g_m \mathcal{G})}{1 + 1/(g_m \bar{G})}$$

$$= -\frac{R_L}{r_s} \frac{1}{1 + \dfrac{1}{g_m r_o}\left[1 + \dfrac{r_o + r_d + R_L}{r_s}\right]} \tag{6.65}$$

The denominator $D(s)$

Since we have a fifth-order system, the denominator is given by:

$$D(s) = 1 + \sum_{n=1}^{5} a_n s^n \tag{6.66}$$

In determining the coefficients of the denominator, we refer to the reference circuit in Fig. 6.13[†] in which:

$$\left. \begin{array}{l} R_1 = R_s + r_g \\ R = R_L + r_d \end{array} \right\} \tag{6.67}$$

The EET for the dependent current source g_m will be used extensively when determining a driving-point impedance looking into a port:

$$R^{(j)}_{(.)} = R^{(j)}_{(.)} \bigg|_{g_m = 0} \frac{1 + g_m \mathcal{G}}{1 + g_m \bar{G}} \tag{6.68}$$

In this equation $\mathcal{G}$ is the inverse gain with port (j) short and $\bar{G}$ is the inverse gain with port (j) open. These are explicitly defined as:

$$\left. \begin{array}{l} \mathcal{G} = \dfrac{v_{gs}}{i_T} \bigg|_{(j) \to short} \\[4mm] \bar{G} = \dfrac{v_{gs}}{i_T} \bigg|_{(j) \to open} \end{array} \right\} \tag{6.69a, b}$$

in which:

$$\begin{array}{ll} i_T \equiv & \text{independent current source replacing } g_m v_{gs} \\ & \text{and pointing in the opposite direction.} \end{array} \tag{6.70}$$

The coefficient a_1: This is given by:

$$a_1 = \Sigma \tau^{(i)} \tag{6.71}$$

in which the summation is taken over the seven ports for which we determine the following resistances by inspecting Fig. 6.13.

$R^{(1)}$: This is clearly seen to be infinite so that:

$$\tau^{(1)} = \frac{L_1}{R^{(1)}} = 0 \tag{6.72}$$

$R^{(2)}$: This is determined by application of the EET for g_m so that we have:

[†] *Note*: Since we are going to refer to this circuit throughout the remainder of this section, it would be convenient to make a copy of it for handy reference.

$$R^{(2)} = R^{(2)}\big|_{g_m = 0} \frac{1 + g_m\mathscr{G}}{1 + g_m\bar{G}} \tag{6.73}$$

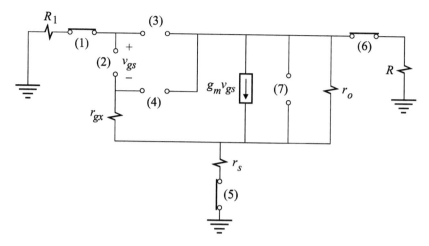

Figure 6.13

With $g_m = 0$, $R^{(2)}$ is seen to be given by:

$$R^{(2)}\big|_{g_m = 0} = R_1 + r_{gx} + r_s \| (r_o + R) \tag{6.74}$$

The inverse gain with respect to $g_m v_{gs}$ with port (2) short is clearly zero so that $\mathscr{G} = 0$ while the inverse gain with port (2) open is the same as in Eq. (6.64), discussed earlier in the determination of A_o. Hence, we have:

$$R^{(2)} = \frac{R_1 + r_{gx} + r_s \| (r_o + R)}{1 + \dfrac{g_m r_o}{1 + \dfrac{r_o + R}{r_s}}} \tag{6.75}$$

$R^{(3)}$: Following the same argument in $R^{(2)}$, we have for $R^{(3)}$:

$$R^{(3)} = R^{(3)}\big|_{g_m = 0} \frac{1 + g_m\mathscr{G}}{1 + g_m\bar{G}} \tag{6.76}$$

With $g_m = 0$, $R^{(3)}$ is given by:

$$R^{(3)}\big|_{g_m = 0} = R_1 + R \| (r_o + r_s) \tag{6.77}$$

The inverse gain $\bar{G}$ is still the same as the one given in Eq. (6.64) but $\mathscr{G}$ is different. In this case, when port (3) is shorted, R_1 becomes parallel with R and v_{gs} appears across r_o. It follows that:

$$\mathscr{G} = r_o \| (r_s + R \| R_1) \tag{6.78}$$

Substituting these results in Eq. (6.76), we obtain:

$$R^{(3)} = [R_1 + R \parallel (r_o + r_s)]\frac{1 + g_m r_o \parallel (r_s + R \parallel R_1)}{1 + \dfrac{g_m r_o}{1 + \dfrac{r_o + R}{r_s}}}$$
(6.79)

$R^{(4)}$: This is given by:

$$R^{(4)} = R^{(4)}\Big|_{g_m=0}\frac{1 + g_m\mathscr{G}}{1 + g_m\bar{G}}$$
(6.80)

in which we can see that:

$$R^{(4)}\Big|_{g_m=0} = r_{gx} + r_o \parallel (r_s + R)$$
(6.81)

The inverse gain $\bar{G}$ is still the same as before, but $\mathscr{G}$ is different and somewhat tricky! Observe carefully that when port (4) is shorted, v_{gs} appears across R with its positive polarity on the ground terminal. Therefore, the inverse gain with port (4) shorted is seen (after taking the Thevenin equivalent of i_T and $r_o \parallel r_{gx}$) to be given by:

$$\mathscr{G} = -r_o \parallel r_{gx}\frac{R}{R + r_s + r_o \parallel r_{gx}}$$
(6.82)

Substituting these results in Eq. (6.80), we obtain:

$$R^{(4)} = [r_{gx} + r_o \parallel (r_s + R)]\frac{1 - g_m\dfrac{r_o \parallel r_{gx}R}{R + r_s + r_o \parallel r_{gx}}}{1 + \dfrac{g_m r_o}{1 + \dfrac{r_o + R}{r_s}}}$$
(6.83)

$R^{(5)}$: This is given by:

$$R^{(5)} = R^{(5)}\Big|_{g_m=0}\frac{1 + g_m\mathscr{G}}{1 + g_m\bar{G}}$$
(6.84)

in which we can easily verify that:

$$R^{(5)}\Big|_{g_m=0} = r_s + r_o + R$$
(6.85)

The inverse gain $\mathscr{G}$ in Eq. (6.84) is determined with port (5) short, which is the same as $\bar{G}$ determined in Eq. (6.64), so that we have:

$$\mathscr{G} = \frac{r_o r_s}{r_s + R_o + R}$$
(6.86)

The inverse gain $\bar{G}$ in Eq. (6.84) is determined with port (5) open so that it can be obtained from $\mathcal{G}$ in Eq. (6.86) simply by letting $r_s \to \infty$. Hence we have:

$$\bar{G} = \frac{r_o r_s}{r_s + r_o + R}\bigg|_{rs \to \infty} = r_o \tag{6.87}$$

Substituting these results in Eq. (6.84), we obtain:

$$R^{(5)} = (r_s + r_o + R)\frac{1 + g_m\dfrac{r_o r_s}{r_s + r_o + R}}{1 + g_m r_o}$$

$$= r_s + \frac{r_o + R}{1 + g_m r_o} \tag{6.88}$$

$R^{(6)}$: This is given by:

$$R^{(6)} = R^{(6)}\big|_{g_m = 0}\frac{1 + g_m\mathcal{G}}{1 + g_m\bar{G}} \tag{6.89}$$

in which:

$$R^{(6)}\big|_{g_m = 0} = R + r_o + r_s$$

As in $R^{(5)}$, the inverse gain $\mathcal{G}$ is given by Eq. (6.86). The inverse gain $\bar{G}$ is zero because when port (6) is opened there is no contribution from i_T (defined in Eq. (6.70)) to v_{gs}. This is so because there is no current flow in r_s and both sides of v_{gs} are at ground potential. Hence, $v_{gs} = 0$, $\bar{G} = 0$ and $R^{(6)}$ is given by:

$$R^{(6)} = (R + r_o + r_s)\left[1 + g_m\frac{r_o r_s}{r_s + r_o + R}\right]$$

$$= r_o + R + r_s(1 + g_m r_o) \tag{6.90}$$

$R^{(7)}$: This is given by r_o in parallel with whatever comes after r_o, which we call $R^{(7')}$, and write it as:

$$R^{(7')} = R^{(7')}\big|_{g_m = 0}\frac{1 + g_m\mathcal{G}}{1 + g_m\bar{G}} \tag{6.91}$$

Clearly, $\mathcal{G} = 0$ because when port (7) is shorted, the independent current source i_T will be shorted and cannot contribute anything to v_{gs}. The inverse gain with port (7) open is simply r_s because i_T passes through r_s which causes v_{gs} to appear directly across r_s. We can also see that with $g_m = 0$, $R^{(7')}$ is simply equal to $R + r_s$. Hence we have:

$$R^{(7')} = \frac{R + r_s}{1 + g_m r_s} \tag{6.92}$$

It follows that:

$$R^{(7)} = r_o \left\| \frac{R + r_s}{1 + g_m r_s} \right.$$

$$= \left. \frac{r_o \| (R + r_s)}{1 + \dfrac{g_m r_o}{1 + \dfrac{r_o + R}{r_s}}} \right\} \qquad (6.93a, b)$$

The time constants which enter into the expression of a_1 in Eq. (6.71) are now summarized:

$$\tau^{(1)} = 0$$

$$\tau^{(2)} = R^{(2)}C_2 = C_2 \frac{R_1 + r_{gx} + r_s \| (r_o + R)}{1 + \dfrac{g_m r_o}{1 + \dfrac{r_o + R}{r_s}}}$$

$$\tau^{(3)} = R^{(3)}C_3 = C_3[R_1 + R \| (r_o + r_s)] \frac{1 + g_m r_o \| (r_s + R \| R_1)}{1 + \dfrac{g_m r_o}{1 + \dfrac{r_o + R}{r_s}}}$$

$$\tau^{(4)} = R^{(4)}C_4 = C_4[r_{gx} + r_o \| (r_s + R)] \frac{1 - \dfrac{g_m R}{1 + \dfrac{R + r_s}{r_o \| r_{gx}}}}{1 + \dfrac{g_m r_o}{1 + \dfrac{R + r_o}{r_s}}} \qquad (6.94a\text{--}g)$$

$$\tau^{(5)} = \frac{L_5}{R^{(5)}} = \frac{L_5}{r_s + \dfrac{r_o + R}{1 + g_m r_o}}$$

$$\tau^{(6)} = \frac{L_6}{R^{(6)}} = \frac{L_6}{r_o + R + r_s(1 + g_m r_o)}$$

$$\tau^{(7)} = R^{(7)}C_7 = C_7 r_o \left\| \frac{R + r_s}{1 + g_m r_s} \right.$$

The coefficient a_1 can now be written explicitly as:

$$a_1 = \frac{L_6 + L_5(1 + g_m r_o)}{r_o + R + r_s(1 + g_m r_o)}$$

$$+ \frac{1}{1 + \dfrac{g_m r_o}{1 + \dfrac{r_o + R}{r_s}}} \left\{ C_2[R_1 + r_{gx} + r_s \| (r_o + R)] \right.$$

$$+ C_3[R_1 + R \| (r_o + r_s)][1 + g_m r_o \| (r_s + R \| R_1)] \tag{6.95}$$

$$+ C_4[r_{gx} + r_o \| (r_s + R)] \left(1 - \frac{g_m R}{1 + \dfrac{r_s + R}{r_o \| r_{gx}}}\right) + \left. C_7 r_o \| (r_o + R) \right\}$$

If we assume that the numerical values in Fig. 6.11 are typical, we can approximate a_1 (see Problem 6.2):

$$a_1 = C_2(R_1 + r_{gx}) + C_3(R_1 + R \| r_o)(1 + g_m r_o \| R \| R_1)$$

$$+ C_4(r_{gx} + r_o \| R)(1 - g_m R \| r_{gx}) + C_7 r_o \| R \tag{6.96}$$

$$+ \frac{L_6 + L_5(1 + g_m r_o)}{r_o + R}$$

The coefficient a_2: This is given by:

$$a_2 = \Sigma\Sigma\tau^{(i)}\tau_{(i)}^{(j)} \tag{6.97}$$

There are $\binom{7}{2} = 21$ terms in a_2 of which, as we shall see, five are zero. We shall use the compact notation (i, j) to denote the pair $\tau^{(i)}\tau_{(i)}^{(j)}$ and proceed starting with $(1, 2)$, $(1, 3)$, *etc.* If an indeterminacy arises, we will reverse the order or add a dummy resistance and let it vanish later in the product.

$(1, 2)$: This is given by $\tau^{(1)}C_2 R_{(1)}^{(2)}$, but since $\tau^{(1)} = 0$ and $R_{(1)}^{(2)} \to \infty$ we have an indeterminacy. Hence, we reverse the order and consider $(2, 1)$ which is given by $\tau^{(2)}L_1/R_{(2)}^{(1)}$. Now, when we determine $R_{(2)}^{(1)}$, we short port (2) and obtain immediately $R_{(2)}^{(1)} = R_1 + r_{gx} + r_s \| (r_o + R)$ so that we have:

$$(2, 1) = \tau^{(2)}\frac{L_1}{R_{(2)}^{(1)}}$$

$$= \frac{L_1 C_2}{1 + \dfrac{g_m r_o}{1 + \dfrac{r_o + R}{r_s}}} \tag{6.98}$$

(1, 3): This is also indeterminate so that we must reverse the order and consider $\tau^{(1)}L_1/R_{(3)}^{(1)}$ instead. Now, when we short port (3), v_{gs} appears directly across $g_m v_{gs}$ which in turn acts like a simple conductance g_m in parallel with r_o. Hence, we see that $R_{(3)}^{(1)} = R_1 + R \parallel (r_s + r_o \parallel g_m^{-1})$ so that we have:

$$(3, 1) = \tau^{(3)} \frac{L_1}{R_{(3)}^{(1)}}$$

$$= L_1 C_3 \frac{1 + \dfrac{R \parallel (r_o + r_s)}{R_1}}{1 + \dfrac{R \parallel (r_s + r_o \parallel g_m^{-1})}{R_1}} \cdot \frac{1 + g_m r_o \parallel (r_s + R \parallel R_1)}{1 + \dfrac{g_m r_o}{1 + \dfrac{r_o + R}{r_s}}} \tag{6.99}$$

Although this is a good expression, it looks like it can use some simplification. This can be performed either manually or by removing the indeterminacy in $(1, 3)$ using a different technique. The simplest way to remove an indeterminacy is to add a dummy resistor which prevents a resistance from becoming zero or infinite. Later, when the final expression is obtained, this dummy resistance is allowed to vanish. In our circuit, we add a resistor r_{dum} across port (2) and determine $R^{(1)}$:

$$R^{(1)} = [R_1 + r_{dum} + r_{gx} + r_s \parallel (r_o + R)] \frac{1 + g_m \mathcal{G}}{1 + g_m \overline{G}} \tag{6.100}$$

in which $\overline{G} = 0$ because when port (1) is open there can be no current flow through r_{dum} and $v_{gs} = 0$. We do not have to determine $\mathcal{G}$ at this point because we are going to deal with it when the product term $(1, 3)$ is formed. Next, we determine $R_{(1)}^{(3)}$:

$$R_{(1)}^{(3)} = [r_{dum} + r_{gx} + (r_s + R) \parallel r_o] \frac{1 + g_m \mathcal{G}'}{1 + g_m \overline{G}'} \tag{6.101}$$

in which $\overline{G}' = 0$ because when ports (3) and (1) are open, there is no current flow through r_{dum} and $v_{gs} = 0$. Substituting these in $(1, 3)$ we obtain:

$$L_1 C_3 \frac{R_{(1)}^{(3)}}{R^{(1)}} = \frac{r_{dum} + r_{gx} + (r_s + R) \parallel r_o}{R_1 + r_{dum} + r_{gx} + r_s \parallel (r_o + R)} \cdot \frac{1 + g_m \mathcal{G}'}{1 + g_m \mathcal{G}} \tag{6.102}$$

Taking the limit $r_{dum} \to \infty$ in this expression we obtain:

$$\left.\mathcal{G}'\right|_{r_{dum} \to \infty} = r_o \parallel (r_s + R)$$

$$\left.\mathcal{G}\right|_{r_{dum} \to \infty} = \frac{r_o r_s}{r_o + r_s + R} \tag{6.103a–c}$$

$$\left.\frac{r_{dum} + r_{gx} + (r_s + R) \parallel r_o}{R_1 + r_{dum} + r_{gx} + r_s \parallel (r_o + R)}\right|_{r_{dum} \to \infty} = 1$$

It follows that:

$$(1,3) = L_1 C_3 \frac{1 + g_m r_o \| (r_s + R)}{1 + g_m \dfrac{r_o r_s}{r_o + r_s + R}} \tag{6.104}$$

(1, 4): This is given by $\tau^{(1)} C_4 R_{(1)}^{(4)}$. Since $R_{(1)}^{(4)}$ is seen to be finite (we do not need to calculate it) and $\tau^{(1)} = 0$, we have:

$$(1,4) = 0 \tag{6.105}$$

(1, 5), (1, 6), (1, 7): These are all zero because $\tau^{(1)} = 0$ and there is no indeterminacy:

$$(1,5) = (1,6) = (1,7) = 0 \tag{6.106}$$

(2, 3): This is given by $\tau^{(2)} C_3 R_{(2)}^{(3)}$. When port (2) is shorted, $g_m v_{gs}$ vanishes so that looking into port (3) we see a resistive bridge circuit in which r_s is the bridge element. We can solve for the bridge resistance quite easily as we did in Chapter 1 by applying the EET to r_s. Since the expression of the bridge resistance is somewhat long, combining it with the expression of $\tau^{(2)}$ may result in an even longer expression for $(2, 3)$. Hence, we investigate reversing the order of the ports to $(3, 2)$ to see if we can obtain a simpler answer using $\tau^{(3)} C_2 R_{(3)}^{(2)}$. To determine $R_{(3)}^{(2)}$, first we look into port (2) with port (3) short and $g_m = 0$ and find:

$$R_{(3)}^{(2)}\big|_{g_m=0} = r_{gx} + r_o \| (r_s + R \| R_1) \tag{6.107}$$

The inverse gain with respect to $g_m v_{gs}$ with port (2) short is zero, i.e. $\mathcal{G} = 0$. With port (2) open (and port (3) short) v_{gs} is simply the voltage across the independent current source i_T so that $\bar{G} = r_o \| (r_s + R \| R_1)$. Hence we have:

$$R_{(3)}^{(2)} = R_{(3)}^{(2)}\big|_{g_m=0} \frac{1 + g_m \mathcal{G}}{1 + g_m \bar{G}}$$

$$= \frac{r_{gx} + r_o \| (r_s + R \| R_1)}{1 + g_m r_o \| (r_s + R \| R_1)} \tag{6.108}$$

Combining this with the expression of $\tau^{(2)}$ in Eq. (6.94b), we obtain:

$$(3,2) = \frac{C_2 C_3}{1 + \dfrac{g_m r_o}{1 + \dfrac{r_o + R}{r_s}}} [R_1 + R \| (r_o + r_s)][r_{gx} + r_o \| (r_s + R \| R_1)] \tag{6.109}$$

(2, 4): This is given by $\tau^{(2)} C_4 R_{(2)}^{(4)}$. It can be seen that $R_{(2)}^{(4)}$ is the same as $R_{(2)}^{(3)}$. Hence, $(2, 4)$ can be deduced from $(2, 3)$ or $(3, 2)$ simply by changing C_3 to C_4 in Eq. (6.109):

$$(2,4) = \cfrac{C_2 C_4}{1 + \cfrac{g_m r_o}{1 + \cfrac{r_o + R}{r_s}}} [R_1 + R \,\|\, (r_o + r_s)][r_{gx} + r_o \,\|\, (r_s + R \,\|\, R_1)] \tag{6.110}$$

(2,5): This is given by $\tau^{(2)} L_5 / R_{(2)}^{(5)}$. In this case we see that $\tau^{(5)}$ has a simpler expression than $\tau^{(2)}$ so that we reverse the order to $(5, 2)$ with the hope of obtaining a simpler expression using $\tau^{(5)} C_2 R_{(5)}^{(2)}$. Already we see that we have a free bonus because we can deduce $R_{(5)}^{(2)}$ from $R^{(2)}$ by letting $r_s \to \infty$ since opening port (5) and letting $r_s \to \infty$ are the same things. Hence we have:

$$R_{(5)}^{(2)} = R^{(2)} \big|_{r_s \to \infty}$$

$$= \frac{R_1 + r_{gx} + r_o + R}{1 + g_m r_o} \tag{6.111}$$

Combining this with the expression of $\tau^{(5)}$ in Eq. (6.94e), we obtain:

$$(5,2) = L_5 C_2 \frac{R_1 + r_{gx} + r_o + R}{r_o + R + (1 + g_m r_o) r_s} \tag{6.112}$$

(2,6): This is given by $\tau^{(2)} L_6 / R_{(2)}^{(6)}$. If we reverse the order, as we did in $(2, 5)$, we can once again write down the answer immediately:

$$(6,2) = \tau^{(6)} \tau_{(6)}^{(2)}$$

$$= \tau^{(6)} (\tau^{(2)} \big|_{R \to \infty}) \tag{6.113}$$

$$= \frac{L_6}{r_o + R + r_s (1 + g_m r_o)} C_2 (r_s + r_{gx} + R_1)$$

In the derivation of $\tau_{(6)}^{(2)}$, we have made use of the fact that opening port (6) is the same thing as letting $R \to \infty$.

(2,7): This is given by $\tau^{(2)} C_7 R_{(2)}^{(7)}$. As in the previous two cases, we reverse the order for the same reason and determine:

$$(7,2) = \tau^{(7)} \tau_{(7)}^{(2)}$$

$$= \tau^{(7)} \tau^{(2)} \big|_{r_o \to 0} \tag{6.114}$$

$$= C_7 r_o \left\| \frac{R + r_s}{1 + g_m r_s} \right. C_2 (R_1 + r_{gx} + r_s \,\|\, R)$$

In the derivation of $\tau_{(7)}^{(2)}$, we have made use of the fact that shorting port (7) is the same thing as letting $r_o \to 0$.

(3,4): This is given by $\tau^{(3)} C_4 R_{(3)}^{(4)}$. Applying the EET to $R_{(3)}^{(4)}$, we can write:

$$R_{(3)}^{(4)} = R_{(3)}^{(4)}|_{g_m = 0} \frac{1 + g_m \mathcal{G}}{1 + g_m \bar{G}} \tag{6.115}$$

In Fig. 6.13 we can see that:

$$R_{(3)}^{(4)}|_{g_m = 0} = r_{gx} + r_o \| (r_s + R \| R_1) \tag{6.116}$$

We can also see that $\mathcal{G} = 0$ because shorting port (4) with port (3) short causes $v_{gs} = 0$. With port (3) and port (4) open, v_{gs} appears directly across r_o so that $\bar{G} = r_o \| (r_s + R_1 \| R)$. Substituting these results in Eq. (6.115) we obtain:

$$R_{(3)}^{(4)} = \frac{r_{gx} + r_o \| (r_s + R \| R_1)}{1 + g_m r_o \| (r_s + R_1 \| R)} \tag{6.117}$$

Combining this with $\tau^{(3)}$ in Eq. (6.94c), we obtain:

$$(3,4) = C_4 C_3 [R_1 + R \| (r_o + r_s)] \frac{r_{gx} + r_o \| (r_s + R \| R_1)}{1 + \dfrac{g_m r_o}{1 + \dfrac{r_o + R}{r_s}}} \tag{6.118}$$

$(3,5)$: This is given by $\tau^{(3)} L_5 / R_{(3)}^{(5)}$. We realize, however, that $R_{(5)}^{(3)}$, which is simply $R_1 + R$, is much simpler than $R_{(3)}^{(5)}$. Hence, we reverse the order and write:

$$(5,3) = \tau^{(5)} C_3 R_{(5)}^{(3)}$$

$$= L_5 C_3 \frac{R_1 + R}{r_s + \dfrac{r_o + R}{1 + g_m r_o}} \tag{6.119}$$

$(3,6)$: This is given by $\tau^{(3)} \tau_{(3)}^{(6)}$ but since $\tau^{(6)}$ is simpler than $\tau^{(3)}$, we reverse the order and, instead, determine $\tau_{(6)}^{(3)}$ which we can obtain from $\tau^{(3)}$ by letting $R \to \infty$:

$$\begin{aligned} \tau_{(6)}^{(3)} &= \tau^{(3)}|_{R \to \infty} \\ &= C_3 (R_1 + r_o + r_s)(1 + g_m r_o \| (r_s + R_1)) \\ &= C_3 [r_o + (1 + g_m r_o)(r_s + R_1)] \end{aligned} \tag{6.120a–c}$$

Combining this with $\tau^{(6)}$ in Eq. (6.94f), we obtain:

$$(6,3) = L_6 C_3 \frac{r_o + (1 + g_m r_o)(r_s + R_1)}{r_o + R + r_s(1 + g_m r_o)} \tag{6.121}$$

$(3,7)$: For the same reason as in $(3,6)$, we reverse the order and determine easily $R_{(7)}^{(3)} = R_1 + r_s \| R$. Hence, using $\tau^{(7)}$ in Eq. (6.94g) we obtain:

$$(7,3) = \tau^{(7)}\tau^{(3)}_{(7)}$$

$$= C_7 C_3 (R_1 + r_s \| R) r_o \left\| \frac{R + r_s}{1 + g_m r_s} \right. \tag{6.122}$$

(4, 5): It is simpler to determine $(5, 4)$, which is given by $\tau^{(5)} C_4 R^{(4)}_{(5)}$, because $R^{(4)}_{(5)}$ can be derived from $R^{(4)}$ in Eq. (6.83) by letting $r_s \to \infty$:

$$R^{(4)}_{(5)} = R^{(4)}|_{r_s \to \infty} = (r_{gx} + r_o) \frac{1 + g_m \cdot 0}{1 + g_m r_o} \tag{6.123}$$

Hence, using $\tau^{(5)}$ in Eq. (6.94e), we obtain:

$$(5, 4) = \tau^{(5)} C_4 R^{(4)}_{(5)}$$

$$= L_5 C_4 \frac{r_{gx} + r_o}{1 + g_m r_o} \frac{1}{r_s + \dfrac{r_o + R}{1 + g_m r_o}} \tag{6.124}$$

$$= L_5 C_4 \frac{r_o + r_{gx}}{r_o + R + (1 + g_m r_o) r_s}$$

(6, 4): This is given by $\tau^{(6)} C_4 R^{(4)}_{(6)}$ in which $R^{(4)}_{(6)}$ can be obtained from $R^{(4)}$ in Eq. (6.83) by letting $R \to \infty$:

$$R^{(4)}_{(6)} = R^{(4)}|_{R \to \infty} = (r_o + r_{gx}) \frac{1 - g_m r_o \| r_{gx}}{1 + g_m \cdot 0} \tag{6.125}$$

Hence, using $\tau^{(6)}$ in Eq. (6.94f), we obtain:

$$(6, 4) = \tau^{(6)} C_4 R^{(4)}_{(6)}$$

$$= \frac{L_6 C_4 (r_o + r_{gx})(1 - g_m r_o \| r_{gx})}{r_o + R + r_s(1 + g_m r_o)} \tag{6.126}$$

$$= L_6 C_4 \frac{r_o + r_{gx}(1 - g_m r_o)}{r_o + R + r_s(1 + g_m r_o)}$$

(7, 4): This is given by $\tau^{(7)} C_4 R^{(4)}_{(7)}$ in which $R^{(4)}_{(7)}$ is immediately verified to be r_{gx}. Using $\tau^{(7)}$ in Eq. (6.94g), we have:

$$(7, 4) = C_7 C_4 r_o \left\| \frac{R + r_s}{1 + g_m r_s} r_{gx} \right. \tag{6.127}$$

(5, 6): This is given by $\tau^{(5)} L_6 / R^{(6)}_{(5)}$ in which we see that $R^{(6)}_{(5)} \to \infty$ so that:

$$(5, 6) = 0 \tag{6.128}$$

(7,5): This is given by $\tau^{(7)}L_5/R_{(7)}^{(5)}$ in which $R_{(7)}^{(5)}$ is immedately verified to be $r_s + R$ so that we have:

$$
(7,5) = C_7L_5\dfrac{r_o\left\|\dfrac{R+r_s}{1+g_mr_s}\right.}{r_s+R}
$$

$$
= \dfrac{C_7L_5r_o}{r_o + R + (1+g_mr_o)r_s}
$$

(6.129a, b)

(7,6): This is given by $\tau^{(7)}L_6/R_{(7)}^{(6)}$ in which $R_{(7)}^{(6)}$ is immediately verified to be $r_s + R$ so that we have:

$$
(7,6) = C_7L_6\dfrac{r_o\left\|\dfrac{R+r_s}{1+g_mr_s}\right.}{r_s+R}
$$

$$
= \dfrac{C_7L_6r_o}{r_o + R + (1+g_mr_o)r_s}
$$

(6.130a, b)

One possible way of collecting all the terms above to obtain the coefficient a_2 is:

$$
a_2 = \dfrac{L_1}{1+\dfrac{g_mr_o}{1+\dfrac{r_o+R}{r_s}}}[C_2 + C_3(1+g_mr_o \| r_s + R)]
$$

$$
+ \dfrac{L_5}{R+r_o+r_s(1+g_mr_o)}[C_2(R + R_1 + r_o + r_{gx})
$$

$$
+ C_3(1+g_mr_o)(R+R_1) + C_4(r_o + r_{gx}) + C_7r_o]
$$

$$
+ \dfrac{L_6}{R+r_o+r_s(1+g_mr_o)}\{C_2(r_s + r_{gx} + R_1)
$$

(6.131)

$$
+ C_3[r_o + (1+g_mr_o)(r_s + R)] + C_4[r_o + r_{gx}(1 - g_mr_o)] + C_7r_o\}
$$

$$
+ \dfrac{C_2C_3C_4}{C_2 \| C_3 \| C_4}[R_1 + R \| (r_o + r_s)]\dfrac{r_{gx} + r_o \| (r_s + R \| R_1)}{1+\dfrac{g_mr_o}{1+\dfrac{r_o+R}{r_s}}}
$$

$$
+ C_7r_o\left\|\dfrac{r_s+R}{1+g_mr_s}\right.[C_2(r_{gx} + R_1 + r_s \| R) + C_3(R_1 + r_s \| R) + C_4r_{gx}]
$$

The coefficient a_3: This is given by:

$$a_3 = \Sigma\Sigma\Sigma \tau^{(i)} \tau^{(j)}_{(i)} \tau^{(k)}_{(i,j)} \tag{6.132}$$

There are $\binom{7}{3} = 35$ terms in a_3 of which, as we shall see, nine are zero. As before, we shall use the compact notation (i, j, k) to denote the triplet $\tau^{(i)} \tau^{(j)}_{(i)} \tau^{(k)}_{(i,j)}$ and proceed beginning with $(1, 2, 3), (1, 2, 4)$, etc. In each triplet there is only one new calculation to be made, which is the port resistance $R^{(k)}_{(i,j)}$. In all cases $R^{(k)}_{(i,j)}$ is determined either by a simple inspection of the reference circuit in Fig. 6.13 or deduced from a previous calculation.

(1, 2, 3): This is given by $(1, 2)C_3 R^{(3)}_{(1,2)}$ in which:

$$R^{(3)}_{(1,2)} = r_{gx} + r_o \| (r_s + R) \tag{6.133}$$

Combining this with $(1, 2)$ in Eq. (6.98), we obtain:

$$(1, 2, 3) = L_1 C_2 C_3 \frac{r_{gx} + r_o \| (r_s + R)}{1 + \dfrac{g_m r_o}{1 + \dfrac{r_o + R}{r_s}}} \tag{6.134}$$

(1, 2, 4): This is given by $(1, 2)C_4 R^{(4)}_{(1,2)}$ in which:

$$R^{(4)}_{(1,2)} = r_{gx} + r_o \| (r_s + R) \tag{6.135}$$

Combining this with $(1, 2)$ in Eq. (6.98), we obtain:

$$(1, 2, 4) = L_1 C_2 C_4 \frac{r_{gx} + r_o \| (r_s + R)}{1 + \dfrac{g_m r_o}{1 + \dfrac{r_o + R}{r_s}}} \tag{6.136}$$

(1, 2, 5): This is given by $(1, 2)L_5/R^{(5)}_{(1,2)}$ in which:

$$R^{(5)}_{(1,2)} = r_s + r_o + R \tag{6.137}$$

Combining this with $(1, 2)$ in Eq. (6.98), we obtain:

$$(1, 2, 5) = \frac{L_1 C_2 L_5}{\left(1 + g_m \dfrac{r_o r_s}{r_s + r_o + R}\right)(r_s + r_o + R)} \left.\begin{array}{c} \\ \\ \end{array}\right\}$$
$$= \frac{L_1 C_2 L_5}{r_o + R + (1 + g_m r_o)r_s} \tag{6.138a, b}$$

(1, 2, 6): This is given by $(1, 2)L_6/R^{(6)}_{(1,2)}$ in which:

$$R^{(6)}_{(1,2)} = r_s + r_o + R \tag{6.138}$$

Combining this with $(1, 2)$ in Eq. (6.98), we obtain:

$$(1, 2, 6) = \frac{L_1 C_2 L_6}{r_o + R + (1 + g_m r_o) r_s} \tag{6.139}$$

$(1, 2, 7)$: This is given by $(1, 2) C_7 R^{(7)}_{(1,2)}$ in which:

$$R^{(7)}_{(1,2)} = r_o \| (r_s + R) \tag{6.140}$$

Combining this with $(1, 2)$ in Eq. (6.98), we obtain:

$$(1, 2, 7) = L_1 C_2 C_7 \frac{r_o \| (r_s + R)}{1 + \dfrac{g_m r_o}{1 + \dfrac{r_o + R}{r_s}}} \tag{6.141}$$

$(1, 3, 4)$: This is given by $(1, 3) C_4 R^{(4)}_{(1,3)}$ in which $R^{(4)}_{(1,3)}$ can be deduced from $R^{(4)}_{(3)}$ in Eq. (6.117):

$$R^{(4)}_{(1,3)} = R^{(4)}_{(3)} |_{R_1 \to \infty}$$

$$= \frac{r_{gx} + r_o \| (r_s + R)}{1 + g_m r_o \| (r_s + R)} \tag{6.142}$$

This is combined with the result obtained for $(1, 3)$ in Eq. (6.99) to yield:

$$(1, 3, 4) = L_1 C_3 C_4 \frac{r_{gx} + r_o \| (r_s + R)}{1 + \dfrac{g_m r_o}{1 + \dfrac{r_o + R}{r_s}}} \tag{6.143}$$

$(1, 3, 5)$: This is given by $(1, 3) L_5 / R^{(5)}_{(1,3)}$ in which:

$$R^{(5)}_{(1,3)} = r_s + r_o \| g_m^{-1} + R \tag{6.144}$$

This is combined with $(1, 3)$ in Eq. (6.99) to yield:

$$(1, 3, 5) = \frac{L_1 C_5 C_3}{1 + \dfrac{g_m r_o}{1 + \dfrac{r_o + R}{r_s}}} \frac{1 + g_m r_o \| (r_s + R)}{r_s + r_o \| g_m^{-1} + R} \tag{6.145}$$

It is apparent that the second ratio in this equation can be simplified further. As usual, whenever we suspect it is possible to obtain a simpler result, we change the order of the ports. In this case, if we take a quick look at $(3, 5, 1)$ we see that $(3, 5)$, given in Eq. (6.119), is slightly simpler than $(1, 3)$ while $R^{(1)}_{(3,5)} = R + R_1$. Now we have:

$$(5,3,1) = (5,3)\frac{L_1}{R^{(1)}_{(3,5)}}$$

$$= \frac{L_1 L_5 C_3}{r_s + \dfrac{r_o + R}{1 + g_m r_o}} \tag{6.146}$$

We would have obtained the same result had we simplified the result in Eq. (6.145).

$(1,3,6)$: This is given by $(1,3)L_6/R^{(6)}_{(1,3)}$. We can see that $R^{(6)}_{(1,3)}$ is the same as $R^{(5)}_{(1,3)}$ so that we should be able to deduce $(1,3,6)$ from $(1,3,5)$ simply by replacing L_5 with L_6 in Eq. (6.146):

$$(6,3,1) = \frac{L_1 L_6 C_3}{r_s + \dfrac{r_o + R}{1 + g_m r_o}} \tag{6.147}$$

$(1,3,7)$: This is given by $(1,3)C_7 R^{(7)}_{(1,3)}$. As in $(1,3,5)$, we change the order to $(3,7,1)$ in which:

$$R^{(1)}_{(3,7)} = R_1 + r_s \| R \tag{6.148}$$

Combining this with $(3,7)$ or $(7,3)$ in Eq. (6.122), we obtain:

$$(7,3,1) = L_1 C_3 C_7 r_o \left\| \frac{R + r_s}{1 + g_m r_s} \right. \tag{6.149}$$

$(1,4,5), (1,4,6), (1,4,7)$: Since $(1,4) = 0$ and $R^{(5)}_{(1,4)}$, $R^{(6)}_{(1,4)}$ and $R^{(7)}_{(1,4)}$ are all finite, we have:

$$\left.\begin{array}{l} (1,4,5) = 0 \\ (1,4,6) = 0 \\ (1,4,7) = 0 \end{array}\right\} \tag{6.150a–c}$$

$(1,5,6), (1,5,7)$: Since $(1,5) = 0$ and $R^{(6)}_{(1,5)}$ and $R^{(7)}_{(1,5)}$ are both finite, we have:

$$\left.\begin{array}{l} (1,5,6) = 0 \\ (1,5,7) = 0 \end{array}\right\} \tag{6.151a, b}$$

$(1,6,7)$: Since $(1,6) = 0$ and $R^{(7)}_{(1,6)} = 0$ we have:

$$(1,6,7) = 0 \tag{6.152}$$

$(2,3,4)$: Since $R^{(4)}_{(3,2)} = 0$ we have:

$$(2,3,4) = 0 \tag{6.153}$$

(2, 3, 5): This is given by $(2,3)L_5/R^{(5)}_{(3,2)}$. Changing the order to $(5,3,2)$ we have:

$$R^{(2)}_{(3,5)} = R^{(2)}_{(3)}\big|_{r_s \to \infty} = \frac{r_{gx} + r_o}{1 + g_m r_o} \tag{6.154}$$

Combining this with $(5,3)$ given in Eq. (6.119), we obtain:

$$
\begin{aligned}
(5,3,2) &= L_5 C_3 C_2 \frac{R_1 + R}{r_s + \dfrac{r_o + R}{1 + g_m r_o}} \frac{r_{gx} + r_o}{1 + g_m r_o} \\[2ex]
&= L_5 C_3 C_2 \frac{R_1 + R}{1 + g_m r_s + \dfrac{R + r_s}{r_o}} \frac{r_{gx}}{r_o \| r_{gx}}
\end{aligned}
\tag{6.155a, b}
$$

(2, 3, 6): This is given by $(2,3)L_6/R^{(6)}_{(2,3)}$. We can change the order either to $(2,6,3)$ or $(3,6,2)$. The former looks a little simpler because if we look into port (3) and short port (2), we get rid of $g_m v_{gs}$ and obtain:

$$R^{(3)}_{(2,6)} = r_o + r_{gx} \| (r_s + R_1) \tag{6.156}$$

Combining this with $(6,2)$ given in Eq. (6.113), we obtain:

$$(6,2,3) = L_6 C_2 C_3 \frac{(r_s + r_{gx} + R_1)[r_o + r_{gx} \| (r_s + R_1)]}{r_o + R + r_s(1 + g_m r_o)} \tag{6.157}$$

This expression can be simplified to make it look like some of the other expressions we had obtained earlier. Hence, r_o and $r_{gx} \| (r_s + R_1)$ are factored out in the numerator:

$$
\begin{aligned}
(6,2,3) &= L_6 C_2 C_3 \frac{(r_s + r_{gx} + R_1)r_o r_{gx} \| (r_s + R_1)\left(\dfrac{1}{r_o} + \dfrac{1}{r_{gx}} + \dfrac{1}{r_s + R_1}\right)}{r_o + R + r_s(1 + g_m r_o)} \\[3ex]
&= L_6 C_2 C_3 \frac{r_{gx}(r_s + R_1)\left(\dfrac{1}{r_o} + \dfrac{1}{r_{gx}} + \dfrac{1}{r_s + R_1}\right)}{1 + g_m r_s + \dfrac{R + r_s}{r_o}} \\[3ex]
&= L_6 C_2 C_3 r_{gx} \frac{1 + \dfrac{r_s + R_1}{r_o \| r_{gx}}}{1 + g_m r_s + \dfrac{R + r_s}{r_o}}
\end{aligned}
\tag{6.158}
$$

(2, 3, 7): This is given by $(2,3)C_7 R^{(7)}_{(2,3)}$. We can change the order either to $(2,7,3,)$ or to $(3, 7, 2)$. The former requires the determination of $R^{(3)}_{(2,7)}$ which is given by $r_{gx} \parallel (R_1 + r_s \parallel R)$, while the latter requires the determination of $R^{(2)}_{(3,7)}$ which is given by r_{gx}. Hence, using the latter, in which $(3, 7)$ is given by Eq. (6.122), we obtain:

$$(7, 3, 2) = C_2 C_3 C_7 r_{gx}(R_1 + r_s \parallel R) r_o \left\| \frac{R + r_s}{1 + g_m r_s} \right. \tag{6.159}$$

(2, 4, 5): This is given by $(2, 4)L_5/R^{(5)}_{(2,4)}$. We change the order to $(2, 5, 4)$ and determine:

$$R^{(4)}_{(2,5)} = (R_1 + R) \parallel (r_{gx} + r_o) \tag{6.160}$$

Combining this result with $(5, 2)$ given in Eq. (6.112), we obtain:

$$(5, 2, 4) = L_5 C_2 C_4 \frac{(R_1 + r_{gx} + r_o + R)(R_1 + R) \parallel (r_{gx} + r_o)}{r_o + R + (1 + g_m r_o) r_s}$$

$$= L_5 C_2 C_4 \frac{(R_1 + R)(r_{gx} + r_o)}{r_o + R + (1 + g_m r_o) r_s} \tag{6.161}$$

$$= L_5 C_2 C_4 \frac{R_1 + R}{1 + g_m r_s + \dfrac{R + r_s}{r_o}} \frac{r_{gx}}{r_{gx} \parallel r_o}$$

(2, 4, 6): This is given by $(2, 4)L_6/R^{(6)}_{(2,4)}$. We change the order to $(2, 6, 4)$ and realize that $R^{(4)}_{(2,6)} = R^{(3)}_{(2,6)}$. It follows that $(2, 6, 4)$ can be deduced from $(2, 3, 6)$ simply by replacing C_3 with C_4 in Eq. (6.158):

$$(6, 2, 4) = L_6 C_2 C_4 r_{gx} \frac{1 + \dfrac{r_s + R_1}{r_o \parallel r_{gx}}}{1 + g_m r_s + \dfrac{R + r_s}{r_o}} \tag{6.162}$$

(2, 4, 7): This is given by $(2, 4)C_7 R^{(7)}_{(2,4)}$. We change the order to $(4, 7, 2)$ and determine:

$$R^{(2)}_{(4,7)} = R_1 + r_s \parallel R \tag{6.163}$$

Combining this with $(7, 4)$ in Eq. (6.127), we obtain:

$$(7, 4, 2) = C_2 C_4 C_7 r_{gx}(R_1 + r_s \parallel R) r_o \left\| \frac{r_s + R}{1 + g_m r_s} \right. \tag{6.164}$$

(2, 5, 6): This is given by $(2, 5)L_6/R^{(6)}_{(2,5)}$ in which:

$$R^{(6)}_{(2,5)} = R + r_o + r_{gx} + R_1 \tag{6.165}$$

Combining this with $(5, 2)$ in Eq. (6.112), we obtain:

$$(5, 2, 6) = \frac{L_6 L_5 C_2}{r_o + R + (1 + g_m r_o) r_s} \tag{6.166}$$

$(2, 5, 7)$: This is given by $(2, 5) C_7 R^{(7)}_{(2,5)}$ in which:

$$R^{(7)}_{(2,5)} = r_o \,\|\, (R + r_{gx} + R_1) \tag{6.167}$$

Combining this with $(5, 2)$ in Eq. (6.112), we obtain:

$$(5, 2, 7) = L_5 C_2 C_7 \frac{R + R_1 + r_{gx}}{1 + g_m r_s + \dfrac{R + r_s}{r_o}} \tag{6.168}$$

$(2, 6, 7)$: This is given by $(2, 6) C_7 R^{(7)}_{(2,6)}$ in which:

$$R^{(7)}_{(2,6)} = r_o \tag{6.169}$$

Combining this with $(6, 2)$ in Eq. (6.113), we obtain:

$$(6, 2, 7) = L_6 C_2 C_7 \frac{r_s + r_{gx} + R_1}{1 + g_m r_s + \dfrac{R + r_s}{r_o}} \tag{6.170}$$

$(3, 4, 5)$: This is given by $(3, 4) L_5 / R^{(5)}_{(3,4)}$. We change the order to $(4, 5, 3)$ and determine:

$$R^{(3)}_{(4,5)} = R_1 + R \tag{6.171}$$

Combining this with $(5, 4)$ in Eq. (6.124), we obtain:

$$\left.\begin{aligned}
(5, 4, 3) &= L_5 C_4 C_3 (R_1 + R) \frac{r_o + r_{gx}}{r_o + R + r_s(1 + g_m r_o)} \\[2mm]
&= L_5 C_4 C_3 \frac{R_1 + R}{1 + g_m r_s + \dfrac{R + r_s}{r_o}} \frac{r_{gx}}{r_{gx} \,\|\, r_o}
\end{aligned}\right\} \tag{6.172a, b}$$

$(3, 4, 6)$: This is given by $(3, 4) L_6 / R^{(6)}_{(3,4)}$. We change the order to $(3, 6, 4)$ and deduce $R^{(4)}_{(3,6)}$ from $R^{(4)}_{(3)}$ in Eq. (6.117):

$$R^{(4)}_{(3,6)} = R^{(4)}_{(3)}\big|_{R \to \infty} = \frac{r_{gx} + r_o \,\|\, (r_s + R_1)}{1 + g_m r_o \,\|\, (r_s + R_1)} \tag{6.173}$$

Combining this with $(6, 3)$ in Eq. (6.121), we obtain:

$$(6, 3, 4) = \frac{C_3 L_6 C_4 (R_1 + r_o + r_s)[r_{gx} + r_o \,\|\, (r_s + R_1)]}{r_o + R + r_s(1 + g_m r_o)} \tag{6.174}$$

which after some simplification, reduces to:

$$(6,3,4) = C_3 L_6 C_4 r_{gx} \frac{1 + \dfrac{r_s + R_1}{r_o \parallel r_{gx}}}{1 + g_m r_s + \dfrac{R + r_s}{r_o}} \tag{6.175}$$

(3,4,7): This is given by $(3,4)C_7 R^{(7)}_{(3,4)}$. We change the order to $(3,7,4)$ and determine:

$$R^{(4)}_{(3,7)} = r_{gx} \tag{6.176}$$

Combining this with $(7,3)$ in Eq. (6.122), we obtain:

$$(7,3,4) = C_4 C_3 C_7 r_{gx} (R_1 + r_s \parallel R) r_o \left\| \frac{R + r_s}{1 + g_m r_s} \right. \tag{6.177}$$

This can also be written after some manipulation as:

$$(7,3,4) = C_4 C_3 C_7 \frac{r_{gx}(R + R_1)}{1 + g_m r_s + \dfrac{r_s + R}{r_o}} (r_s + R_1 \parallel R) \tag{6.178}$$

(3,5,6): This is given by $(3,5)L_6/R^{(6)}_{(3,5)}$ in which:

$$R^{(6)}_{(3,5)} = R + R_1 \tag{6.179}$$

Combining this with $(5,3)$, in Eq. (6.119) we obtain:

$$(5,3,6) = \frac{L_6 L_5 C_3}{r_s + \dfrac{r_o + R}{1 + g_m r_o}} \tag{6.180}$$

(3,5,7): This is given by $(3,5)C_7 R^{(7)}_{(3,5)}$ in which:

$$R^{(7)}_{(3,5)} = r_o \parallel g_m^{-1} \tag{6.181}$$

Combining this with $(5,3)$, in Eq. (6.119) we obtain:

$$\left. \begin{aligned} (5,3,7) &= L_5 C_3 C_7 \frac{R_1 + R}{r_s + \dfrac{r_o + R}{1 + g_m r_o}} r_o \parallel g_m^{-1} \\[2em] &= L_5 C_3 C_7 \frac{R_1 + R}{1 + g_m r_s + \dfrac{r_s + R}{r_o}} \end{aligned} \right\} \tag{6.182a, b}$$

(3,6,7): This is given by $(3,6)C_7 R^{(7)}_{(3,6)}$. We change the order to $(6,7,3)$ and determine readily:

$$R^{(3)}_{(6,7)} = R_1 + r_s \qquad (6.183)$$

Combining this with $(7, 6)$, in Eq. $(6.130b)$ we obtain:

$$(7, 6, 3) = L_6 C_3 C_7 \frac{r_s + R_1}{1 + g_m r_s + \dfrac{r_s + R}{r_o}} \qquad (6.184)$$

(4, 5, 6): Since $R^{(6)}_{(5,4)} \to \infty$ and $(4, 5)$ is finite, we have:

$$(4, 5, 6) = 0 \qquad (6.185)$$

(4, 5, 7): This is given by $(4, 5)C_7 R^{(7)}_{(4,5)}$ in which:

$$R^{(7)}_{(4,5)} = r_o \| r_{gx} \qquad (6.186)$$

Combining this with $(5, 4)$, in Eq. (6.124) we obtain:

$$
\left.
\begin{aligned}
(4, 5, 7) &= L_5 C_4 C_7 \frac{1 + \dfrac{r_{gx}}{r_o}}{1 + \dfrac{R + r_s(1 + g_m r_o)}{r_o}} r_o \| r_{gx} \\[2em]
&= L_5 C_4 C_7 \frac{r_{gx}}{1 + g_m r_s + \dfrac{R + r_s}{r_o}}
\end{aligned}
\right\} \qquad (6.187a, b)
$$

(4, 6, 7): This is given by $(4, 6)C_7 R^{(7)}_{(4,6)}$. We change the order to $(6, 7, 4)$ and determine:

$$R^{(4)}_{(6,7)} = r_{gx} \qquad (6.188)$$

Combining this with $(6, 7)$ in Eq. $(6.130b)$, we obtain:

$$(4, 6, 7) = L_6 C_7 C_4 \frac{r_{gx}}{1 + g_m r_s + \dfrac{R + r_s}{r_o}} \qquad (6.189)$$

(5, 6, 7): Since $(5, 6) = 0$ as determined earlier in Eq. (6.128) and $R^{(7)}_{(5,6)}$ is finite, we have:

$$(5, 6, 7) = 0 \qquad (6.190)$$

One possible way of collecting all the terms above to obtain the coefficient a_3 is:

$$a_3 = \frac{L_1 C_2}{1 + \dfrac{g_m r_o}{1 + \dfrac{r_o + R}{r_s}}} \left\{ (C_3 + C_4)[r_{gx} + r_o \| (r_s + R)] + C_7 r_o \| (r_s + R) \right.$$

$$+ \frac{L_5 + L_6}{r_o + r_s + R} \bigg\} + L_1 C_3 \left[C_7 r_o \left\| \frac{R + r_s}{1 + g_m r_s} + \frac{L_5 + L_6}{r_s + \dfrac{r_o + R}{1 + g_m r_o}} \right. \right.$$

$$+ C_4 \frac{r_{gx} + r_o \| (r_s + R)}{1 + \dfrac{g_m r_o}{1 + \dfrac{r_o + R}{r_s}}} \left. \right]$$

$$+ \frac{1}{1 + g_m r_s + \dfrac{r_s + R}{r_o}} \left\{ \frac{C_2 C_3 C_4}{C_2 \| C_3 \| C_4} r_{gx}(R + R_1) \left[C_7(r_s + R_1 \| R) \right. \right. \tag{6.191}$$

$$+ \frac{L_5}{r_o \| r_{gx}} + \frac{L_6}{R + R_1} \left(1 + \frac{r_s + R_1}{r_o \| r_{gx}} \right) \right] + C_2 L_5 \left[\frac{L_6}{r_o} + C_7(r_{gx} + R_1 + R) \right]$$

$$+ C_7 L_6 [C_2(r_s + r_{gx} + R_1) + C_3(r_s + R_1) + C_4 r_{gx}]$$

$$+ C_3 L_5 \left[\frac{L_6(1 + g_m r_o)}{r_o} + C_7(R_1 + R) \right] + C_4 L_5 C_7 r_{gx} \bigg\}$$

The coefficient a_4: This is given by:

$$a_4 = \Sigma\Sigma\Sigma\Sigma \tau^{(i)} \tau_{(i)}^{(j)} \tau_{(i,j)}^{(k)} \tau_{(i,j,k)}^{(l)} \tag{6.192}$$

There are $\binom{7}{4} = 35$ terms to determine of which ten are zero.

$(1,2,3,4)$: $R_{(1,2,3)}^{(4)} = 0$ so that:

$$(1,2,3,4) = 0 \tag{6.193}$$

$(1,2,3,5)$: This is given by $(1,2,3)L_5/R_{(1,2,3)}^{(5)}$. Before going ahead with the determination of $R_{(1,2,3)}^{(5)}$, it is always a good idea to check if we can get a simpler answer by changing the order of the ports. Hence, taking a quick glance at $(1,2,3,5)$ versus $(1,2,5,3)$, we see that $(1,2,5)$ is simpler than $(1,2,3)$ and that $R_{(1,2,5)}^{(3)}$, which is equal to $r_{gx} + r_o$, is simpler than $R_{(1,2,3)}^{(5)}$, which is equal to $r_s + r_{gx} \| r_o + R$. Therefore, we change the order to $(1,2,5,3)$ and obtain:

$$(1,2,5,3) = (1,2,5)C_3 R_{(1,2,5)}^{(3)}$$

$$= (1, 2, 5)C_3(r_{gx} + r_o) \tag{6.194}$$

Combining this result with $(1, 2, 5)$ given in Eq. (6.138), we obtain:

$$(1, 2, 5, 3) = L_1 L_5 C_2 C_3 \frac{r_{gx} + r_o}{r_o + R + (1 + g_m r_o)r_s} \tag{6.195}$$

$$= \frac{L_1 L_5 C_2 C_3}{1 + g_m r_s + \dfrac{r_s + R}{r_o}} \frac{r_{gx}}{r_o \| r_{gx}} \tag{6.195}$$

$\underline{(1, 2, 3, 6)}$: For the same reason as in $(1, 2, 3, 5)$, we change the order to $(1, 2, 6, 3)$ and determine:

$$R^{(3)}_{(1,2,6)} = r_{gx} + r_o \tag{6.196}$$

which is the same as $R^{(3)}_{(1,2,5)}$ above. Combining this with $(1, 2, 6)$ given in Eq. (6.139), we obtain:

$$(1, 2, 6, 3) = \frac{L_1 L_6 C_2 C_3}{1 + g_m r_s + \dfrac{r_s + R}{r_o}} \frac{r_{gx}}{r_o \| r_{gx}} \tag{6.197}$$

$\underline{(1, 2, 3, 7)}$: Changing the order to $(1, 2, 7, 3)$ we determine:

$$(1, 2, 7, 3) = (1, 2, 7)C_3 R^{(3)}_{(1,2,7)} \tag{6.198a}$$

in which:

$$R^{(3)}_{(1,2,7)} = r_{gx} \tag{6.198b}$$

Combining these with $(1, 2, 7)$ given in Eq. (6.141), we obtain:

$$(1, 2, 7, 3) = L_1 C_2 C_7 C_3 r_{gx} \frac{r_o \| (r_s + R)}{1 + \dfrac{g_m r_o}{1 + \dfrac{r_o + R}{r_s}}} \tag{6.199}$$

$\underline{(1, 2, 4, 5)}$: We change the order to $(1, 2, 5, 4)$ and determine:

$$(1, 2, 5, 4) = (1, 2, 5)C_4 R^{(4)}_{(1,2,5)} \tag{6.200a}$$

in which:

$$R^{(4)}_{(1,2,5)} = r_{gx} + r_o \tag{6.200b}$$

Combining this result with $(1, 2, 5)$ given in Eq. (6.138), we obtain:

$$(1,2,5,4) = \frac{L_1 L_5 C_2 C_4}{1 + g_m r_s + \dfrac{r_s + R}{r_o} r_o \| r_{gx}} \frac{r_{gx}}{}$$

(6.201)

(1, 2, 4, 6): We change the order to $(1, 2, 6, 4)$ and determine:

$$(1,2,6,4) = (1,2,6)C_4 R^{(4)}_{(1,2,6)}$$

(6.202a)

in which:

$$R^{(4)}_{(1,2,6)} = r_{gx} + r_o$$

(6.202b)

Combining these with $(1, 2, 6)$ given in Eq. (6.139), we obtain:

$$(1,2,6,4) = \frac{L_1 L_6 C_2 C_4}{1 + g_m r_s + \dfrac{r_s + R}{r_o} R_o \| r_{gx}} \frac{r_{gx}}{}$$

(6.203)

(1, 2, 4, 7): We change the order to $(1, 2, 7, 4)$ and determine:

$$(1,2,7,4) = (1,2,7)C_4 R^{(4)}_{(1,2,7)}$$

(6.204a)

in which:

$$R^{(4)}_{(1,2,7)} = r_{gx}$$

(6.204b)

Combining these with $(1, 2, 7)$ given in Eq. (6.141), we obtain:

$$(1,2,7,4) = L_1 C_2 C_7 C_4 \frac{r_{gx} r_o \| (r_s + R)}{1 + \dfrac{g_m r_o}{1 + \dfrac{r_o + R}{r_s}}}$$

$$= L_1 C_2 C_7 C_4 \frac{r_{gx}(r_s + R)}{1 + g_m r_s + \dfrac{R + r_s}{r_o}}$$

(6.205)

(1, 2, 5, 6): Since $R^{(6)}_{(1,2,5)} \to \infty$ and $(1, 2, 5)$ is finite, we have:

$$(1,2,5,6) = (1,2,5) \frac{L_6}{R^{(6)}_{(1,2,5)}} = 0$$

(6.206)

(1, 2, 5, 7): This is given by:

$$(1,2,5,7) = (1,2,5)C_7 R^{(7)}_{(1,2,5)}$$

(6.207)

in which:

$$R^{(7)}_{(1,2,5)} = r_o$$

(6.208)

Combining these with $(1, 2, 5)$ given in Eq. (6.138), we obtain:

$$(1, 2, 5, 7) = \frac{L_1 C_2 L_5 C_7}{1 + g_m r_s + \dfrac{R + r_s}{r_o}} \tag{6.209}$$

(1, 2, 6, 7): This is given by:

$$(1, 2, 6, 7) = (1, 2, 6) C_7 R_{(1,2,6)}^{(7)} \tag{6.210a}$$

in which:

$$R_{(1,2,6)}^{(7)} = r_o \tag{6.210b}$$

Combining these with $(1, 2, 6)$ given in Eq. (6.139), we obtain:

$$(1, 2, 6, 7) = \frac{L_1 C_2 L_6 C_7}{1 + g_m r_s + \dfrac{R + r_s}{r_o}} \tag{6.211}$$

(1, 3, 4, 5): We change the order to $(3, 4, 5, 1)$ and determine:

$$(3, 4, 5, 1) = (3, 4, 5) \frac{L_1}{R_{(3,4,5)}^{(1)}} \tag{6.212a}$$

in which:

$$R_{(3,4,5)}^{(1)} = R + R_1 \tag{6.212b}$$

Combining these with $(3, 4, 5)$ given in Eq. (6.172b), we obtain:

$$(1, 3, 4, 5) = \frac{L_1 L_5 C_3 C_4}{1 + g_m r_s + \dfrac{R + r_s}{r_o}} \frac{r_{gx}}{r_o \| r_{gx}} \tag{6.213}$$

(1, 3, 4, 6): We can see that $R_{(1,3,4)}^{(6)}$ and $R_{(1,3,4)}^{(5)}$ are equal. We also have:

$$\left. \begin{aligned} (1, 3, 4, 6) &= (1, 3, 4) L_6 / R_{(1,3,4)}^{(6)} \\ (1, 3, 4, 5) &= (1, 3, 4) L_5 / R_{(1,3,4)}^{(5)} \end{aligned} \right\} \tag{6.214a, b}$$

Hence, $(1, 3, 4, 6)$ can be easily deduced from $(1, 3, 4, 5)$ simply by replacing L_5 with L_6 in Eq. (6.213). Hence we have:

$$(1, 3, 4, 6) = \frac{L_1 L_6 C_3 C_4}{1 + g_m r_s + \dfrac{R + r_s}{r_o}} \frac{r_{gx}}{r_o \| r_{gx}} \tag{6.215}$$

(1, 3, 4, 7): This is given by:

$$(1, 3, 4, 7) = (1, 3, 4)C_7 R^{(7)}_{(1,3,4)} \tag{6.216a}$$

in which:

$$R^{(7)}_{(1,3,4)} = r_{gx} \| r_o \| (r_s + R) \tag{6.216b}$$

Combining these with $(1, 3, 4)$ given in Eq. (6.143), we obtain:

$$(1, 3, 4, 7) = L_1 C_3 C_4 C_7 \frac{r_{gx} + r_o \| (r_s + R)}{1 + \dfrac{g_m r_o}{1 + \dfrac{r_o + R}{r_s}}} r_{gx} \| r_o \| (r_s + R)$$

$$= L_1 C_3 C_4 C_7 \frac{r_{gx} r_o \| (r_s + R)}{1 + \dfrac{g_m r_o r_s}{r_s + r_o + R}} \tag{6.217}$$

$$= L_1 C_3 C_7 C_4 \frac{r_{gx}(r_s + R)}{1 + g_m r_s + \dfrac{R + r_s}{r_o}}$$

$(1, 3, 5, 6)$: We can see that $R^{(6)}_{(1,3,5)} \to \infty$ while $(1, 3, 5)$ is finite. Hence we have:

$$(1, 3, 5, 6) = (1, 3, 5)\frac{L_6}{R^{(6)}_{(1,3,5)}} = 0 \tag{6.218}$$

$(1, 3, 5, 7)$: We change the order to $(1, 3, 7, 5)$ and determine:

$$(1, 3, 7, 5) = (1, 3, 7)\frac{L_5}{R^{(5)}_{(1,3,7)}} \tag{6.219a}$$

in which we determine:

$$R^{(5)}_{(1,3,7)} = r_s + R \tag{6.219b}$$

Combining these with $(1, 3, 7)$ given Eq. (6.149), we obtain:

$$\left.\begin{aligned}(1, 3, 7, 5) &= \frac{L_1 L_5 C_3 C_7}{r_s + R} r_o \left\| \frac{R + r_s}{1 + g_m r_s} \right. \\[2ex] &= \frac{L_1 L_5 C_3 C_7}{1 + g_m r_s + \dfrac{R + r_s}{r_o}}\end{aligned}\right\} \tag{6.220a, b}$$

$(1, 3, 6, 7)$: Change the order to $(1, 3, 7, 6)$ and determine:

$$(1, 3, 7, 6) = (1, 3, 7)\frac{L_6}{R^{(6)}_{(1,3,7)}} \tag{6.221}$$

Here, we see that $R^{(6)}_{(1,3,7)}$ is the same as $R^{(5)}_{(1,3,7)}$ so that $(1,3,7,6)$ can be deduced from $(1,3,7,5)$ in Eq. (6.220b) simply by replacing L_5 with L_6:

$$(1,3,7,6) = \frac{L_1 L_6 C_3 C_7}{1 + g_m r_s + \dfrac{R + r_s}{r_o}} \tag{6.222}$$

$(1,4,5,6)$: We can see that $R^{(6)}_{(1,4,5)} \to \infty$ while $(1,4,5) = 0$, as determined earlier in Eq. (6.150a), so that we have:

$$(1,4,5,6) = (1,4,5)\frac{L_6}{R^{(6)}_{(1,4,5)}} = 0 \tag{6.223}$$

$(1,4,5,7)$: Since $(1,4,5) = 0$ and $R^{(7)}_{(1,4,5)}$ is finite, we have:

$$(1,4,5,7) = 0 \tag{6.224}$$

$(1,4,6,7)$: Since $(1,4,6) = 0$ and $R^{(7)}_{(1,4,6)}$ is finite, we have:

$$(1,4,6,7) = 0 \tag{6.225}$$

$(1,5,6,7)$: Since $(1,5,6) = 0$ and $R^{(7)}_{(1,5,6)}$ is finite, we have:

$$(1,5,6,7) = 0 \tag{6.226}$$

$(2,3,4,5)$, $(2,3,4,6)$, $(2,3,4,7)$: Since $(2,3,4) = 0$ and $R^{(5)}_{(2,3,4)}$, $R^{(6)}_{(2,3,4)}$ and $R^{(7)}_{(2,3,4)}$ are finite, we have:

$$\left.\begin{array}{l}(2,3,4,5) = 0 \\[4pt] (2,3,4,6) = 0 \\[4pt] (2,3,4,7) = 0\end{array}\right\} \tag{6.227a–c}$$

$(2,3,5,6)$: This is given by:

$$(2,3,5,6) = (2,3,5)\frac{L_6}{R^{(6)}_{(2,3,5)}} \tag{6.228}$$

in which we determine:

$$R^{(6)}_{(2,3,5)} = R + R_1 \tag{6.229}$$

Combining these with $(2,3,5)$ given in Eq. (6.155b), we obtain:

$$(2,3,5,6) = \frac{C_2 C_3 L_5 L_6}{1 + g_m r_s + \dfrac{R + r_s}{r_o}} \cdot \frac{r_{gx}}{r_o \| r_{gx}} \tag{6.230}$$

$(2,3,5,7)$: This is given by:

$$(2, 3, 5, 7) = (2, 3, 5)C_7 R_{(2,3,5)}^{(7)} \tag{6.231}$$

in which:

$$R_{(2,3,5)}^{(7)} = r_o \| r_{gx} \tag{6.232a}$$

Combining these with $(2, 3, 5)$ in Eq. $(6.155b)$, we obtain:

$$(2, 3, 5, 7) = C_2 C_3 C_7 L_5 \frac{r_{gx}(R + R_1)}{1 + g_m r_s + \dfrac{R + r_s}{r_o}} \tag{6.232b}$$

$(\mathbf{2, 3, 6, 7})$: We change the order and determine:

$$(3, 6, 7, 2) = (3, 6, 7)C_2 R_{(3,6,7)}^{(2)} \tag{6.233a}$$

in which we see that:

$$R_{(3,6,7)}^{(2)} = r_{gx} \tag{6.233b}$$

Combining these with $(3, 6, 7)$ given in Eq. (6.184), we obtain:

$$(2, 3, 6, 7) = L_6 C_3 C_7 C_2 \frac{r_{gx}(r_s + R_1)}{1 + g_m r_s + \dfrac{r_s + R}{r_o}} \tag{6.234}$$

$(\mathbf{2, 4, 5, 6})$: This is given by:

$$(2, 4, 5, 6) = (2, 4, 5)\frac{L_6}{R_{(2,4,5)}^{(6)}} \tag{6.235a}$$

in which we determine:

$$R_{(2,4,5)}^{(6)} = R_1 + R \tag{6.235b}$$

Combining these with $(2, 4, 5)$ given in Eq. (6.161), we obtain:

$$(2, 4, 5, 6) = \frac{C_2 C_4 L_5 L_6}{1 + g_m r_s + \dfrac{R + r_s}{r_o}} \frac{r_{gx}}{r_o \| r_{gx}} \tag{6.236}$$

$(\mathbf{2, 4, 5, 7})$: This is given by:

$$(2, 4, 5, 7) = (2, 4, 5)C_7 R_{(2,4,5)}^{(7)} \tag{6.237a}$$

in which we determine:

$$R_{(2,4,5)}^{(7)} = r_o \| r_{gx} \tag{6.237b}$$

Combining these with $(2, 4, 5)$ given in Eq. (6.161), we obtain:

$$(2,4,5,7) = C_2C_4C_7L_5 \frac{r_{gx}(R + R_1)}{1 + g_m r_s + \dfrac{R + r_s}{r_o}} \tag{6.238}$$

(2, 4, 6, 7): This is given by:

$$(2,4,6,7) = (2,4,6)C_7 R^{(7)}_{(2,4,6)} \tag{6.239a}$$

in which we determine:

$$R^{(7)}_{(2,4,6)} = r_o \| r_{gx} \| (r_s + R_1) \tag{6.239b}$$

Combining these with $(2, 4, 6)$ given in Eq. (6.162), we obtain:

$$(2,4,6,7) = L_6C_2C_4C_7 r_{gx} \frac{\left(1 + \dfrac{r_s + R_1}{r_o \| r_{gx}}\right) r_o \| r_{gx} \| (r_s + R_1)}{1 + g_m r_s + \dfrac{R + r_s}{r_o}}$$

$$= L_6C_2C_4C_7 \frac{r_{gx}(r_s + R_1)}{1 + g_m r_s + \dfrac{R + r_s}{r_o}} \tag{6.240}$$

(2, 5, 6, 7): This is given by:

$$(2,5,6,7) = (2,5,6)C_7 R^{(7)}_{(2,5,6)} \tag{6.241a}$$

in which we determine:

$$R^{(7)}_{(2,5,6)} = r_o \tag{6.241b}$$

Combining these with $(2, 5, 6)$ given in Eq. (6.166), we obtain:

$$(2,5,6,7) = \frac{L_6L_5C_2C_7}{1 + g_m r_s + \dfrac{r_s + R}{r_o}} \tag{6.242}$$

(3, 4, 5, 6): This is given by:

$$(3,4,5,6) = (3,4,5) \frac{L_6}{R^{(6)}_{(3,4,5)}} \tag{6.243a}$$

in which we determine:

$$R^{(6)}_{(3,4,5)} = R_1 + R \tag{6.243b}$$

Combining these with $(3, 4, 5)$ given in Eq. (6.172b), we obtain:

$$(3,4,5,6) = \frac{C_3 C_4 L_5 L_6}{1 + g_m r_s + \dfrac{R + r_s}{r_o} r_{gx} \| r_o} \frac{r_{gx}}{r_{gx} \| r_o} \tag{6.244}$$

$(\mathbf{3,4,5,7})$: This is given by:

$$(3,4,5,7) = (3,4,5)C_7 R^{(7)}_{(3,4,5)} \tag{6.245a}$$

in which we determine:

$$R^{(7)}_{(3,4,5)} = r_{gx} \| r_o \tag{6.245b}$$

Combining these with $(3,4,5)$ given in Eq. $(6.172b)$, we obtain:

$$(3,4,5,7) = C_3 C_4 C_7 L_5 \frac{r_{gx}(R + R_1)}{1 + g_m r_s + \dfrac{R + r_s}{r_o}} \tag{6.246}$$

$(\mathbf{3,4,6,7})$: This is given by:

$$(3,4,6,7) = (3,4,6)C_7 R^{(7)}_{(3,4,6)} \tag{6.247a}$$

in which we determine:

$$R^{(7)}_{(3,4,6)} = r_{gx} \| r_o \| (r_s + R_1) \tag{6.247b}$$

Combining these with $(3,4,6)$ given in Eq. (6.175), we obtain in the same manner as in Eq. (6.240):

$$(3,4,6,7) = C_4 C_3 C_7 L_6 \frac{r_{gx}(r_s + R_1)}{1 + g_m r_s + \dfrac{R + r_s}{r_o}} \tag{6.248}$$

$(\mathbf{3,5,6,7})$: This is given by:

$$(3,5,6,7) = (3,5,6)C_7 R^{(7)}_{(3,5,6)} \tag{6.249a}$$

in which we determine:

$$R^{(7)}_{(3,5,6)} = r_o \| g_m^{-1} \tag{6.249b}$$

Combining these with $(3,5,6)$ given in Eq. (6.180), we obtain:

$$(3,5,6,7) = L_6 L_5 C_3 C_7 \frac{r_o \| g_m^{-1}}{r_s + \dfrac{r_o + R}{1 + g_m r_o}}$$

$$= \frac{L_6 L_5 C_3 C_7}{1 + g_m r_s + \dfrac{R + r_s}{r_o}} \tag{6.250}$$

The coefficient a_4 is given by:

$$a_4 = \frac{C_2C_3 + C_2C_4 + C_3C_4}{1 + g_m r_s + \dfrac{R + r_s}{r_o}} C_7 r_{gx}[(L_1 + L_6)(r_s + R)$$

$$+ L_5(R + R_1)] + \frac{L_1L_6 + L_1L_5 + L_5L_6}{1 + g_m r_s + \dfrac{R + r_s}{r_o}} \left[C_7(C_2 + C_3) \right.$$ (6.251)

$$\left. + \frac{r_{gx}}{r_{gx} \| r_o}(C_2C_3 + C_2C_4 + C_3C_4) \right]$$

Some of the terms in Eq. (6.251) can be collected to yield a more compact expression for a_4:

$$a_4 = \frac{1}{1 + g_m r_s + \dfrac{R + r_s}{r_o}} \left\{ \frac{L_1L_5L_6}{L_1 \| L_5 \| L_6} \left[C_7(C_2 + C_3) + \frac{r_{gx}}{r_{gx} \| r_o} \frac{C_2C_3C_4}{C_2 \| C_3 \| C_4} \right] \right.$$

$$\left. + \frac{C_2C_3C_4 r_{gx}(r_s + R_1)}{C_2 \| C_3 \| C_4} C_7 \left[L_1 + L_6 + L_5 \frac{R_1 + R}{r_s + R} \right] \right\}$$ (6.252)

The coefficient a_5: This is given by:

$$a_5 = \Sigma\Sigma\Sigma\Sigma\Sigma \tau^{(i)} \tau_{(i)}^{(j)} \tau_{(i,j)}^{(k)} \tau_{(i,j,k)}^{(l)} \tau_{(i,j,k,l)}^{(m)}$$ (6.253)

There are $\binom{7}{5} = 21$ terms to determine, of which twelve are zero. In the surviving nine terms, all the new port resistances $R_{(i,j,k,l)}^{(m)}$ are equal to $r_{gx} \| r_o$.

$(1, 2, 3, 4, n)$: Since $(1, 2, 3, 4) = 0$ as determined earlier in Eq. (6.193), $(1, 2, 3, 4, n) = 0$ for $n = 5, 6, 7$ (no indeterminate terms). Hence, we have:

$$\left. \begin{aligned} (1, 2, 3, 4, 5) &= 0 \\ (1, 2, 3, 4, 6) &= 0 \\ (1, 2, 3, 4, 7) &= 0 \end{aligned} \right\}$$ (6.254a–c)

$(1, 2, 3, 5, 6)$: Since $R_{(1,2,3,5)}^{(6)} \to \infty$ and $(1, 2, 3, 5)$ is finite, we have:

$$(1, 2, 3, 5, 6) = (1, 2, 3, 5)\frac{L_6}{R_{(1,2,3,5)}^{(6)}} = 0$$ (6.255)

$(1, 2, 3, 5, 7)$: This is given by:

$$(1, 2, 3, 5, 7) = (1, 2, 3, 5)C_7 R_{(1,2,3,5)}^{(7)}$$ (6.256a)

in which:

$$R^{(7)}_{(1,2,3,5)} = r_o \parallel r_{gx} \tag{6.256b}$$

Combining these with $(1, 2, 3, 5)$ in Eq. (6.195), we obtain:

$$(1, 2, 3, 5, 7) = \frac{L_1 L_5 C_2 C_3 C_7}{1 + g_m r_s + \dfrac{r_s + R}{r_o}} r_{gx} \tag{6.257}$$

$\underline{(1, 2, 3, 6, 7)}$: This is given by:

$$(1, 2, 3, 6, 7) = (1, 2, 3, 6) C_7 R^{(7)}_{(1,2,3,6)} \tag{6.258a}$$

in which:

$$R^{(7)}_{(1,2,3,6)} = r_o \parallel r_{gx} \tag{6.258b}$$

Combining these with $(1, 2, 3, 6)$ in Eq. (6.197), we obtain:

$$(1, 2, 3, 6, 7) = \frac{L_1 L_6 C_2 C_3 C_7}{1 + g_m r_s + \dfrac{r_s + R}{r_o}} r_{gx} \tag{6.259}$$

$\underline{(1, 2, 4, 5, 6)}$: $R^{(6)}_{(1,2,4,5)} \to \infty$ and $(1, 2, 4, 5)$ is finite so that:

$$(1, 2, 4, 5, 6) = (1, 2, 4, 5) \frac{L_6}{R^{(6)}_{(1,2,4,5)}} = 0 \tag{6.260}$$

$\underline{(1, 2, 4, 5, 7)}$: This is given by:

$$(1, 2, 4, 5, 7) = (1, 2, 4, 5) C_7 R^{(7)}_{(1,2,4,5)} \tag{6.261a}$$

in which:

$$R^{(7)}_{(1,2,4,5)} = r_o \parallel r_{gx} \tag{6.261b}$$

Combining these with $(1, 2, 4, 5)$ in Eq. (6.201), we obtain:

$$(1, 2, 4, 5, 7) = \frac{L_1 C_2 C_4 L_5 C_7}{1 + g_m r_s + \dfrac{r_s + R}{r_o}} r_{gx} \tag{6.262}$$

$\underline{(1, 2, 4, 6, 7)}$: This is given by:

$$(1, 2, 4, 6, 7) = (1, 2, 4, 6) C_7 R^{(7)}_{(1,2,4,6)} \tag{6.263a}$$

in which:

$$R^{(7)}_{(1,2,4,6)} = r_o \parallel r_{gx} \tag{6.263b}$$

Combining these with $(1, 2, 4, 6)$ in Eq. (6.203), we obtain:

$$(1,2,4,6,7) = \frac{L_1 C_2 C_4 L_6 C_7}{1 + g_m r_s + \dfrac{r_s + R}{r_o}} r_{gx} \tag{6.264}$$

$\underline{(1, 2, 5, 6, 7)}$: Since $(1, 2, 5, 6) = 0$ as determined in Eq. (6.206) and $R_{(1,2,5,6)}^{(7)} = r_o$, we have:

$$(1, 2, 5, 6, 7) = 0 \tag{6.265}$$

$\underline{(1, 3, 4, 5, 6)}$: $R_{(1,3,4,5)}^{(6)} \to \infty$ and $(1, 3, 4, 5)$ is finite so that:

$$(1, 3, 4, 5, 6) = (1, 3, 4, 5) \frac{L_6}{R_{(1,2,4,5)}^{(6)}} = 0 \tag{6.266}$$

$\underline{(1, 3, 4, 5, 7)}$: This is given by:

$$(1, 3, 4, 5, 7) = (1, 3, 4, 5) C_7 R_{(1,3,4,5)}^{(7)} \tag{6.267a}$$

in which we determine:

$$R_{(1,3,4,5)}^{(7)} = r_{gx} \| r_o \tag{6.267b}$$

Combining these with $(1, 3, 4, 5)$ given in Eq. (6.213), we obtain:

$$(1, 3, 4, 5, 7) = \frac{L_1 L_5 C_3 C_4 C_7}{1 + g_m r_s + \dfrac{r_s + R}{r_o}} r_{gx} \tag{6.268}$$

$\underline{(1, 3, 4, 6, 7)}$: This is given by:

$$(1, 3, 4, 6, 7) = (1, 3, 4, 6) C_7 R_{(1,3,4,6)}^{(7)} \tag{6.269a}$$

in which we determine:

$$R_{(1,3,4,6)}^{(7)} = r_{gx} \| r_o \tag{6.269b}$$

Combining these with $(1, 3, 4, 6)$ given in Eq. (6.215), we obtain

$$(1, 3, 4, 6, 7) = \frac{L_1 L_6 C_3 C_4 C_7}{1 + g_m r_s + \dfrac{r_s + R}{r_o}} r_{gx} \tag{6.270}$$

$\underline{(1, 3, 5, 6, 7)}$: Since $(1, 3, 5, 6) = 0$ as determined in Eq. (6.218) and $R_{(1,3,5,6)}^{(7)}$ is finite, we have:

$$(1, 3, 5, 6, 7) = 0 \tag{6.271}$$

$\underline{(1, 4, 5, 6, 7)}$: Since $(1, 4, 5, 6) = 0$ as determined in Eq. (6.223) and $R_{(1,4,5,6)}^{(7)}$ is finite, we have:

$$(1, 4, 5, 6, 7) = 0 \tag{6.272}$$

(2, 3, 4, 5, 6), **(2, 3, 4, 5, 7)**, **(2, 3, 4, 6, 7)**: Since $(2, 3, 4, 5) = 0$ and $(2, 3, 4, 6) = 0$ as determined in Eqs. (6.227a, b) and no indeterminacy occurs when ports (6) and (7) are considered, we have:

$$\left.\begin{array}{l} (2, 3, 4, 5, 6) = 0 \\ (2, 3, 4, 5, 7) = 0 \\ (2, 3, 4, 6, 7) = 0 \end{array}\right\} \tag{6.273a-c}$$

(2, 3, 5, 6, 7): This is given by:

$$(2, 3, 5, 6, 7) = (2, 3, 5, 6)R^{(7)}_{(2,3,5,6)} \tag{6.274a}$$

in which we determine:

$$R^{(7)}_{(2,3,5,6)} = r_o \,\|\, r_{gx} \tag{6.274b}$$

Combining these with $(2, 3, 5, 6)$ given in Eq. (6.230), we obtain:

$$(2, 3, 5, 6, 7) = \frac{C_2 C_3 C_7 L_5 L_6}{1 + g_m r_s + \dfrac{R + r_s}{r_o}} r_{gx} \tag{6.275}$$

(2, 4, 5, 6, 7): This is given by:

$$(2, 4, 5, 6, 7) = (2, 4, 5, 6)C_7 R^{(7)}_{(2,4,5,6)} \tag{6.276a}$$

in which we determine:

$$R^{(7)}_{(2,4,5,6)} = r_o \,\|\, r_{gx} \tag{6.276b}$$

Combining these with $(2, 4, 5, 6)$ given in Eq. (6.236), we determine:

$$(2, 4, 5, 6, 7) = \frac{C_2 C_4 C_7 L_5 L_6}{1 + g_m r_s + \dfrac{R + r_s}{r_o}} r_{gx} \tag{6.277}$$

(3, 4, 5, 6, 7): This is given by:

$$(3, 4, 5, 6, 7) = (3, 4, 5, 6)C_7 R^{(7)}_{(3,4,5,6)} \tag{6.278a}$$

in which we determine:

$$R^{(7)}_{(3,4,5,6)} = r_o \,\|\, r_{gx} \tag{6.278b}$$

Combining these with $(3, 4, 5, 6)$ given in Eq. (6.244), we determine:

$$(3, 4, 5, 6, 7) = \frac{C_4 C_3 C_7 L_5 L_6}{1 + g_m r_s + \dfrac{R + r_s}{r_o}} r_{gx} \tag{6.279}$$

The coefficient a_5 is given by:

$$a_5 = \frac{L_1 L_5 L_6 C_2 C_3 C_4}{L_1 \| L_5 \| L_6 C_2 \| C_3 \| C_4} \frac{C_7 r_{gx}}{1 + g_m r_s + \dfrac{R + r_s}{r_o}} \tag{6.280}$$

The numerator $N(s)$

The numerator is given by:

$$N(s) = 1 + \sum_{n=1}^{4} b_i s^n \tag{6.281}$$

in which the coefficients b_n are determined using null double injection.

The coefficient b_1: This is given by:

$$b_i = \sum_{i=1}^{7} t^{(i)} = \Sigma C_i \mathcal{R}^{(i)} + \Sigma \frac{L_i}{\mathcal{R}^{(i)}} \tag{6.282}$$

in which $t^{(i)}$ is the time constant formed at port (i) by the reactive element at that port and the null resistance $\mathcal{R}^{(i)}$ looking into that port using null double injection. These are determined as follows.

$\mathcal{R}^{(1)}$: This is infinite since ports (2) and (3) are open. Hence, we have:

$$\left. \begin{array}{l} \mathcal{R}^{(1)} \to \infty \\[2mm] t^{(1)} = \dfrac{L_1}{\mathcal{R}^{(1)}} = 0 \end{array} \right\} \tag{6.283a, b}$$

$\mathcal{R}^{(2)}$: We see in Fig. 6.14 that when the output voltage and hence the output current are both nulled, the voltage drop across r_s, given by $I_T r_s$, is equal and opposite to the voltage drop across r_o which is given by $g_m r_o V_T$. Hence, we have:

$$g_m V_T r_o = -I_T r_s \tag{6.284}$$

It follows that:

$$\left. \begin{array}{l} \mathcal{R}^{(2)} = \dfrac{V_T}{I_T} = -\dfrac{r_s}{g_m r_o} \\[4mm] t^{(2)} = C_2 \mathcal{R}^{(2)} = -\dfrac{C_2 \, r_s}{g_m \, r_o} \end{array} \right\} \tag{6.285a, b}$$

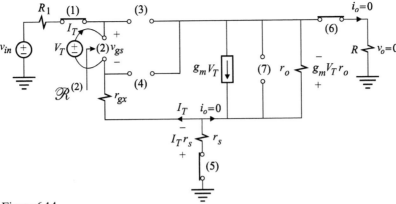

Figure 6.14

$\mathscr{R}^{(3)}$: We see in Fig. 6.15 that with the output nulled, v_{gs} is given by:

$$v_{gs} = -(V_T + I_T r_s) \tag{6.286}$$

We can also see that the voltage drop across r_s, given by $I_T r_s$, is equal to the voltage drop across r_o which is given by $g_m v_{gs} r_o$. Hence, we have:

$$I_T r_s = g_m v_{gs} r_o$$
$$= -g_m r_o (V_T + I_T r_s) \tag{6.287}$$

It follows that:

$$\left. \begin{array}{l} \mathscr{R}^{(3)} = -r_s \left(1 + \dfrac{1}{g_m r_o \parallel r_s} \right) \\[3mm] t^{(3)} = -C_3 r_s \left(1 + \dfrac{1}{g_m r_o \parallel r_s} \right) \end{array} \right\} \tag{6.288a, b}$$

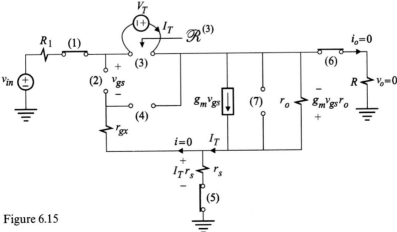

Figure 6.15

$\mathcal{R}^{(4)}$: We see from Fig. 6.16 that when the output is nulled, the current through r_s and the voltage across it are both zero so that the voltage across r_{gx} is V_T. Hence, we have:

$$\left.\begin{aligned}\mathcal{R}^{(4)} &= r_{gx}\\[4pt]t^{(4)} &= C_4 r_{gx}\end{aligned}\right\}\qquad\text{(6.289a, b)}$$

$\mathcal{R}^{(5)}$: The test current looking into port (5) is the same as the output current which when nulled causes the test current to be zero and $\mathcal{R}^{(5)}$ to be infinite. Hence, we have:

$$\left.\begin{aligned}\mathcal{R}^{(5)} &\to \infty\\[4pt]t^{(5)} &= \frac{L_5}{\mathcal{R}^{(5)}} = 0\end{aligned}\right\}\qquad\text{(6.290a, b)}$$

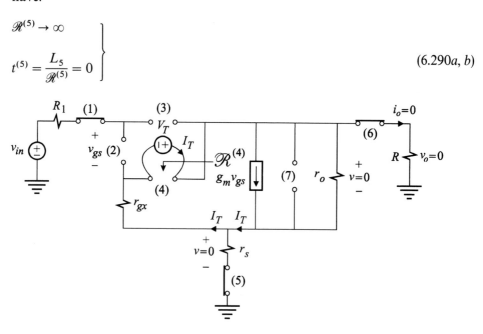

Figure 6.16

$\mathcal{R}^{(6)}$: The test current looking into port (6) with the output nulled is clearly zero so that we have:

$$\left.\begin{aligned}\mathcal{R}^{(6)} &\to \infty\\[4pt]t^{(6)} &= \frac{L_5}{\mathcal{R}^{(5)}} = 0\end{aligned}\right\}\qquad\text{(6.291a, b)}$$

$\mathcal{R}^{(7)}$: We see in Fig. 6.17 that when the output is nulled, the voltage across r_s is zero. It follows that the voltage across port (7) is zero. Hence, a test current source connected at port (7) under this null condition, will develop zero volts across it so that the null resistance looking into port (7) is zero:

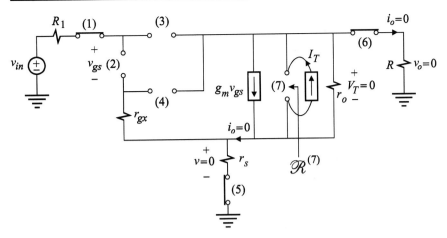

Figure 6.17

$$\mathcal{R}^{(7)} = 0$$
$$t^{(7)} = C_7 \mathcal{R}^{(7)} = 0$$

$$(6.292a, b)$$

The coefficient b_1 is given by:

$$b_1 = -\frac{C_2}{g_m}\frac{r_s}{r_o} - C_3 r_s \left(1 + \frac{1}{g_m r_o \| r_s}\right) + C_4 r_{gx}$$

$$(6.293)$$

The coefficient b_2: This is given by:

$$b_2 = \Sigma\Sigma t^{(i)} t_{(i)}^{(j)}$$

$$(6.294)$$

As in the analysis of the denominator, we shall use (i, j) to denote the pair $t^{(i)} t_{(i)}^{(j)}$ and determine these as follows.

(1, n): Since $t^{(1)} = 0$ and there are no indeterminate forms in $t_{(1)}^{(n)}$, we have:

$$(1, n) = 0 \quad n = 1, 2, \ldots, 7$$

$$(6.295)$$

(2, 3): We can verify in Fig. 6.18 that:

$$V_T = I_T r_o + r_{gx}\left(I_T + \frac{I_T r_o}{r_s}\right)$$

$$(6.296)$$

It follows that:

$$\mathcal{R}_{(2)}^{(3)} = \frac{r_o r_{gx}}{r_o \| r_{gx} \| r_s}$$
$$(2, 3) = -C_2 C_3 \frac{r_s r_{gx}}{g_m r_o \| r_{gx} \| r_s}$$

$$(6.297a, b)$$

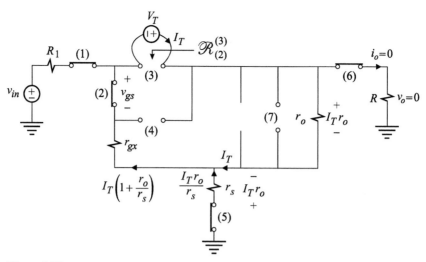

Figure 6.18

(2, 4): This is identical to (2, 3) so that we have:

$$(2,4) = -C_2 C_3 \frac{r_s r_{gx}}{g_m r_o \parallel r_{gx} \parallel r_s} \tag{6.298}$$

(2, 5): When port (2) is shorted, $g_m v_{gs} = 0$ and the current through r_o becomes the same as the output current. It follows that when the output voltage and, hence, the output current are both nulled, the voltage drops across r_o and R are both zero. Hence, the upper side of r_s is at virtual ground and the null resistance looking into port (5) is simply r_s. Hence, we have:

$$\left. \begin{aligned} \mathscr{R}_{(2)}^{(5)} &= r_s \\[2mm] (2,5) &= -\frac{C_2 L_5}{g_m r_o} \end{aligned} \right\} \tag{6.299a, b}$$

(2, 6): Since $\mathscr{R}_{(2)}^{(6)} \to \infty$ and $t^{(2)}$ is finite, we have:

$$(2,6) = t^{(2)} \frac{L_6}{\mathscr{R}_{(2)}^{(6)}} = 0 \tag{6.300}$$

(2, 7): When port (2) is shorted, $g_m v_{gs}$ vanishes. Hence, looking into port (7) with the output current nulled we see r_o:

$$\mathscr{R}_{(2)}^{(7)} = r_o \tag{6.301}$$

It follows that:

$$(2,7) = -C_7 C_2 \frac{r_s}{g_m} \tag{6.302}$$

(3, 4): Since $t^{(4)}$ is simpler than $t^{(3)}$, we change the order to (4, 3) and determine $\mathcal{R}_{(4)}^{(3)}$ as shown in Fig. 6.19. Since the output voltage and current are nulled, the voltage across r_s is $I_T r_s$ and it appears across r_{gx}. It follows that:

$$I_T = -\frac{I_T r_s}{r_o \| r_{gx}} - g_m V_T \tag{6.303}$$

Hence, we have:

$$\left. \begin{aligned} \mathcal{R}_{(4)}^{(3)} &= -\frac{1}{g_m} \frac{r_s}{r_s \| r_o \| r_{gx}} \\[2mm] (4, 3) &= -C_3 C_4 \frac{r_s r_{gx}}{g_m r_o \| r_{gx} \| r_s} \end{aligned} \right\} \tag{6.304a, b}$$

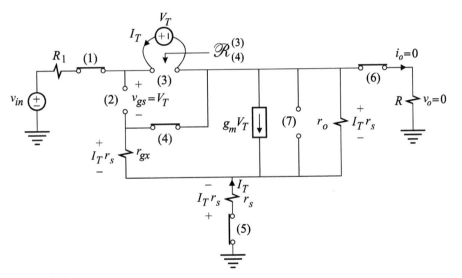

Figure 6.19

(3, 5): When port (3) is shorted, the dependent source $g_m v_{gs}$ acts like a simple conductance g_m in parallel with r_o. When the output voltage is nulled, the upper side of $r_o \| g_m^{-1}$ is at virtual ground so that the null resistance looking into port (5) is:

$$\mathcal{R}_{(3)}^{(5)} = r_s + r_o \| g_m^{-1} \tag{6.305}$$

It follows that:

$$(3, 5) = -C_3 L_5 \frac{r_s \left(1 + \dfrac{1}{g_m r_o \| r_s}\right)}{r_s + r_o \| g_m^{-1}}$$

$$= -C_3 L_5 \left(1 + \frac{1}{g_m r_o} \right)$$

(6.306)

(3, 7): Looking into port (7) with port (3) short and the output voltage nulled, we have according to Fig. 6.20:

$$I_T = \frac{V_T}{r_s} + \frac{V_T}{r_o} + g_m V_T$$

(6.307)

It follows that:

$$\left. \begin{array}{l} \mathscr{R}_{(3)}^{(7)} = r_s \| r_o \| g_m^{-1} \\[2mm] (3, 7) = -C_3 C_7 \dfrac{r_s}{g_m} \end{array} \right\}$$

(6.308a, b)

The reader can easily verify that all the remaining pairs are zero so that b_2 is given by:

$$b_2 = -\frac{C_2 C_3 C_4}{C_2 \| C_3 \| C_4} \frac{r_s r_{gx}}{r_s \| r_o \| r_{gx}} - C_7 (C_3 + C_2) \frac{r_s}{g_m}$$

$$- L_5 \left[C_3 \left(1 + \frac{1}{g_m r_o} \right) + \frac{C_2}{g_m r_o} \right]$$

(6.309)

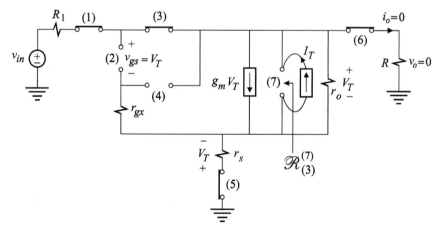

Figure 6.20

The coefficient a_3: This is given by:

$$a_3 = \Sigma\Sigma\Sigma t^{(i)} t_{(i)}^{(j)} t_{(i,j)}^{(k)}$$

(6.310)

In what follows, only the nonzero triplets will be given. The reader can verify that all other triplets are zero.

(2, 3, 5): When ports (2) and (3) are shorted, the dependent source $g_m v_{gs}$ vanishes and r_{gx} and r_o appear in parallel. Now, when the output voltage is nulled, the upper node of $r_{gx} \| r_o$ is at virtual ground so that the null resistance looking into port (5) is given by:

$$\mathscr{R}^{(5)}_{(2,3)} = r_s + r_{gx} \| r_o \tag{6.311}$$

It follows that:

$$(2, 3, 5) = -C_2 C_3 L_5 \frac{r_{gx}}{g_m r_{gx} \| r_o} \tag{6.312}$$

(2, 3, 7): The determination of $\mathscr{R}^{(7)}_{(2,3)}$ is very similar to $\mathscr{R}^{(7)}_{(3)}$ shown in Fig. 6.20. In this case, when port (2) is shorted, r_{gx} appears in parallel with r_o and $g_m v_{gs}$ vanishes so that we have:

$$I_T = \frac{V_T}{r_s} + \frac{V_T}{r_o \| r_{gx}} \tag{6.313}$$

It follows that:

$$\left.\begin{aligned} \mathscr{R}^{(7)}_{(2,3)} &= r_s \| r_o \| r_{gx} \\[2mm] (2, 3, 7) &= -C_2 C_3 C_7 \frac{r_s r_{gx}}{g_m} \end{aligned}\right\} \tag{6.314a, b}$$

(2, 4, 5): This is similar to $(2, 3, 5)$ in which C_3 is replaced with C_4. Hence, we have:

$$(2, 4, 5) = -C_2 C_4 L_5 \frac{r_{gx}}{g_m r_{gx} \| r_o} \tag{6.315}$$

(2, 4, 7): This is similar to $(2, 3, 7)$ in which C_3 is replaced with C_4. Hence, we have:

$$(2, 4, 7) = -C_2 C_4 C_7 \frac{r_s r_{gx}}{g_m} \tag{6.316}$$

(2, 5, 7): The determination of $\mathscr{R}^{(7)}_{(2,5)}$ is similar to that of $\mathscr{R}^{(7)}_{(2)}$ in Eq. (6.301) so that we have:

$$\mathscr{R}^{(7)}_{(2,5)} = r_o \tag{6.317}$$

It follows that:

$$(2, 5, 7) = -\frac{L_5 C_2 C_7}{g_m} \tag{6.318}$$

(3, 4, 5): The null resistance $\mathscr{R}^{(5)}_{(3,4)}$ is the same as $\mathscr{R}^{(5)}_{(2,3)}$ in Eq. (6.311) so that we have:

$$(3,4,5) = -\frac{L_5 C_4 C_3}{g_m} \frac{r_{gx}}{r_{gx} \| r_o} \tag{6.319}$$

(3, 4, 7): The null resistance $\mathscr{R}^{(7)}_{(3,4)}$ is the same as $\mathscr{R}^{(7)}_{(2,3)}$ in Eq. (6.314) so that we have:

$$(3,4,7) = -C_3 C_4 C_7 \frac{r_{gx} r_s}{g_m} \tag{6.320}$$

(3, 5, 7): The resistance looking into port (7) with port (3) short and port (5) open, regardless of whether the output is nulled or not, is given by:

$$\mathscr{R}^{(7)}_{(3,5)} = r_o \| g_m^{-1} \tag{6.321}$$

It follows that:

$$(3,5,7) = -\frac{L_5 C_3 C_7}{g_m} \tag{6.322}$$

Grouping these results, we obtain for b_3:

$$b_3 = -\left(\frac{L_5}{r_{gx} \| r_o} + C_7 r_s\right) \frac{C_2 C_3 C_4}{C_2 \| C_3 \| C_4} \frac{r_{gx}}{g_m} - (C_2 + C_3)\frac{L_5 C_7}{g_m} \tag{6.323}$$

The coefficient b_4: This is given by:

$$b_4 = \Sigma\Sigma\Sigma\Sigma t^{(i)} t^{(j)}_{(i)} t^{(k)}_{(i,j)} t^{(l)}_{(i,j,k)} \tag{6.324}$$

in which the nonzero terms are determined as follows.

(2, 3, 5, 7): The resistance looking into port (7) with ports (2) and (3) short and port (5) open, regardless of whether the output is nulled or not, is simply given by:

$$\mathscr{R}^{(7)}_{(2,3,5)} = r_o \| r_{gx} \tag{6.325}$$

It follows that:

$$(2,3,5,7) = -L_5 C_2 C_3 C_7 \frac{r_{gx}}{g_m} \tag{6.326}$$

(2, 4, 5, 7), **(3, 4, 5, 7)**: As in Eq. (6.325), we have by inspection:

$$\mathscr{R}^{(7)}_{(2,4,5)} = \mathscr{R}^{(7)}_{(3,4,5)} = r_o \| r_{gx} \tag{6.327}$$

It follows that:

$$\left.\begin{aligned}
(2,4,5,7) &= -L_5 C_2 C_4 C_7 \frac{r_{gx}}{g_m} \\
\\
(3,4,5,7) &= -L_5 C_3 C_4 C_7 \frac{r_{gx}}{g_m}
\end{aligned}\right\} \tag{6.328a, b}$$

Grouping the results above, we obtain:

$$b_4 = -L_5 C_7 \frac{C_2 C_3 C_4}{C_2 \| C_3 \| C_4} \frac{r_{gx}}{g_m} \tag{6.329}$$

Example 6.3   Using the element values in Fig. 6.11 we obtain the following values for the voltage gain transfer function:

$$
\left.
\begin{aligned}
A_o &= 2.256 \ (7.0673 \ \text{dB}) \\
b_1 &= 6.017752 \times 10^{-13} \ \text{s} \\
b_2 &= 1.922437 \times 10^{-23} \ \text{s}^2 \\
b_3 &= 3.432182 \times 10^{-34} \ \text{s}^3 \\
b_4 &= 9.042019 \times 10^{-47} \ \text{s}^4 \\
a_1 &= 5.284670 \times 10^{-11} \ \text{s} \\
a_2 &= 6.288061 \times 10^{-12} \ \text{s}^2 \\
a_3 &= 3.877904 \times 10^{-33} \ \text{s}^3 \\
a_4 &= 9.129621 \times 10^{-45} \ \text{s}^4 \\
a_5 &= 2.177460 \times 10^{-57} \ \text{s}^5
\end{aligned}
\right\} \tag{6.330a–j}
$$

Upon determining the roots of the numerator and the denominator, the transfer function can be written in the following factored form:

$$A(s) = A_o \frac{\left(1 - \dfrac{s}{\omega_{oz} Q_{oz}} + \dfrac{s^2}{\omega_{oz}^2}\right)\left(1 + \dfrac{s}{\omega_{z1}}\right)\left(1 + \dfrac{s}{\omega_{z2}}\right)}{\left(1 + \dfrac{s}{\omega_{p1}}\right)\left(1 + \dfrac{s}{\omega_o Q_o} + \dfrac{s^2}{\omega_o^2}\right)\left(1 + \dfrac{s}{\omega_{h1}}\right)\left(1 + \dfrac{s}{\omega_{h1}}\right)} \tag{6.331}$$

in which:

$$
\left.
\begin{aligned}
\omega_{p1} &= (2\pi) \, 4.41 \times 10^9 \ \text{rad/s} \\
\omega_o &= (2\pi) \, 22.11 \times 10^9 \ \text{rad/s} \\
Q_o &= 0.766 \\
\omega_{h1} &= (2\pi) \, 39.7 \times 10^9 \ \text{rad/s} \\
\omega_{h2} &= (2\pi) \, 594.6 \times 10^9 \ \text{rad/s} \\
\omega_{oz} &= (2\pi) \, 21.47 \times 10^9 \ \text{rad/s} \\
Q_{oz} &= 1.274 \\
\omega_{z1} &= (2\pi) \, 25.86 \times 10^9 \ \text{rad/s} \\
\omega_{z2} &= (2\pi) \, 595.1 \times 10^9 \ \text{rad/s}
\end{aligned}
\right\} \tag{6.332a–i}
$$

The magnitude and phase of the voltage gain are shown in Fig. 6.21. The bandwidth is seen to be 4.14 GHz and is dictated by the dominant pole, f_{p1}, as given above in Eq. (6.332a). If the dominant pole is approximated by the first term a_1, then we obtain for the bandwidth:

$$f_{p1} \approx \frac{1}{2\pi a_1} = 3.01\,\text{GHz} \tag{6.333}$$

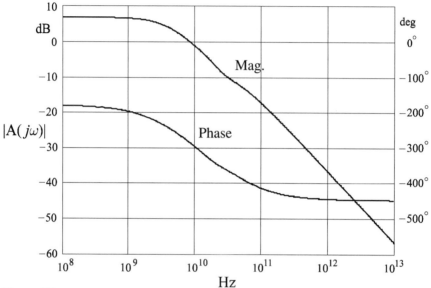

Figure 6.21

This is not a very good approximation because the roots of the denominator are not well separated. A more accurate approximation is given by:

$$f_{p1} \approx \frac{1}{2\pi \left(a_1 - \dfrac{a_2}{a_1} \right)} = 3.89\,\text{GHz} \tag{6.334}$$

If a reasonable analytical expression of the dominant pole is desired, we must use the first approximation in which the approximate expression of a_1 given by Eq. (6.96) can be used. ☐

6.5   Review

The task of analyzing the frequency response of a complicated circuit can be greatly simplified by removing *all* the reactive elements and analyzing a set of purely resistive circuits which are generated by opening and shorting the reactive ports according to a straightforward permutation scheme required by the NEET. If need be, the resistive circuits can be simplified by further application of the NEET. If an indeterminacy arises whenever the product of two port resistances is taken, either the order of the ports can be reversed or a dummy resistor can be introduced and allowed to vanish later after the product is formed. When the product of several port resistances is taken, one can assess rather quickly the order of the ports which results in the simplest expression. Remarkably, we can see that the algebra never gets out of hand since each component of the final answer is determined either by inspection or by a line or two of algebra.

Problems

6.1 Approximate factors of a polynomial with well-separated roots. (*a*) Show that when the roots of a third-order polynomial are well separated and have the same sign, they can be factored as shown in the denominator of Eq. (6.39). (*b*) Generalize the result in part (*a*) to an *n*th-order polynomial. (*c*) Verify the analytical expression of the dominant and second pole in Eqs. (6.42) and (6.43). (*d*) As an exercise, compare the numerical values of the approximate and exact values of the third pole.

6.2 Approximate expression of the dominant pole of a MESFET amplifier. Using the numerical values of the MESFET amplifier in Fig. 6.11, determine the value of A_o in Eq. (6.65) and verify the approximation for a_1, and hence of the dominant pole, or bandwidth, in Eq. (6.96).

6.3 Frequency response of a video amplifier. A two-stage video amplifier[3] with a midband gain of 5 and a bandwidth of about 70 MHz is shown in Fig. 6.22a. The amplifier is capable of driving a load of 75 Ω with 1 V peak-to-peak as shown in Fig. 6.22b and has an input impedance of about 75 Ω. All the simulation results shown are obtained by Or CAD/Pspice.

The complete frequency response is shown in Fig. 6.22c and is given by:

$$\frac{v_o(s)}{v_{in}(s)} = A_m \frac{N(s)}{D_L(s)D_H(s)} \tag{6.335}$$

in which A_m is the midband gain, $D_L(s)$ is the low-frequency behavior and $D_H(s)$ is the high-frequency behavior. The small-signal parameters of the transistors corresponding to the model in Fig. 6.22d are evaluated at the dc operating point and given in the table.

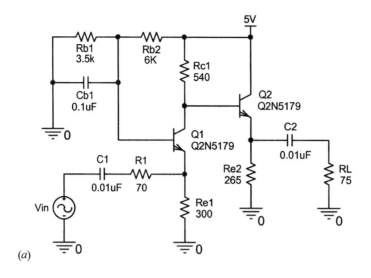

(a)

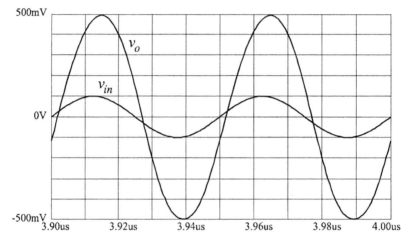

(b)

Figure 6.22

Table

	Q_Q1	Q_Q2
MODEL	Q2N5179	Q2N5179
IB	4.21E − 05	1.33E − 04
IC	3.07E − 03	9.00E − 03
VBE	8.16E − 01	8.50E − 01
VBC	− 1.52E + 00	− 1.73E + 00
VCE	2.34E + 00	2.58E + 00
BETADC	7.28E + 01	6.76E + 01

Table (*cont.*)

	Q_Q1	Q_Q2
GM	1.06E – 01	2.70E – 01
RPI	6.87E + 02	2.16E + 02
RX	1.00E + 01	1.00E + 01
RO	3.31E + 04	1.13E + 04
CBE	1.68E – 11	4.67E – 11
CBC	6.40E – 13	6.25E – 13
BETAAC	7.27E + 01	5.84E + 01
FT/FT2	9.65E + 08	9.10E + 08

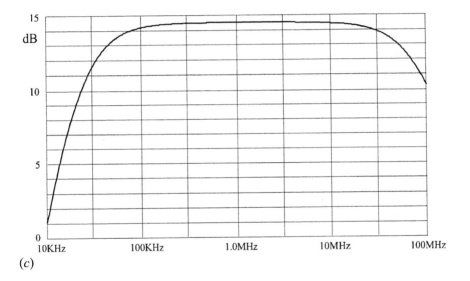

(*c*)

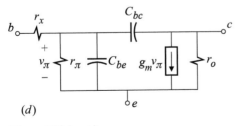

(*d*)

Figure 6.22 (*cont.*)

(*a*) Determine the midband gain using the equivalent circuit model of the amplifier in Fig. 6.22*e* and show that it is given by:

$$A_m = \frac{R_{c1}}{R_1} \frac{1}{1 + \dfrac{R_{c1} + r_{x2} + r_{\pi2}}{(1 + \beta_2)R_L \| r_{o2} \| R_{e2}}}$$

(6.336)

$$\times \frac{1 + \dfrac{r_{\pi1} + r_{x1}}{\beta_1 r_{o1}}}{1 + \dfrac{r_{\pi1} + r_{x1}}{\beta_1 [R_{e1} \| R_1 \| (r_{\pi1} + r_{x1})] \left[1 + \dfrac{R_{c1} \| R_{in2} + R_{e1} \| R_1 \| (r_{x1} + r_{\pi1})}{r_{o1}} \right]}}$$

in which:

$$R_{in2} = r_{x2} + r_{\pi2} + (1 + \beta_2)(R_{e2} \| R_L \| r_{o2})$$

(6.337)

Using numerical values, obtain a few different approximations for A_m with different degrees of accuracies.

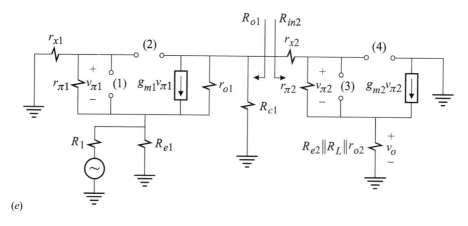

(e)

Figure 6.22 (*cont.*)

(b) Determine the coefficients a_i in the high-frequency response $D_H(s)$ using the 4-EET and the reference circuit in Fig. 6.22e with $v_{in} = 0$:

$$D_H(s) = 1 + \sum_{n=1}^{4} a_n s^n$$

(6.338)

Hint: (i) The coefficient a_1 is given by:

$$a_1 = \sum_{i=1}^{4} C_i R^{(i)}$$

(6.339)

in which $R^{(i)}$ are given by:

$$R^{(1)} = r_{\pi1} \| r_{x1} \frac{1 + \dfrac{R_{e1} \| R_1}{r_{x1} \| (r_{o1} + R_{c1} \| R_{in2})}}{1 + \dfrac{R_{e1} \| R_1}{r_{x1} + r_{\pi1}} \dfrac{1 + \beta_1 + \dfrac{r_{x1} + r_{\pi1} + R_{c1} \| R_{in2}}{r_{o1}}}{1 + \dfrac{R_{c1} \| R_{in2}}{r_{o1}}}} \tag{6.340}$$

$$R^{(2)} = R_{c1} \| R_{in2} \left(1 + \frac{r_{x1}}{R_{c1} \| R_{in2} \| R_{e1} \| R_1} \right)$$
$$\times \frac{1 + \dfrac{1}{g_{m1}} \dfrac{1}{r_{o1} \| r_{\pi1} \| (r_{x1} \| R_{c1} \| R_{in2} + R_{e1} \| R_1)}}{1 + \dfrac{1}{\beta_1} \dfrac{r_{\pi1} + r_{x1}}{r_{o1} \| [R_{c1} \| R_{in2} + R_{e1} \| R_1 \| (r_{\pi1} + r_{x1})]}} \tag{6.341}$$

$$R^{(3)} = r_{\pi2} \| (r_{x2} + R_{o1}) \frac{1 + \dfrac{R_L \| R_{e2} \| r_{o2}}{r_{x2} + R_{o1}}}{1 + \dfrac{R_L \| R_{e2} \| r_{o2}}{r_{x2} + R_{o1} + r_{\pi2}} (1 + \beta_2)} \tag{6.342}$$

$$R^{(4)} = (r_{x2} + R_{o1}) \| r_{\pi2} \frac{1 + \dfrac{(1 + \beta_2)(R_L \| R_{e2} \| r_{o2})}{r_{\pi2}}}{1 + \dfrac{(1 + \beta_2)(R_L \| R_{e2} \| r_{o2})}{r_{\pi2} + r_{x2} + R_{o1}}} \tag{6.343}$$

in which:

$$R_{o1} = R_{c1} \left\| \left(1 + \frac{\beta_1}{1 + \dfrac{r_{x1} + r_{\pi1}}{r_{o1} \| R_1 \| R_{e1}}} \right) [r_{o1} + (r_{x1} + r_{\pi1}) \| R_{e1} \| R_1] \right. \tag{6.344}$$

(ii) When determining product terms in the higher coefficients, try to recognize short cuts like:

$$R^{(n)}_{(1,2)} = R^{(n)} \Big|_{\substack{r_{\pi1} \to 0 \\ g_{m1} \to 0}} \tag{6.345}$$

(c) Show that $N(s)$ in Eq. (6.335) is given by:

$$N(s) = \left(1 + \frac{s}{\omega_{oz} Q_{oz}} + \frac{s^2}{\omega_{oz}^2} \right) \left(1 - s \frac{C_{cb2}}{g_{m2}} \right) \tag{6.346}$$

in which:

$$\omega_{oz} = \cfrac{1}{\sqrt{C_{bc1}C_{be1}r_{o1}r_{\pi 1} \left\| \cfrac{r_{x1}}{1 + g_{m1}r_{o1}} \right.}}$$

$$\approx \cfrac{1}{\sqrt{C_{bc1}C_{be1} \cfrac{r_{x1}}{g_{m1}}}}$$

(6.347)

$$Q_{oz} = \cfrac{\sqrt{C_{bc1}C_{be1} \cfrac{r_{o1}}{r_{\pi 1} \left\| \cfrac{r_{x1}}{1 + g_{m1}r_{o1}} \right.}}}{C_{be1} + C_{cb1}\left[1 + \cfrac{r_{o1}}{r_{\pi 1}}(1 + \beta_1)\right]}$$

$$\approx \cfrac{\sqrt{C_{bc1}C_{be1} \cfrac{g_{m1}}{r_{x1}}}}{C_{be1} + C_{cb1}g_{m1}r_{o1}}$$

(6.348)

(d) Using the numerical values in Fig. 6.22a and in the table, evaluate your analytical results and obtain a good approximation of the dominant pole or the bandwidth using:

$$f_{p1} = \frac{1}{2\pi a_1}$$

(6.349)

REFERENCES

1. P. R. Gray and R. G. Meyer, *Analysis and Design of Analog Integrated Circuits*, Second Edition, John Wiley & Sons, Chapter 12.
2. K. Benboudjema, M. Boukadoun, G. Vasilescu and G. Alquié, "Symbolic Analysis of Linear Microwave Circuits by Extension of the Polynomial Interpolation Method", *IEEE Transaction on Circuits and Systems I: Fundamental Theory and Applications*, Vol. 45, No. 9, Sept. 1998, pp. 936–944.
3. P. Horowitz and W. Hill, *The Art of Electronics*, Second Edition, Cambridge University Press, Chapter 13.

7 Passive filters

Where inductors and transformers still get respect

7.1 Introduction

Before the advent and wide-spread use of operational amplifiers, much of network theory was concerned with the analysis and synthesis of passive filters. Since vacuum tube amplifiers were bulky, required power supplies and consumed power, feedback techniques were not investigated to synthesize filters. Their potential to do so, however, was well known. Since there is a very great number of books on network synthesis, we shall limit our discussion here only to the analysis of a few interesting passive filters mainly to demonstrate the unique techniques of this book. Today, passive filters are mostly used in analog communication circuits and switching power converter circuits. Switching power converter circuits will be discussed in Chapter 8. Chapter 7, however, concludes with a section on special infinite networks in which some thoughts are presented on unifying all three linear elements, R, L and C, into a single element using fractional calculus.

7.2 *RC* filters with gain

The possibility of obtaining any voltage or current gain[1] from a purely RC network may seem counterintuitive at first, simply because such a network can neither have a flat gain larger than unity nor exhibit any resonance. After a little thought, we consider if it is at all possible, starting from a flat gain, A_o, to have a zero, ω_z, before a pole, ω_p, so that $A_o(\omega_p/\omega_z) > 1$. For a ladder network it can be shown that this is not possible (see Problem 7.1) but for other types of networks, such as the one illustrated in Fig. 7.1a, it is indeed possible to obtain a gain larger than unity. We shall analyze this circuit using the 2-EET, whereby the two capacitors are designated as extra elements and removed as shown in Fig. 7.1b, in which we see that the low-frequency gain is unity, i.e. $A_o = 1$.

With the input voltage source shorted we can verify the following driving-point impedances by inspection of Fig. 7.1b:

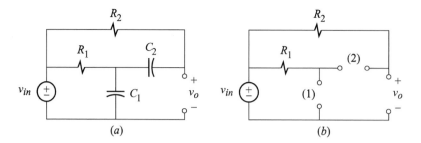

Figure 7.1

$$R^{(1)} = R_1$$

$$R^{(2)} = R_1 + R_2$$

$$R^{(2)}_{(1)} = R_2$$

(7.1a–c)

The denominator is thus given by:

$$D(s) = 1 + s(C_1 R^{(1)} + C_2 R^{(2)}) + s^2 C_1 C_2 R^{(1)} R^{(2)}_{(1)}$$

$$= 1 + s[C_1 R_1 + C_2(R_1 + R_2)] + s^2 C_1 C_2 R_1 R_2$$

(7.2)

The null driving-point impedances are shown in Fig. 7.2b, whence we can verify the following by inspection:

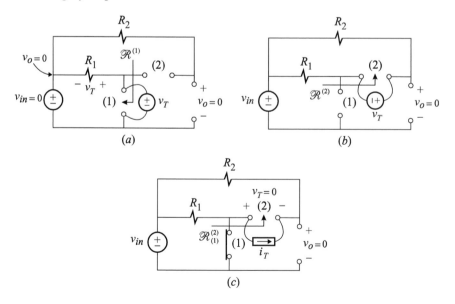

Figure 7.2

$$\left.\begin{array}{l} \mathscr{R}^{(1)} = R_1 \\[6pt] \mathscr{R}^{(2)} = R_1 + R_2 \\[6pt] \mathscr{R}_{(1)}^{(2)} = 0 \end{array}\right\} \tag{7.3a-c}$$

Substituting these in the numerator:

$$N(s) = 1 + s(C_1\mathscr{R}^{(1)} + C_2\mathscr{R}^{(2)}) + s^2 C_1 C_2 \mathscr{R}^{(1)}\mathscr{R}_{(1)}^{(2)}$$

$$= 1 + s[C_1 R_1 + C_2(R_1 + R_2)] \tag{7.4}$$

The transfer function is now given by:

$$\frac{v_o(s)}{v_{in}(s)} = A_o \frac{1 + a_1 s}{1 + a_1 s + a_2 s^2} \tag{7.5}$$

in which:

$$\left.\begin{array}{l} A_o = 1 \\[6pt] a_1 = C_1 R_1 + C_2(R_1 + R_2) \\[6pt] a_2 = C_1 C_2 R_1 R_2 \end{array}\right\} \tag{7.6a-c}$$

If the poles in Eq. (7.5) are well separated, then they can be factored approximately as $(1 + a_1 s)(1 + sa_2/a_1)$. This will cause the zero to cancel with the first pole preventing the magnitude of the transfer function from exceeding unity. If on the other hand the poles coalesce, then the denominator factors as a perfect square and the transfer function can be written as:

$$\left.\begin{array}{l} A(s) = \dfrac{1 + a_1 s}{\left(1 + \dfrac{sa_1}{2}\right)^2} \\[24pt] \qquad = \dfrac{1 + \dfrac{s}{\omega_1}}{\left(1 + \dfrac{s}{2\omega_1}\right)^2} \end{array}\right\} \tag{7.7a, b}$$

It can be seen in Eq. (7.7b) that the zero occurs before the double pole so that the magnitude will exceed unity and peak in the neighborhood of the pole. The peak can be estimated by the magnitude of the transfer function at $2\omega_1$:

$$A_{max} \approx \frac{\sqrt{1 + 2^2}}{1 + 1} \approx 1.12 \tag{7.8}$$

A comparison of the denominators in Eqs. (7.5) and (7.7a) yields:

$$a_1 = 2\sqrt{a_2} \tag{7.9}$$

This is the relationship which the circuit elements must satisfy in order for the denominator to factor as in Eq. (7.7a). Substituting Eqs. (7.6b, c) in (7.9) we obtain:

$$R_1C_1 + R_2C_2 + R_1C_2 = 2\sqrt{R_1C_1R_2C_2} \tag{7.10}$$

which can be written as:

$$\left(\sqrt{R_1C_1} - \sqrt{R_2C_2}\right)^2 + R_1C_2 = 0 \tag{7.11}$$

which clearly *cannot* be satisfied exactly for any choice of elements with positive values. To see how this equation can be satisfied approximately, we factor out $\sqrt{R_1C_1}$ and divide it out to obtain:

$$\left(1 - \sqrt{\frac{R_2C_2}{R_1C_1}}\right)^2 + \frac{C_2}{C_1} = 0 \tag{7.12}$$

which can be satisfied approximately if the elements are chosen as follows:

$$\left.\begin{array}{l} R_2C_2 = R_1C_1 \\[2mm] \dfrac{C_2}{C_1} = \dfrac{R_1}{R_2} \ll 1 \end{array}\right\} \tag{7.13a, b}$$

With this choice of elements, the transfer function can be written as:

$$A(s) = \frac{1 + s/\omega_z}{(1 + s/\omega_p)^2} \tag{7.14}$$

in which:

$$\omega_p \approx 2\omega_z = \frac{1}{R_1C_1} \tag{7.15}$$

For the numerical values $R_1 = 1\,\text{K}$, $R_2 = 100\,\text{K}$, $C_1 = 1\,\text{nF}$ and $C_2 = 10\,\text{pF}$, the magnitude response is shown in Fig. 7.3. The maximum value of the transfer function is $A_{max} = 1.153$ and occurs at $f_{max} \approx 113\,\text{kHz}$.

Using higher-order versions of the circuit in Fig. 7.1 it is possible to obtain higher gains (see Problem 7.2).

The third-order, twin-T, band-pass filter shown in Fig. 7.4a is another *RC* circuit which can produce a gain larger than unity. The three capacitors are designated as the extra elements and the reference circuit is shown in Fig. 7.4b in which the reference gain is seen to be unity, i.e. $A_o = 1$.

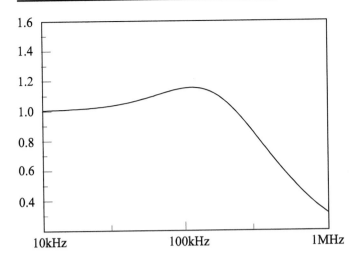

Figure 7.3

With the input voltage source shorted, we can determine the following driving-point impedances by a simple inspection of Fig. 7.4b:

$$R^{(1)} = R_1$$

$$R^{(2)} = R_{(1)}^{(2)} = R_2$$

$$R^{(3)} = R_2 + R_3$$

$$R_{(1)}^{(3)} = R_1 + R_2 + R_3$$

$$R_{(2)}^{(3)} = R_3$$

$$R_{(1,2)}^{(3)} = R_1 + R_3$$

(7.16a–f)

The denominator according to the 3-EET is given by:

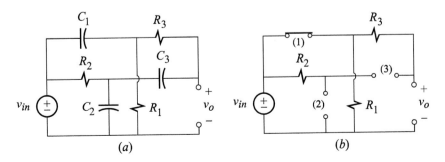

(a)   (b)

Figure 7.4

$$D(s) = 1 + \frac{1}{sC_1R_1} + sC_2R_2 + sC_3(R_2 + R_3)$$

$$+ \frac{1}{sC_1R_1}sC_2R_2 + \frac{1}{sC_1R_1}sC_3(R_1 + R_2 + R_3) \qquad (7.17)$$

$$+ sC_2R_2C_3R_3 + \frac{1}{sC_1R_1}sC_2R_2sC_3(R_1 + R_3)$$

The numerator is given by:

$$N(s) = 1 + \frac{1}{sC_1\mathscr{R}^{(1)}} + sC_2\mathscr{R}^{(2)} + sC_3\mathscr{R}^{(3)}$$

$$+ \frac{1}{sC_1\mathscr{R}^{(1)}}sC_2\mathscr{R}^{(2)}_{(1)} + \frac{1}{sC_1\mathscr{R}^{(1)}}sC_3\mathscr{R}^{(3)}_{(1)} + \frac{1}{sC_2\mathscr{R}^{(2)}}sC_3\mathscr{R}^{(3)}_{(2)} \qquad (7.18)$$

$$+ \frac{1}{sC_1\mathscr{R}^{(1)}}sC_2\mathscr{R}^{(2)}_{(1)}sC_3\mathscr{R}^{(3)}_{(1,2)}$$

The null driving-point impedances are determined next. In Fig. 7.5a, $\mathscr{R}^{(1)}$ is seen to be infinite because when the output voltage is nulled, i.e. $v_o = 0$, the current through R_3 and, hence, the voltage drop across R_1 are both zero so that the total current through R_1 and R_2, which is equal to i_T, is also zero. Hence, we have:

$$\mathscr{R}^{(1)} \to \infty \qquad (7.19)$$

In Fig. 7.5b, a null in the output voltage corresponds to a zero voltage at the input so that the test voltage source appears across R_2 and we have:

$$\mathscr{R}^{(2)} = R_2 \qquad (7.20)$$

In Fig. 7.5c, since the sum of the voltage drops across R_2 and R_3 is equal to v_T while the current through each of them is i_T, we have:

$$\mathscr{R}^{(3)} = R_2 + R_3 \qquad (7.21)$$

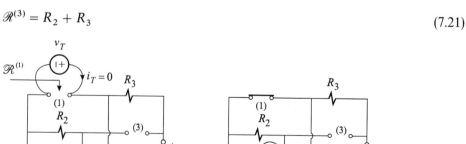

(a)   (b)

Figure 7.5

In Fig. 7.5d, $\mathscr{R}_{(1)}^{(2)}$ may seem confusing because of the presence of v_{in}, but if we connect a resistance across port (1), which eventually can be made infinite, then we can see that a null in the output is caused by $v_{in} = 0$, just as in Fig. 7.5b, so that $\mathscr{R}_{(1)}^{(2)} = \mathscr{R}^{(2)}$. Hence, we have:

$$\mathscr{R}_{(1)}^{(2)} = R_2 \tag{7.22}$$

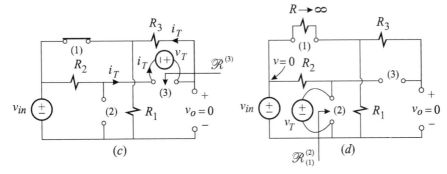

(c)   (d)

Figure 7.5 (*cont.*)

In Fig. 7.5e, a null in the output voltage causes the voltage across $R_1 + R_3$ to be zero so that the current i_T is zero too. It follows that $\mathscr{R}_{(1)}^{(3)} \to \infty$ and the product $(1/sC_1\mathscr{R}^{(1)})sC_3\mathscr{R}_{(1)}^{(3)}$ in Eq. (7.18) is indeterminate. To remove this indeterminacy, we interchange the order of the ports and consider the product $sC\mathscr{R}^{(3)}(1/sC_1\mathscr{R}_{(3)}^{(1)})$ in which $\mathscr{R}_{(3)}^{(1)}$ is shown in Fig. 7.5f. In this figure we have:

$$
\left.
\begin{aligned}
v_T &= i_3(R_3 + R_2) \\
i_T &= i_1 + i_3 \\
i_3 R_3 &= i_1 R_1
\end{aligned}
\right\} \tag{7.23a–c}
$$

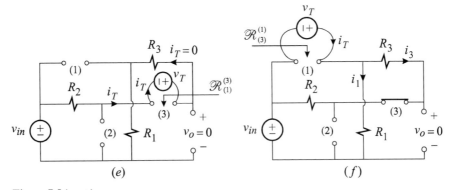

(e)   (f)

Figure 7.5 (*cont.*)

in which the last equation follows from the null in the output voltage. Simultaneous solution of these equations yields:

$$\mathscr{R}^{(1)}_{(3)} = \frac{R_2 + R_3}{1 + R_3/R_1} \tag{7.24}$$

In Fig. 7.5g, we can see that:

$$\mathscr{R}^{(3)}_{(2)} = 0 \tag{7.25a}$$

In the last term of the numerator, we shall change the order of the ports to 1–3–2 in order to avoid an indeterminacy ($\mathscr{R}^{(3)}_{(1,2)}$ is infinite) and determine in Fig. 7.5h:

$$\mathscr{R}^{(2)}_{(1,3)} = 0 \tag{7.25b}$$

The numerator is now given by:

$$N(s) = 1 + s[C_2 R_2 + C_3(R_2 + R_3)] + \frac{C_3}{C_1}\left(1 + \frac{R_3}{R_1}\right) \tag{7.26}$$

The numerator can also be determined using the method discussed in Section 2.4 (see Problem 7.3). The transfer function is now given by:

$$A(s) = \frac{N(s)}{D(s)}$$

$$= A_o \frac{(s/\omega_o)(1 + s/\omega'_o)}{1 + a_1(s/\omega_o) + a_2(s/\omega_o)^2 + (s/\omega_o)^3} \tag{7.27}$$

in which:

$$\left.\begin{aligned}
\omega'_o &= \frac{1 + (1 + R_3/R_1)C_3/C_1}{C_2 R_2 + C_3(R_3 + R_2)} \\[2mm]
\omega_o &= \frac{1}{\sqrt[3]{R_1 C_1 R_2 C_2 R_3 C_3}} \\[2mm]
A_o &= \omega_o[R_1 C_1 + (R_1 + R_3)C_3] \\[2mm]
a_1 &= \omega_o[R_1 C_1 + R_2 C_2 + (R_1 + R_2 + R_3)C_3] \\[2mm]
a_2 &= \omega_o^2[R_1 C_1 R_2 C_2 + R_1 C_1(R_2 + R_3)C_3 + R_2 C_2(R_1 + R_3)C_3]
\end{aligned}\right\} \tag{7.28a–e}$$

In order to have band-pass characteristics, $A(s)$ must be symmetrical with respect to ω_o on the log-frequency axis, which in turn requires that:

$$A(s/\omega_o) = A(\omega_o/s) \tag{7.29}$$

Letting $s/\omega_o \rightarrow \omega_o/s$ in Eq. (7.27) we get:

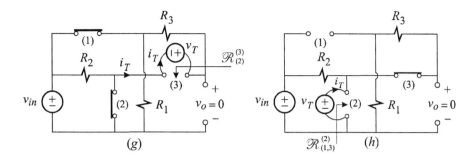

Figure 7.5 (*cont.*)

$$A(\omega_o/s) = A_o \frac{(\omega_o/s)(1 + \omega_o^2/s\omega_o')}{1 + a_1(\omega_o/s) + a_2(\omega_o/s)^2 + (\omega_o/s)^3}$$

$$= A_o \frac{(s/\omega_o)(s/\omega_o + \omega_o/\omega_o')}{(s/\omega_o)^3 + a_1(s/\omega_o)^2 + a_2(s/\omega_o) + 1}$$

(7.30a, b)

Equating Eqs. (7.30b) and (7.27) yields:

$$\left. \begin{aligned} \omega_o' &= \omega_o \\ a_1 &= a_2 \equiv a \end{aligned} \right\}$$

(7.31a, b)

When Eqs. (7.31a, b) are substituted in (7.27), then $s = -\omega_o$ will be a root of the denominator which will cancel with the zero to yield:

$$A(s) = A_o \frac{s/\omega_o}{1 + (a-1)(s/\omega_o) + (s/\omega_o)^2}$$

$$= \frac{A_o}{a-1} \frac{1}{1 + \frac{1}{a-1}\left(\frac{s}{\omega_o} + \frac{\omega_o}{s}\right)}$$

(7.32)

At $s = j\omega_o$, the magnitude peaks at $A_o/(a-1)$ and the phase passes through zero. In what follows, we would like to see how the circuit values are to be chosen so that Eqs. (7.31a, b) are satisfied and $A_o/(a-1) > 1$. In Eqs. (7.28a, b) ω_o' and ω_o can be rewritten as:

$$\omega_o' = \frac{1}{R_2 C_1} \frac{1 + \dfrac{C_1}{C_3} + \dfrac{R_3}{R_1}}{1 + \dfrac{R_3}{R_2} + \dfrac{C_2}{C_3}}$$

$$\omega_o = \frac{1}{(R_2 C_1)^{\frac{1}{3}}(R_1 C_2 R_3 C_3)^{\frac{1}{3}}}$$

(7.33a, b)

So that if $\omega_o = \omega_o'$ for all R_i and C_i, we must have:

$$R_2C_1 = R_1C_2 = R_3C_3 = \frac{1}{\omega_o} \tag{7.34}$$

When Eq. (7.34) is substituted in the expression for a we obtain for the peak value in Eq. (7.32):

$$\frac{A_o}{a-1} = \frac{1 + \dfrac{C_1}{C_2} + \dfrac{C_3}{C_2}}{\dfrac{C_1}{C_2} + \dfrac{C_3}{C_2} + \dfrac{C_2}{C_1} + \dfrac{C_3}{C_1}} \tag{7.35}$$

in which we have two independent ratios:

$$\left.\begin{aligned}\alpha &= \frac{C_1}{C_2} = \frac{R_1}{R_2} \\[2mm] \beta &= \frac{C_3}{C_2} = \frac{R_1}{R_3}\end{aligned}\right\} \tag{7.36a, b}$$

Substituting these in Eq. (7.35) we get:

$$\frac{A_o}{a-1} = \frac{\alpha(1 + \alpha + \beta)}{\alpha(\alpha + \beta) + 1 + \beta} \tag{7.37}$$

The peak is plotted in Fig. 7.6 for positive values of α and β, where it can be seen that the maximum value of the peak occurs for $\beta = 0$. This maximum is larger than unity and can be determined by setting the derivative of Eq. (7.37) with respect to α to zero while $\beta \to 0$. For practical design, β can be made small and does not have to be zero. This yields:

$$\left.\begin{aligned}\left(\frac{A_o}{a-1}\right)_{max} &= \frac{4 + 3\sqrt{2}}{4 + \sqrt{2}} = 1.2 \\[2mm] \alpha_{max} &= 1 + \sqrt{2}\end{aligned}\right\} \tag{7.38a, b}$$

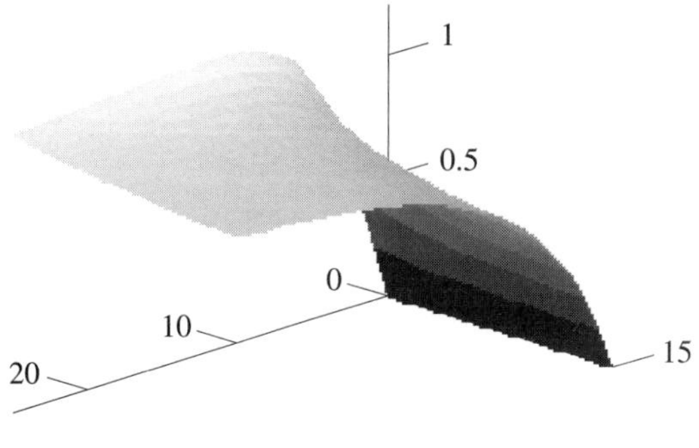

Figure 7.6

7.3　Lattice filters

Lattice filters[2] are bridge circuits with identical opposite branches as shown in Fig. 7.7a. A schematic abbreviation is shown in Fig. 7.7b in which only a pair of opposite branches is shown. These filters are generally difficult to build and tune because they require components with tight tolerances. There are two other disadvantages to a symmetrical bridge circuit. First, the output is not referenced to the same point as the input voltage. Second, the number of reactive elements required is twice the order of the transfer function because the bridge is ideally balanced. These disadvantages can be eliminated by using any of the three circuits in Figs. 7.8b–d, which are entirely equivalent to the symmetrical lattice in Fig. 7.8a. One way to show the equivalence between these networks is to prove that they have the same two-port parameters (see Problem 7.4).

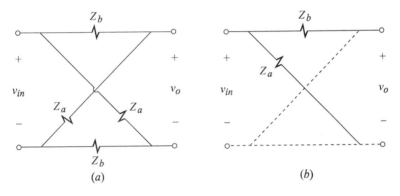

(a)　　　　　　(b)

Figure 7.7

Example 7.1　For a particular choice of element values, the circuit in Fig. 7.9 acts as a second-order delay equalizer. We shall first determine its frequency response and then its time-domain response to a pulse waveform. Note that the 1:1 ideal inverting transformer and the inductor, L_b, are formed by winding a primary and secondary on a single magnetic core. To achieve tight coupling ($k \geq 0.999$), the primary and secondary wires are first twisted together and then wound around the magnetic core.

It is quite easy to see how this circuit can be designed to operate as a delay equalizer with unity gain. At very low frequencies, L_b shorts out the inverting transformer while C_a acts as an open circuit. Hence the voltage gain at low frequencies is simply given by the voltage divider R_s and R_L. If the series and parallel resonant branches are designed to have the same resonant frequency, then

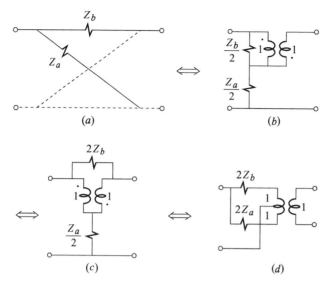

(a)   (b)

(c)   (d)

Figure 7.8

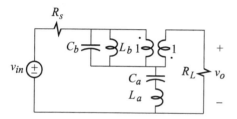

Figure 7.9

at resonance the parallel branch acts as an open circuit while the series branch acts as a short circuit. Therefore, at resonance the magnitude of the output voltage does not change but appears inverted, or shifted 180° in phase, because it coincides with the secondary voltage of the inverting transformer. As the frequency is increased beyond resonance, C_b shorts out the inverting transformer while L_a acts as an open circuit and once again the output becomes in phase with the input. Hence, over the entire frequency range the magnitude of the output voltage remains constant while its phase is shifted by 360°.

Since the circuit is relatively simple, each resonant branch can be treated as a single impedance element and the 2-EET can be used to analyze the circuit. The reference circuit is shown in Fig. 7.10, whence we see that the low-frequency gain is given by:

$$A_o = \frac{R_L}{R_L + R_s}$$

(7.39)

The transfer function is of the form:

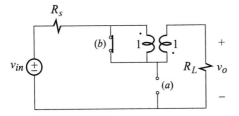

Figure 7.10

$$A(s) = A_o \frac{N(s)}{D(s)} \tag{7.40}$$

in which $N(0) = D(0) = 1$. The numerator can be determined by inspecting the transform network for null conditions in the response as discussed in Chapter 2. According to Fig. 7.11, a null in $v_o(s)$ is accompanied by a null in $i_o(s)$ so that $i_b(s) = i_a(s) = i(s)$ and:

$$-i(s)Z_b(s) + i(s)Z_a(s) = 0 \tag{7.41}$$

It follows that the zeros of $Z_a(s) - Z_b(s)$ are the same as the zeros of $N(s)$. This yields:

$$Z_a(s) - Z_b(s) = sL_a + \frac{1}{sC_a} - \frac{L_b/C_b}{sL_b + 1/sC_b} = 0 \tag{7.42}$$

Multiplying out in Eq. (7.42) and normalizing the leading constant to unity yields:

$$N(s) = 1 + s^2(L_aC_a + L_bC_b - C_aL_b) + s^4L_aC_aL_bC_b \tag{7.43}$$

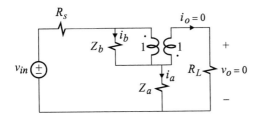

Figure 7.11

As explained earlier, in order for this circuit to behave as a delay equalizer, both branches must have the same resonant frequency so that we have:

$$L_aC_a = L_bC_b = \frac{1}{\omega_o^2} \tag{7.44}$$

With the intention of substituting Eq. (7.44) in (7.43), let us manipulate C_aL_b in (7.43):

$$C_a L_b = C_a L_b \frac{L_a}{L_a} = \frac{C_a L_a}{L_a/L_b} = \frac{1}{(Q\omega_o)^2} \qquad (7.45)$$

in which Q is defined as:

$$Q = \sqrt{\frac{L_a}{L_b}} = \sqrt{\frac{C_b}{C_a}} \qquad (7.46)$$

Substituting Eqs. (7.44) and (7.45) in (7.43) we obtain:

$$N(s) = 1 + \left(\frac{s}{\omega_o}\right)^2 \left(2 - \frac{1}{Q^2}\right) + \left(\frac{s}{\omega_o}\right)^4$$

$$= 1 + 2\left(\frac{s}{\omega_o}\right)^2 + \left(\frac{s}{\omega_o}\right)^4 - \left(\frac{s}{\omega_o}\right)^2 \frac{1}{Q^2}$$

$$= \left[1 + \left(\frac{s}{\omega_o}\right)^2\right]^2 - \left(\frac{s}{\omega_o}\right)^2 \frac{1}{Q^2} \qquad (7.47)$$

$$= \left[1 - \left(\frac{s}{\omega_o}\right)\frac{1}{Q} + \left(\frac{s}{\omega_o}\right)^2\right]\left[1 + \left(\frac{s}{\omega_o}\right)\frac{1}{Q} + \left(\frac{s}{\omega_o}\right)^2\right]$$

The reason for factoring the numerator as shown will become clear as soon as the denominator is determined.

The denominator is determined by applying the 2-EET to the circuit in Fig. 7.12 in which $v_{in} = 0$. According to the 2-EET we have:

$$\Delta(s) = 1 + \frac{R^{(a)}}{Z_a(s)} + \frac{Z_b(s)}{R^{(b)}} + \frac{Z_b(s)}{R^{(b)}} \frac{R_{(b)}^{(a)}}{Z_a(s)} \qquad (7.48)$$

The denominator, $D(s)$, in Eq. (7.40) is given by the *numerator* of $\Delta(s)$ above simply because the denominator of $\Delta(s)$ contributes to $N(s)$, which has already been determined and therefore we need not worry about it.

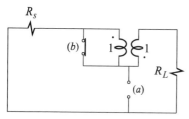

Figure 7.12

The determination of $R^{(a)}$, $R^{(b)}$ and $R_{(b)}^{(a)}$ is shown in Figs. 7.13a–c. In Fig. 7.13a, the voltage which appears across $R_s + R_L$ is $2v_T$ because of the inverting transformer, so that the current i is given by:

$$i = \frac{2v_T}{R_s + R_L}$$
(7.49)

It follows that:

$$R^{(b)} = \frac{v_T}{2i} = \frac{R_s + R_L}{4}$$
(7.50)

In Fig. 7.13c, it can be seen that:

$$R^{(a)} = R_s \| R_L$$
(7.51)

In Fig. 7.13c, the voltages across the primary and secondary of the inverting transformer are related by:

$$-(iR_s - v_T) = iR_L - v_T$$
(7.52)

It follows that:

$$R^{(a)}_{(b)} = \frac{v_T}{2i} = \frac{R_s + R_L}{4}$$
(7.53)

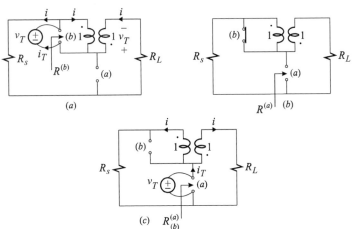

Figure 7.13

Substituting these results in Eq. (7.48) and making use of the definitions of ω_o and Q given in Eqs. (7.44), (7.45) and (7.46) we obtain from the numerator of $A(s)$ the following expression for $D(s)$:

$$D(s) = \left[1 + \left(\frac{s}{\omega_o}\right)^2\right]^2 + \left(\frac{s}{Q\omega_o}\right)^2$$

$$+ \left[1 + \left(\frac{s}{\omega_o}\right)^2\right]\left(\frac{4L_b}{R_s + R_L} + C_a R_s \| R_L\right)s$$
(7.54)

Upon comparing this expression of $D(s)$ to $N(s)$ in Eq. (7.47), we see that if the circuit is designed such that the elements in the last term in Eq. (7.54) are chosen such that:

$$\frac{4L_b}{R_s + R_L} + C_a R_s \parallel R_L = \frac{2}{\omega_o Q} \tag{7.55}$$

then $D(s)$ factors as a perfect square:

$$D(s) = \left[1 + \frac{s}{Q\omega_o} + \left(\frac{s}{\omega_o} \right)^2 \right]^2 \tag{7.56}$$

and the transfer function reduces to the desired form of a second-order delay:

$$A(s) = A_o \frac{1 - \dfrac{s}{Q\omega_o} + \left(\dfrac{s}{\omega_o} \right)^2}{1 + \dfrac{s}{Q\omega_o} + \left(\dfrac{s}{\omega_o} \right)^2} \tag{7.57}$$

The transfer function in Eq. (7.57) has a constant magnitude of A_o and a total phase shift of $-360°$ with a center frequency of ω_o. A numerical example of $A(s)$ will be given shortly.

We continue to study the condition in Eq. (7.55) that the circuit elements must satisfy in order to obtain the second-order delay given by Eq. (7.57). Substituting Eq. (7.45) in (7.55) and dividing out by C_a we obtain:

$$\frac{4}{R_s + R_L} \frac{L_b}{C_a} + R_s \parallel R_L = 2 \sqrt{\frac{L_b}{C_a}} \tag{7.58}$$

Now if we let $x = \sqrt{L_b/C_a}$ in this equation we obtain the following peculiar result:

$$\frac{(x - R_s)(x - R_L)}{R_s + R_L} = 0 \tag{7.59}$$

which states that either R_s must satisfy the following condition:

$$R_s = 2 \sqrt{\frac{L_b}{C_a}} = \frac{2}{Q} \sqrt{\frac{L_a}{C_a}} \tag{7.60}$$

and R_L can be any value, or R_L must satisfy the same condition:

$$R_L = 2 \sqrt{\frac{L_b}{C_a}} = \frac{2}{Q} \sqrt{\frac{L_a}{C_a}} \tag{7.61}$$

and R_s can be any value. Now we turn to Eq. (7.57) to show how Q and ω_o must be chosen for a desired delay, τ_d. If the bandwidth of the signal is bounded from above by ω_H, then ω_o must be chosen larger than ω_H:

$$\omega_o \geq \omega_H \tag{7.62}$$

Since the phase response of an ideal delay circuit is a linear function of frequency, $-\omega\tau_d$, the optimum choice of Q in Eq. (7.57) is 0.707. This yields a phase response which is nearly a linear function of frequency for frequencies below ω_o:

$$\phi(\omega) \approx -\omega\left(\frac{\pi}{\omega_o}\right); \; \omega \leq \omega_o \tag{7.63}$$

so that the delay is given by:

$$\tau_d = \frac{\pi}{\omega_o} = \frac{1}{2f_o} \tag{7.64}$$

The maximum possible delay is therefore given by:

$$\tau_{d_{max}} = \frac{1}{2f_H} \tag{7.65}$$

To obtain the design equations of the circuit elements for a desired delay and a given source resistance, R_s, we substitute Eq. (7.64) in (7.60) and get:

$$\left.\begin{aligned} C_a &= \frac{\tau_d}{R_s}\frac{2}{Q\pi} \\[2mm] L_a &= \tau_d R_s \frac{Q}{2\pi} \end{aligned}\right\} \tag{7.66a, b}$$

The values of L_b and C_b follow from Eq. (7.47):

$$\left.\begin{aligned} L_b &= \frac{L_a}{Q^2} \\[2mm] C_b &= Q^2 C_a \end{aligned}\right\} \tag{7.67a, b}$$

The same design equations apply, with R_s replaced by R_L, when designing for a given load R_L.

For the following numerical values, the magnitude and phase response are shown in Fig. 7.14:

$$L_a = 0.177\,\mu\text{H} \qquad C_a = 56.6\,\text{nF} \qquad R_s = 5\,\Omega$$

$$L_b = 0.354\,\mu\text{H} \qquad C_b = 28.3\,\text{nF} \qquad R_L = 8\,\Omega$$

The following values are computed:

$$f_o = \frac{1}{2\pi\sqrt{L_aC_a}} = \frac{1}{2\pi\sqrt{L_bC_b}} = 1.59\,\text{MHz}$$

$$Q = \sqrt{\frac{L_a}{L_b}} = \sqrt{\frac{C_b}{C_a}} = 0.707$$

$$A_o = \frac{R_L}{R_L + R_s} = \frac{8}{15}$$

$$\tau_d = \frac{1}{2f_o} = 300\,\text{ns}$$

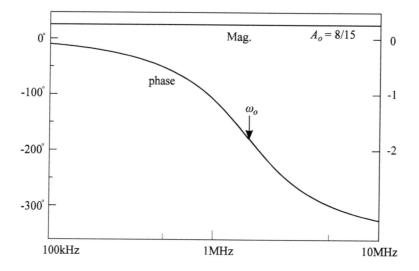

Figure 7.14

The phase response is plotted on a linear frequency axis in Fig. 7.15 to show that it is nearly a linear function of the frequency in the range $f \le f_o$ with a slope equal to the delay of the network.

The time-domain response to a 1-volt pulse with 5-µs rise-and-fall times is shown in Fig. 7.16a. A significant portion of the frequency content of this pulse is well within 1.6 MHz. In this figure the applied pulse is shown scaled with the dc gain of network A_o in order to see the delay clearly. A magnified version of this figure during the initial rise time is shown in Fig. 7.16b.

Delay networks have a wide variety of applications in pulse and microwave networks. □

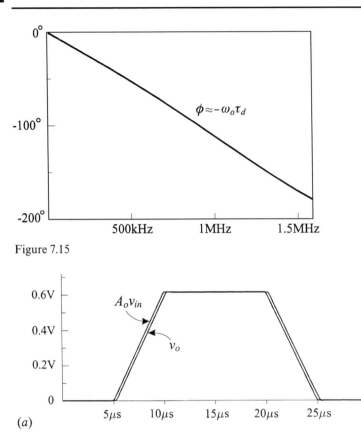

Figure 7.15

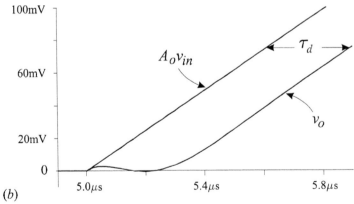

(a)

(b)

Figure 7.16

7.4   Resonant filters

Resonant filters are mainly used as narrowband tuning circuits in radio communication. They are also used as narrowband impedance transformation circuits with properties similar to those of a transformer. In high-frequency switching power converters, resonant circuits can be used to reduce voltage or current stresses on the switching devices by providing zero voltage or zero current transition at the switching instants.

7.4.1   Parallel resonant filters

We shall begin by analyzing the parallel resonant circuit with triple damping shown in Fig. 7.17a and derive approximate results for high-Q or narrowband operation.

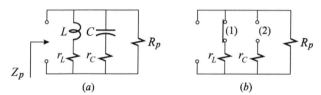

Figure 7.17

Upon inspection of the circuit in Fig. 7.17a we obtain:

$$Z_p(s) = R_p \| r_L \frac{(1 + sL/r_L)(1 + sr_C C)}{1 + a_1 s + a_2 s^2} \tag{7.68}$$

in which the denominator is determined by applying the 2-EET to the reference circuit in Fig. 7.17b whence we have by inspection:

$$\left.\begin{aligned} R^{(1)} &= r_L + R_p \\ R^{(2)} &= r_C + r_L \| R_p \\ R^{(2)}_{(1)} &= r_C + R_p \end{aligned}\right\} \tag{7.69a–c}$$

The coefficients a_1 and a_2 in the denominator are now given by:

$$\left.\begin{aligned} a_1 &= \frac{L}{R^{(1)}} + CR^{(2)} = \frac{L}{r_L + R_p} + C(r_C + r_L \| R_p) \\ a_2 &= \frac{L}{R^{(1)}} CR^{(2)}_{(1)} = LC\frac{r_C + R_p}{r_L + R_p} \end{aligned}\right\} \tag{7.70a, b}$$

The denominator in Eq. (7.68) is obtained by substituting for a_1 and a_2:

$$Z_p(s) = R_p \| r_L \frac{(1 + s/\omega_1)(1 + s/\omega_2)}{1 + s/Q\omega_o + (s/\omega_o)^2} \tag{7.71}$$

in which:

$$\omega_o = \sqrt{\frac{1}{LC}} \sqrt{\frac{1 + r_L/R_p}{1 + r_C/R_p}} \tag{7.72}$$

$$\left. \begin{aligned} \omega_1 &= \frac{r_L}{L} \\[2ex] \omega_2 &= \frac{1}{r_C C} \end{aligned} \right\} \tag{7.73a, b}$$

$$\frac{1}{\omega_o Q} = \frac{L}{r_L + R_p} + C(r_C + r_L \| R_p) \tag{7.74}$$

Since we are interested in narrowband operation, we must determine the conditions that the damping elements should satisfy to obtain high-Q. According to Eq. (7.74) the condition for high-Q is given by:

$$Q = \frac{1}{\dfrac{\omega_o L}{r_L + R_p} + \omega_o C(r_C + r_L \| R_p)} \gg 1 \tag{7.75}$$

Since both factors in the denominator are positive, each factor must be much smaller than unity:

$$\left. \begin{aligned} \frac{\omega_o L}{r_L + R_p} &\ll 1 \\[2ex] \omega_o C(r_C + r_L \| R_p) &\ll 1 \end{aligned} \right\} \tag{7.76a, b}$$

Taking the product of these factors and making use of the expression of the resonant frequency in Eq. (7.72) we obtain:

$$\frac{r_C + r_L \| R_p}{r_C + R_p} \ll 1 \tag{7.77}$$

It can be seen that the condition in Eq. (7.77) can be satisfied if r_C, r_L and R_p satisfy:

$$r_C, r_L \ll R_p \tag{7.78}$$

in light of which the condition in Eq. (7.76a) becomes:

$$q_p = \frac{R_p}{\omega_o L} \gg 1 \tag{7.79}$$

Hence, the damping elements must satisfy the conditions in Eqs. (7.78) and (7.79) for high-Q operation. Next we define the following q-factors for the inductor and the capacitor both of which, according to Eqs. (7.78) and (7.79), can be seen to be much larger than unity:

$$q_L = \frac{\omega_o L}{r_L} = \frac{Z_o}{r_L} \gg 1$$

$$q_C = \frac{1}{\omega_o C r_C} = \frac{Z_o}{r_C} \gg 1$$

$$(7.80a, b)$$

in which Z_o is the characteristic impedance given by:

$$Z_o = \omega_o L = \frac{1}{\omega_o C} = \sqrt{\frac{L}{C}} \qquad (7.81)$$

Next, it will be shown that a high-Q, triple-damped parallel resonant circuit behaves as a simple parallel LCR circuit over a relatively wide range of frequencies. According to Eqs. (7.73a, b) and (7.80):

$$\frac{\omega_1}{\omega_o} = \frac{r_L}{\omega_o L} = \frac{1}{q_L} \ll 1$$

$$\frac{\omega_2}{\omega_o} = \frac{1}{\omega_o r_C C} = q_C \gg 1$$

$$(7.82a, b)$$

It follows that the zeros of the impedance function in Eq. (7.71) are far away from the resonant frequency, as shown in Fig. 7.18, so that it can now be approximated in the frequency range $\omega_1 < \omega < \omega_2$ as:

$$Z_p(s) \approx r_L \frac{s/\omega_1}{1 + s/Q\omega_o + (s/\omega_o)^2}$$

$$\approx \frac{sL}{1 + s/Q\omega_o + (s/\omega_o)^2}$$

$$(7.83)$$

in which:

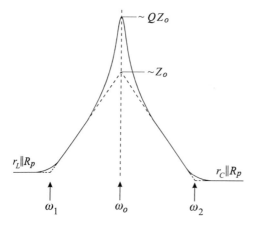

Figure 7.18

$$\omega_o \approx \frac{1}{\sqrt{LC}} \tag{7.84}$$

Equation (7.83) is recognized to be the impedance of a simple parallel LCR circuit in which R is an effective parallel resistance, which is determined by approximating Q in Eq. (7.74) using the conditions in Eqs. (7.78) and (7.79):

$$\frac{1}{Q} \approx \omega_o L \left[\frac{1}{R_p} + \frac{1}{Z_o^2}(r_C + r_L) \right] = \frac{\omega_o L}{R_e} \tag{7.85}$$

in which R_e is the effective parallel resistance given by:

$$R_e = R_p \left\| \frac{Z_o^2}{r_C} \right\| \frac{Z_o^2}{r_L} \tag{7.86}$$

The equivalent simple parallel resonant circuit and its impedance are shown in Figs. 7.19a and b. The Q in Eq. (7.85) can be expressed in terms of the individual q-factors defined in Eqs. (7.79) and (7.80):

$$Q = q_p \| q_L \| q_C \tag{7.87}$$

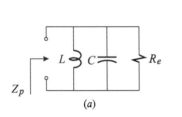
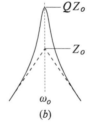

(a)   (b)

Figure 7.19

7.4.2   Tapped parallel resonant filter

The circuit to be considered next is the capacitively tapped parallel resonant circuit,[3] shown in Fig. 7.20. First, we shall derive an exact expression of the impedance, $Z_p(s)$, and then obtain an approximation for it for high-Q.

The low-frequency asymptote of $Z_p(s)$ is given by r_L, while its zeros are given by

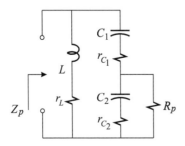

Figure 7.20

the zeros of the impedances of the two parallel branches. The zero of the impedance of the first branch, $r_L + sL$, is the same zero as before and is given by ω_1 in Eq.
(7.73a). The zero of the impedance of the second branch is obtained by applying
the 2-EET to the reference circuit in Fig. 7.21 in which the input port is shorted.
We have by inspection of Fig. 7.21:

$$R^{(1)} = r_{C_1} + R_p$$

$$R^{(2)} = r_{C_2} + R_p \qquad\qquad (7.88a\text{--}c)$$

$$R^{(2)}_{(1)} = r_{C_2} + r_{C_1} \| R_p$$

The second factor in the numerator of $Z_p(s)$ is now given by:

$$1 + s(C_1 R^{(1)} + C_2 R^{(2)}) + s^2 C_1 C_2 R^{(1)} R^{(2)}_{(1)} \qquad (7.89)$$

Substituting Eqs. (7.88) in (7.89) yields:

$$1 + s[C_1(r_{C_1} + R_p) + C_2(r_{C_2} + R_p)] +$$

$$s^2 C_1 C_2 (r_{C_1} + R_p)(r_{C_2} + r_{C_1} \| R_p) \qquad (7.90)$$

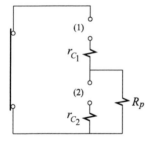

Figure 7.21

Since, for high-Q we should have $r_{C_1}, r_{C_2} \ll R_p$, Eq. (7.90) simplifies to:

$$1 + s(C_1 + C_2)R_p + s^2 C_1 C_2 R_p (r_{C_2} + r_{C_1}) \qquad (7.91)$$

This is a quadratic of the form $1 + a_1 s + a_2 s^2$ in which clearly $a_1 \gg a_2/a_1$ so that it
factors to an excellent approximation as:

$$(1 + s/\omega_2)(1 + s/\omega_3) \qquad (7.92)$$

where:

$$\omega_2 = \frac{1}{(C_1 + C_2)R_p}$$

$$\omega_3 = \frac{1}{C_1 \| C_2(r_{C_2} + r_{C_1})}$$

(7.93a, b)

The results in Eq. (7.93) are not surprising. In fact they can be arrived at by a simple inspection of the impedance of the second branch, whose zeros are nothing more than the poles of its admittance, which are given by the network obtained upon shorting its terminals. With the terminals shorted, we can easily see that at low frequencies the capacitive reactances are high so that r_{C_1} and r_{C_2} can be neglected and C_1 and C_2 appear in parallel together with R_p. It follows that ω_2 is given by Eq. (7.93a). At very high frequencies, as the reactances become very small and comparable to r_{C_1} and r_{C_2}, R_p drops out of the picture and C_1, C_2, r_{C_1} and r_{C_2} appear in series and yield the high-frequency corner at ω_3 in Eq. (7.93b).

The impedance $Z_p(s)$ can now be written as:

$$Z_p(s) = r_L \frac{(1 + s/\omega_1)(1 + s/\omega_2)(1 + s/\omega_3)}{1 + a_1 s + a_2 s^2 + a_3 s^3}$$

(7.94)

in which the coefficients a_i are determined by applying the 3-EET to the reference circuit in Fig. 7.22 according to which we have by inspection:

$$R^{(1)} = r_{C_1} + r_L + R_p \approx R_p$$

$$R^{(2)} = r_{C_2} + R_p \approx R_p$$

$$R^{(3)} \to \infty$$

$$R^{(2)}_{(1)} = r_{C_2} + (r_L + r_{C_1}) \| R_p \approx r_{C_2} + r_L + r_{C_1}$$

$$R^{(3)}_{(1)} = R^{(1)}$$

$$R^{(3)}_{(2)} \to \infty$$

$$R^{(2)}_{(3,1)} = r_{C_2} + R_p \approx R_p$$

(7.95a–g)

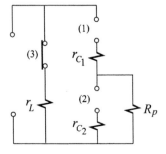

Figure 7.22

The coefficients a_i are given by:

$$
\left.\begin{aligned}
a_1 &= C_1 R^{(1)} + C_2 R^{(2)} + \frac{L}{R^{(3)}} \approx (C_1 + C_2) R_p \\[2mm]
a_2 &= C_1 R^{(1)} C_2 R^{(2)}_{(1)} + C_1 R^{(1)} \frac{L}{R^{(3)}_{(1)}} + C_2 R^{(2)} \frac{L}{R^{(3)}_{(2)}} \approx LC_1 \\[2mm]
a_3 &= C_1 R^{(1)} \frac{L}{R^{(3)}_{(1)}} C_2 R^{(2)}_{(3,1)} \approx LC_1 C_2 R_p
\end{aligned}\right\}
\qquad (7.96a\text{--}c)
$$

Since this is a resonant circuit, the denominator must factor into a real pole and a quadratic with a high-Q. In order to determine whether the real pole occurs below or above the resonant frequency, we examine the circuit in Fig. 7.20 with the reactive elements in their places. At very low frequencies we see that the inductor drops out of the picture while C_1 and C_2 appear in parallel with R_p giving rise to a dominant pole at:

$$
\omega_p = \frac{1}{(C_1 + C_2) R_p}
\qquad (7.97)
$$

which is consistent with the dominant behavior of the denominator given by $1 + a_1 s$. As the frequency is increased, the inductive reactance increases and eventually resonates with the series combination of C_1 and C_2 at:

$$
\omega_o = \frac{1}{\sqrt{LC_1 \parallel C_2}}
\qquad (7.98)
$$

giving rise to the high-frequency quadratic whose Q-factor is all we need to determine. This can be done analytically as follows. The denominator in factored form can be written as:

$$
(1 + s/\omega_p)[1 + s/\omega_o Q + (s/\omega_o)^2]
\qquad (7.99)
$$

which upon expansion yields:

$$
1 + s\left(\frac{1}{\omega_p} + \frac{1}{\omega_o Q}\right) + s^2\left(\frac{1}{\omega_o^2} + \frac{1}{Q\omega_o \omega_p}\right) + \frac{s^3}{\omega_o^2 \omega_p}
\qquad (7.100)
$$

A comparison of the coefficients of s in Eq. (7.100) with the coefficients a_i yields:

$$a_1 = \frac{1}{\omega_p} + \frac{1}{\omega_o Q} \approx \frac{1}{\omega_p}$$

$$a_2 = \frac{1}{\omega_o^2} + \frac{1}{Q\omega_o\omega_p}$$

$$a_3 = \frac{1}{\omega_o^2\omega_p}$$

$$(7.101a\text{--}c)$$

Substituting for a_1 and a_2 from Eqs. (7.96a, c) in Eqs. (7.101a, c) we obtain the dominant pole ω_p and the resonant frequency ω_o given in Eqs. (7.97) and (7.98), respectively. When a_2 in Eq. (7.96b) is substituted in Eq. (7.101b) along with the expressions for ω_o and ω_p, we get:

$$\frac{1}{Q} = \omega_o\omega_p\left(a_2 - \frac{1}{\omega_o^2}\right) = \frac{r_{C_1} + r_{C_2} + r_L}{\omega_o L} + \frac{\omega_o L}{n^2 R_p} \tag{7.102}$$

in which:

$$n = \frac{C_1 + C_2}{C_1} \tag{7.103}$$

One way of rewriting Eq. (7.102) is:

$$\frac{1}{Q} = \omega_o L\left(\frac{1}{Z_o^2/r_{C_1}} + \frac{1}{Z_o^2/r_{C_2}} + \frac{1}{Z_o^2/r_L} + \frac{1}{n^2 R_p}\right) = \frac{\omega_o L}{R_e} \tag{7.104}$$

which suggests that the resonance is damped by an equivalent resistance R_e given by:

$$R_e = (n^2 R_p)\left\|\frac{Z_o^2}{r_{C_1}}\right\|\left\|\frac{Z_o^2}{r_{C_2}}\right\|\left\|\frac{Z_o^2}{r_L}\right\| \tag{7.105}$$

Another way of rewriting Eq. (7.102) is by making use of parallel notation for q-factors:

$$Q = \frac{r_{C_1}}{\omega_o L} + \frac{r_{C_2}}{\omega_o L} + \frac{r_L}{\omega_o L} + \frac{\omega_o L}{n^2 R_p} = q_{C_1} \| q_{C_2} \| q_L \| q_p \tag{7.106}$$

When the factored denominator is substituted in Eq. (7.94), the dominant pole will cancel with the zero at ω_2 to yield:

$$Z_p(s) = r_L \frac{(1 + s/\omega_1)(1 + s/\omega_3)}{1 + s/\omega_o Q + (s/\omega_o)^2} \tag{7.107}$$

Since $\omega_1 \ll \omega_o \ll \omega_3$, the impedance around resonance can be approximated as the impedance of a simple parallel resonant circuit damped by R_e:

$$Z_p(s) \approx \frac{sL}{1 + s/\omega_o Q + (s/\omega_o)^2} \tag{7.108}$$

The significant result obtained here is that the capacitively tapped parallel resonant circuit in Fig. 7.20 acts exactly like an ideal transformer in the resonant region whose effective turns ratio is given by n in Eq. (7.103). This is shown in Fig. 7.23.

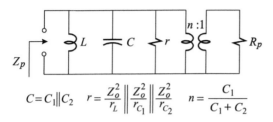

$$C = C_1 \| C_2 \quad r = \frac{Z_o^2}{r_L} \left\| \frac{Z_o^2}{r_{C_1}} \right\| \frac{Z_o^2}{r_{C_2}} \quad n = \frac{C_1}{C_1 + C_2}$$

Figure 7.23

In the preceding analysis, we saw that, except for Q, we were able to determine the denominator completely by inspecting the circuit at low and high frequencies and determining the dominant pole and the resonant frequency in Eqs. (7.97) and (7.98), respectively. A simple way to determine Q is to recognize that it is given by the ratio of the energy stored in the resonant circuit to the energy dissipated per cycle at resonance multiplied by 2π. For simplicity, if the parasitic elements are ignored, the definition of Q yields:

$$Q = \frac{\frac{1}{2} C_1 \| C_2 V^2}{\left(\frac{C_1}{C_1 + C_2} \frac{V}{\sqrt{2}} \right)^2 \frac{1}{R_p}} \omega_o$$

$$= \frac{n^2 R_p}{\omega_o L} \tag{7.109}$$

which agrees with the result derived earlier.

7.4.3   The three-winding transformer

Finally, we consider the three-winding parallel resonant transformer[3-5] shown in Fig. 7.24, which is the basic building block in the antenna, oscillator and IF amplifier stage of a superheterodyne receiver.

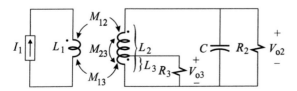

Figure 7.24

When the three-winding transformer is replaced with its equivalent circuit model (see Problem 7.6), we obtain the circuit in Fig. 7.25a in which r_{L_i} are the winding resistances; $L_{\sigma i}$ are the leakage inductances; and n_2 and n_3 are the effective turns ratio between each secondary and the primary. The leakage inductances are given by:

$$
\left.
\begin{aligned}
L_{\sigma 1} &= L_1 \left(1 - \frac{k_{13} k_{12}}{k_{23}} \right) \\[2mm]
L_{\sigma 2} &= L_2 \left(1 - \frac{k_{23} k_{12}}{k_{13}} \right) \\[2mm]
L_{\sigma 3} &= L_3 \left(1 - \frac{k_{23} k_{13}}{k_{13}} \right)
\end{aligned}
\right\}
\qquad (7.110a\text{–}c)
$$

in which L_i is the self-inductance of each winding and k_{ij} is the coupling coefficient between the ith and jth winding. The coupling coefficients and the mutual inductances, M_{ij}, are related by:

$$
k_{ij} = \frac{M_{ij}}{\sqrt{L_i L_j}}
\qquad (7.111)
$$

The effective turns ratio between each secondary and the primary is given by:

$$
\left.
\begin{aligned}
n_2 &= \frac{k_{23}}{k_{13}} \sqrt{\frac{L_2}{L_1}} \\[2mm]
n_3 &= \frac{k_{23}}{k_{12}} \sqrt{\frac{L_3}{L_1}}
\end{aligned}
\right\}
\qquad (7.112a, b)
$$

Finally, the magnetizing inductance referred to the primary side is given by:

$$
M_1 = L_1 \frac{k_{13} k_{12}}{k_{23}}
\qquad (7.113)
$$

The two transfer functions that we shall determine are:

$$
\left.
\begin{aligned}
Z_{12}(s) &= \frac{V_{o2}(s)}{I_1(s)} \\[2mm]
Z_{13}(s) &= \frac{V_{o3}(s)}{I_1(s)}
\end{aligned}
\right\}
\qquad (7.114a, b)
$$

We can work with a simpler circuit if we reflect $L_{\sigma 3} + R_3$ into the primary side as shown in Fig. 7.25b. We shall use the knowledge and analytical experience that we have gained from the past two examples to determine these two transfer functions by inspection.

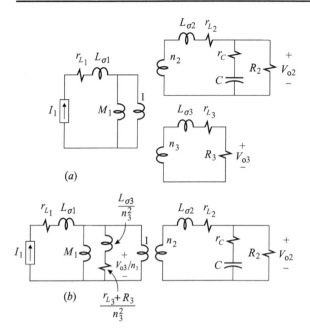

Figure 7.25

Since $r_{L_1} + L_{\sigma 1}$ is in series with the current source, it does not contribute to either transfer function of interest. Also, the three remaining inductors M_1, $L_{\sigma 2}$ and $L_{\sigma 3}/n_3^2$ are not linearly independent so that the denominator is of third-order rather than fourth. Inspection of Fig. 7.25b reveals that $Z_{12}(s)$ has three zeros and is given by:

$$Z_{12}(s) = K\frac{s[1 + sL_{\sigma 3}/(r_{L_3} + R_3)](1 + sr_cC)}{D(s)} \qquad (7.115)$$

in which the zero at the origin will later be combined with the high-Q quadratic factor in the denominator, which is determined by examining the resonance of the circuit in Fig. 7.25b with the current source open. Here, we see that C resonates with an effective inductance $M_1 n_2^2 + L_{\sigma 2} = L_2$, which is nothing more than the self-inductance of the secondary winding, n_2. Hence the resonant frequency is given by:

$$\omega_o = \frac{1}{\sqrt{L_2 C}} \qquad (7.116)$$

This resonance is damped by the four resistances in the circuit. The overall Q-factor is given by the parallel combination of the individual q-factors defined for the resistors:

$$q_3 = \frac{R_3 + r_{L_3}}{Z_o} \frac{n_2^2}{n_3^2}$$

$$q_2 = \frac{Z_o}{r_{L_2}}$$

$$q_C = \frac{Z_o}{r_C} \qquad\qquad\qquad (7.117a\text{–}d)$$

$$q_{R_2} = \frac{R_2}{Z_o}$$

The overall Q-factor is now given by:

$$Q = q_3 \parallel q_2 \parallel q_C \parallel q_{R_2} \qquad\qquad (7.118)$$

To determine whether the real pole of $D(s)$ occurs before or after resonance we examine the circuit below and above resonance. Below resonance the dominating elements are M_1 and R_2, as is expected of the parallel resonance and given by the low-frequency behavior of the quadratic. At frequencies well above resonance, the dominating elements are $L_{\sigma 3}/n_3^2$ and $L_{\sigma 2}$, which appear effectively in series with each other, along with damping resistances, through the $1:n_2$ transformer. When these elements are reflected to the n_2-side of the transformer, the reciprocal of the time constant of the resulting circuit, shown in Fig. 7.26, yields the high-frequency pole:

$$\omega_p = \frac{\dfrac{n_2^2}{n_3^2}(R_3 + r_{L_3}) + r_{L_2} + r_C \parallel R_2}{\dfrac{n_2^2}{n_3^2} L_{\sigma 3} + L_{\sigma 2}} \qquad\qquad (7.119)$$

The denominator in Eq. (7.115) can now be factored:

$$D(s) = \left(1 + \frac{s}{\omega_o Q} + \frac{s^2}{\omega_o^2}\right)\left(1 + \frac{s}{\omega_p}\right) \qquad\qquad (7.120)$$

The zero at the origin in Eq. (7.115) is now combined with the quadratic factor so that the transimpedance function can be written as:

$$Z_{12}(s) = R_{12} \frac{\left(1 + \dfrac{s}{\omega_1}\right)\left(1 + \dfrac{s}{\omega_2}\right)}{\left[1 + Q\left(\dfrac{s}{\omega_o} + \dfrac{\omega_o}{s}\right)\right]\left(1 + \dfrac{s}{\omega_p}\right)} \qquad\qquad (7.121)$$

in which, ω_1 and ω_2 are given by:

$$\omega_1 = \frac{r_{L_3} + R_3}{L_{\sigma 3}} \Bigg\}$$

$$\omega_2 = \frac{1}{r_C C} \Bigg\}$$ (7.122a, b)

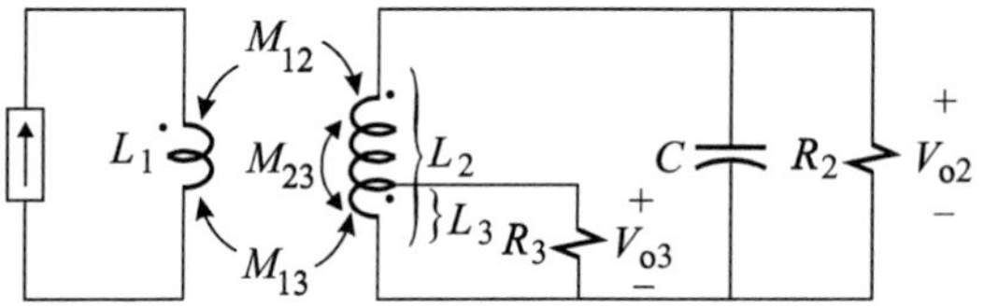

Figure 7.26

In the same equation, R_{12} is the transresistance at resonance, which is simply given by:

$$R_{12} = \frac{R_e}{n_2}$$ (7.123a)

where R_e is the effective parallel resistance given by:

$$R_e = R_2 \parallel R_3 \left(\frac{n_2}{n_3}\right)^2 \left\| \frac{Z_o^2}{r_{L_2}} \right\| \frac{Z_o^2}{r_C}$$ (7.123b)

The numerator of the second transimpedance function, $Z_{13}(s)$, is given by the zeros of the impedance of the branch connected across the secondary of the transformer in Fig. 7.25b. Hence we have:

$$r_{L_2} + s L_{\sigma 2} + R_2 \frac{1 + s C r_C}{1 + s C (r_C + R_2)} = 0$$ (7.124)

Multiplying out by $1 + s C (r_C + R_2)$ and collecting terms, we obtain the numerator of Z_{13}:

$$N(s) = 1 + s \left[\frac{L_{\sigma 2}}{r_{L_2} + R_2} + (r_C + r_{L_2} \parallel R_2) C\right]$$

$$+ s^2 \frac{r_C + R_2}{r_{L_2} + R_2} C L_{\sigma 2}$$ (7.125)

Since the parasitic elements are very small, $N(s)$ in Eq. (7.125) is a sharp resonance which contributes a notch to $Z_{13}(s)$. Hence, we define the frequency of the notch and its Q-factor:

$$\left.\begin{array}{l} \omega_{oz} = \dfrac{1}{\sqrt{CL_{\sigma 2}}} \sqrt{\dfrac{r_{L_2} + R_2}{r_C + R_2}} \approx \dfrac{1}{\sqrt{CL_{\sigma 2}}} \\[4mm] Q_{oz} = \dfrac{r_{L_2} + R_2}{\omega_{oz} L_{\sigma 2}} \left\| \dfrac{1}{\omega_{oz}(r_C + r_{L_2} \| R_2)C} \approx \dfrac{R_2}{\omega_{oz} L_{\sigma 2}} \right. \end{array}\right\}$$

(7.126a, b)

Using these definitions:

$$Z_{13}(s) = R_{13} \dfrac{1 + \dfrac{s}{\omega_{oz} Q_{oz}} + \dfrac{s^2}{\omega_{oz}^2}}{\left[1 + Q\left(\dfrac{s}{\omega_o} + \dfrac{\omega_o}{s}\right)\right]\left(1 + \dfrac{s}{\omega_p}\right)}$$

(7.127)

in which R_{13} is simply given by:

$$R_{13} = R_{12} \dfrac{n_3}{n_2}$$

(7.128)

The approximate transimpedances derived above compare very well with simulation results (see Problem 7.7).

7.5   Infinite scaling networks

Infinite networks are theoretical entities which are primarily useful in obtaining approximate solutions to certain types of large networks. They are also useful in generating theoretical discussion, which can easily degenerate into meaningless discourse, or for serving as a halfway house for, what some would consider, otherwise wayward mathematics. It should be made clear that infinite networks, just like finite networks, are point-like entities and do not have any spatial extent so that one is neither concerned with the time it takes for a signal to traverse the entire network nor with any kind of potential field over it.

7.5.1   Infinite grid

As an example of a useful application of infinite networks, consider determining the resistance between points A and B of the large square grid of identical resistors shown in Fig. 7.27. If we assume the network is infinite, then we can make use of its symmetry and propose to solve for the resistance by arguing that a test current source injected between points A and B will split equally among the four resistors at each node so that the voltage drop across the current source is $2(i_T/4)R = v_T$ and:

$$R_{AB} = \frac{R}{2} \qquad (7.129)$$

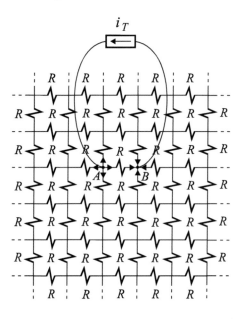

Figure 7.27

This answer is correct, as one can verify by simulating a large grid. In fact it can be shown that, as long as the network is infinite, the resistance between any two points inside a finite neighborhood is $R/2$ (see Problem 7.8).

As an example of a meaningless theoretical argument, consider the infinite network in Fig. 7.28 in which it can be argued that the current i_T can be *any* value without seemingly violating either Ohm's law or Kirchhoff's voltage and current laws. In this figure, according to KVL and Ohm's law, the current in the first parallel branch is given by:

$$i_1 = \frac{v_T - i_T 2R}{R} = \frac{v_T}{R} - 2i_T \qquad (7.130)$$

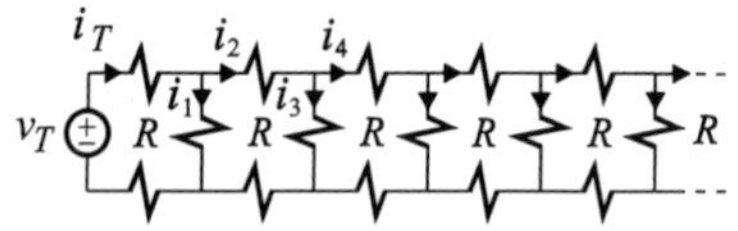

Figure 7.28

According to KVL, i_2 is given by:

$$i_2 = i_T - \left(\frac{v_T}{R} - 2i_T\right) = 3i_T - \frac{v_T}{R} \tag{7.131}$$

Applying KVL and Ohm's law to the second loop, the current i_3 in the second parallel branch is given by:

$$i_3 = \frac{i_1 R - i_2 2R}{R} = \frac{3v_T}{R} - 8i_T \tag{7.132}$$

According to KCL, i_4 is given by:

$$i_4 = i_2 - i_3 = 11i_T - \frac{4v_T}{R} \tag{7.133}$$

We can continue and write a general equation for the current in the nth branch in terms of i_T and v_T and conclude that for a given v_T, we can assign the value of i_T arbitrarily and determine the current and voltage in any branch without violating any circuit law. The reason we are able to get away with this nonsense is that we are postponing the error indefinitely to infinity by showing that, in any finite neighborhood, the voltage and current of any branch are consistent with the voltage and current of any other branch. The easiest way to show that this method is wrong is to show that the power dissipated in the network tends to infinity while the power supplied by the source is finite. Another way of showing that the method is incorrect is to show that it simply does not work whenever the infinite network is truncated. The correct solution to i_T can be obtained in a straight-forward manner by determining the input resistance seen by v_T (see Problem 7.9).

7.5.2   Infinite scaling networks

Certain types of infinite networks with reactive elements exhibit very interesting frequency response characteristics which can model certain physical phenomena. For example, it has been observed that the phase of the impedance of a metal–electrolyte interface tends to a constant value of $<90°$ at low frequencies instead of $90°$ as should be expected of a capacitor. This behavior, known as constant phase angle behavior, is attributed to the roughness of the metal surface which can be modeled[6] by an infinite branching RC network in which the resistors and capacitors in successive branches are related by a scaling factor (see Problem 7.10). Infinite networks of this kind are called scaling networks or fractal networks because their frequency response, under certain assumptions, obeys a scaling law.

    Anomalous eddy current losses in ferromagnetic materials is another physical process whose cumulative effect on the terminal characteristics of an inductor can be easily modeled by a fractal network.[7] These losses are due to fairly complicated processes within the ferromagnetic material whose dependence on the excitation frequency and peak flux density is given by a power law:

$$P_e = P_{eo} \left(\frac{B_m}{B_o}\right)^{D_B} \left(\frac{f}{f_o}\right)^{D_f}$$

(7.134)

in which $D_B, D_f < 2$. A classical analysis yields a power law in which $D_B = D_f = 2$, which is almost never observed experimentally except at frequencies well above the normal rating of the material. The case in which $D_B = D_f = 2$ can be modeled by a parallel LR circuit as shown in Fig. 7.29 in which the resistor accounts for the so-called *classical* eddy current losses. To show that these losses are proportional to the square of the applied B field and frequency of excitation, we express the applied sinusoidal voltage in terms of B and f using Faraday's law:

$$v(t) = \frac{d\lambda}{dt} = NA_c\frac{dB}{dt} = (NA_c\omega B_m)\sin(\omega t) = V_m\sin(2\pi f t)$$

(7.135)

in which:

$\lambda \equiv$ flux linkage

$N \equiv$ number of turns

$A_c \equiv$ area of cross-section   (7.136a–e)

$B \equiv$ flux density

$V_m \equiv NA_c2\pi f B_m$

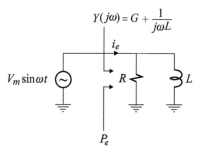

Figure 7.29

According to Eq. (7.136e) the rms value of the applied voltage, $V = V_m/\sqrt{2}$, is proportional to the product of the peak flux density, B_m, and the excitation frequency, f, so that the losses in the resistor are proportional to the square of the frequency and the peak flux density:

$$V^2/R \propto B_m^2 f^2$$

(7.137)

Since the model in Fig. 7.29 is not an adequate representation of the observed losses, we investigate extending the simple parallel LR network into an infinite branching network as shown in Fig. 7.30. Our main objective is to show that the real part of the input admittance of this network has a noninteger power dependence on the frequency of excitation from which the power loss formula in Eq. (7.134) can be deduced. The input admittance in Fig. 7.30 can be expressed as a continued fraction:

$$Y(s) = G + \cfrac{1}{sL + \cfrac{1}{G\dfrac{a}{k} + \cfrac{1}{sL + \cfrac{1}{G\left(\dfrac{a}{k}\right)^2 + \cfrac{1}{sk^2 L + \ldots}}}}} \tag{7.138}$$

in which $G = 1/R$. This equation can be written as:

$$Y(s) = G + \cfrac{1}{sL + \dfrac{k}{a}\cfrac{1}{G + \cfrac{1}{asL + \dfrac{k}{a}\cfrac{1}{G + \cfrac{1}{sa^2 L + \ldots}}}}} \tag{7.139}$$

which can be seen to satisfy the following *functional* equation:

$$\left.\begin{aligned} Y(s) &= G + \cfrac{1}{sL + \dfrac{k}{a}\dfrac{1}{Y(as)}} \\[2em] &= G + \cfrac{Y(as)}{sLY(as) + \dfrac{k}{a}} \end{aligned}\right\} \tag{7.140a, b}$$

This equation can be solved in the limit of low frequencies where $Y(s)$ becomes very large so that G can be ignored and the following assumption, which we shall justify later, is satisfied:

$$\lim_{s \to 0} s\, Y(s) \to 0 \tag{7.141}$$

Hence at low frequencies, Eq. (7.140b) reduces to:

$$Y(s) = \frac{a}{k} Y(as) \tag{7.142}$$

This is a scaling equation which is common to all simple geometric fractals (Koch curve, Sierpinski's triangle, Fournier universe) modeled after the Cantor set in which $a > k > 1$. The solution of Eq. (7.142) is given by:

$$Y(s) = G\left(\frac{\omega_o}{s}\right)^{\eta_f}$$
(7.143)

in which ω_o is a normalizing frequency and η_f is the celebrated fractal dimension of the Cantor set and is given by:

$$\eta_f = \frac{\ln(a/k)}{\ln a} < 1$$
(7.144)

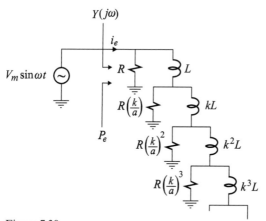

Figure 7.30

Since $\eta_f < 1$, the solution in Eq. (7.143) satisfies the assumption in Eq. (7.141). We shall refer to η_f as the frequency dimension, rather than a fractal dimension, in order to distinguish it from the magnetic dimension, η_B, to be introduced later.

The magnitude and phase of $Y(s)$ are obtained by letting $s = j\omega$ in Eq. (7.143):

$$Y(j\omega) = G\left(\frac{\omega_o}{\omega}\right)^{\eta_f} e - j\frac{\pi}{2}\eta_f$$
(7.145)

The magnitude in decibels is given by:

$$|Y(j\omega)| = -20\eta_f \log\left(\frac{\omega}{\omega_o}\right)$$
(7.146)

which is a line with a slope of $-20\eta_f$ dB/dec. The phase is a constant angle greater than $-90°$ given by:

$$\angle Y(j\omega) = -\frac{\pi}{2}\eta_f$$
(7.147)

Both, the magnitude and phase plots are shown in Fig. 7.31.

Whenever a power law, or a scaling law, like the one in Eq. (7.143) is used to describe a physical process, or quantity, the independent variable must always be normalized in order to avoid *meaningless* units. For example, in Eq. (7.143), ω_o is the normalizing variable without which it takes the following form:

$$Y(s) = Ks^{-n_f} \tag{7.148}$$

Figure 7.31

In this equation K has the "units" of $\Omega^{-1}s^{n_f}$, which has no physical significance because seconds raised to a fractional power cannot be related to a physical measure or scale. This is in contrast to units of seconds raised to an integer power such as "s" or "s²" because these can be easily related to physical quantities such as velocity or acceleration, respectively. Therefore, even though Eq. (7.148) is perfectly acceptable mathematically, it must be normalized properly as in Eq. (7.143) in order to avoid constants with awkward physical units. The physical significance of the normalizing frequency ω_o in Eq. (7.143) is that it is the frequency at which $|Y(j\omega)|$ equals G and beyond which Eq. (7.143) is no longer valid. We can see this from the circuit in Fig. 7.30 in which, as the frequency of the source is increased, $Y(j\omega) \rightarrow G$. Since the inductor in the first branch, L, is the smallest of all the inductors (the others being $k^n L$ with $k > 1$), ω_o is the frequency beyond which the reactance of L is much larger than R. It follows that ω_o is the corner frequency at which $\omega_o L$ and R are equal so that:

$$\omega_o = \frac{R}{L} \tag{7.149}$$

We can now write the complete solution of the infinite network in Fig. 7.30 by combining the solutions for $\omega > \omega_o$ and $\omega < \omega_o$ simply by adding the two together:

$$Y(s) = G\left[1 + \left(\frac{\omega_o}{s}\right)^{n_f}\right] \tag{7.150}$$

This is an approximate solution to the scaling equation in Eq. (7.140). An asymptotic plot of the magnitude in Eq. (7.150) is shown in Fig. 7.32.

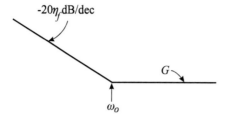

Figure 7.32

Next we determine the power loss in this infinite network according to:

$$P_e = \frac{V_m^2}{2} \mathcal{R}e[Y(j\omega)] \tag{7.151}$$

Substituting Eqs. (7.136e) and (7.145) in Eq. (7.150) we obtain:

$$P_e \propto B_m^2 f^2 \left(\frac{f}{f_o}\right)^{-\eta_f} \tag{7.152}$$

Because of nonlinear effects, the inductance and hence f_o depend on the applied field. This dependence is assumed also to obey a power law:

$$f_o = f_0 \left(\frac{B_m}{B_0}\right)^{\eta_B} \tag{7.153}$$

in which B_0 is a normalizing constant. Substitution of Eq. (7.153) in (7.152) yields the observed power losses in ferromagnetic materials:

$$P_e \propto \left(\frac{B_m}{B_0}\right)^{2-\eta_f\eta_B} \left(\frac{f}{f_0}\right)^{2-\eta_f} \tag{7.154}$$

7.5.3   A generalized linear element and a unified R, L and C model

Finally, it is interesting to note that the expression of the admittance derived in Eq. (7.143) generalizes the relationship between the voltage and current of a two-terminal linear element of which a resistor, inductor and a capacitor are special cases.[7] This is shown in Fig. 7.33 in which:

$$Y(s) = \frac{i(s)}{v(s)} = Y_o \left(\frac{\omega_o}{s}\right)^{-\eta_f} \tag{7.155}$$

where:

$$Y_o = |Y(j\omega_o)| \tag{7.156}$$

If $\eta_f = 0$, then Eq. (7.155) reduces to the conductance of a resistor:

$$Y(s) = Y_o = \frac{1}{R} \tag{7.157}$$

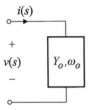

Figure 7.33

If $\eta_f = 1$, then Eq. (7.155) reduces to the admittance of an inductor:

$$Y(s) = Y_o \frac{\omega_o}{s} = \frac{1}{sL} \tag{7.158}$$

in which $L = 1/Y_o\omega_o$. Finally, if $\eta_f = -1$, we get from Eq. (7.155) the admittance of a capacitor:

$$Y(s) = Y_o \frac{s}{\omega_o} = sC \tag{7.159}$$

in which $C = Y_o/\omega_o$. Note that, in each case, $1/Y_o$ is simply a resistance or the magnitude of the inductive or capacitive reactance at ω_o.

The spectral domain operation in Eq. (7.155) corresponds to a time-domain operation which is given by the convolution of $v(t)$ with the inverse Laplace transform of $s^{-\eta_f}$:

$$
\begin{aligned}
i(t) &= G\omega_o^{\eta_f} \mathcal{L}^{-1}\{s^{-\eta_f}\} * v(t) \\
&= G\omega_o^{\eta_f} \int_0^t \frac{(t-\tau)^{\eta_f - 1}}{\Gamma(\eta_f)} v(\tau) d\tau
\end{aligned} \tag{7.160}
$$

in which $\Gamma(\eta_f)$ is the gamma function. The integral in Eq. (7.160) is known as the Riemann–Liouville fractional integral, which reduces to the familiar time-domain relationship between the voltage and current of a resistor, inductor and a capacitor. Using fractional derivatives, we can invert Eq. (7.160) and express $v(\tau)$ in terms of $i(\tau)$:

$$v(\tau) = \frac{1}{G\omega_o^{\eta_f}} \frac{d^{\eta_f}}{dt^{\eta_f}} i(t) \tag{7.161}$$

7.6   Review

A few interesting and representative passive networks are discussed in this chapter. Several RC filters with voltage gain larger than unity are discussed. Although their maximum theoretical gain is not known, a practical design of an RC filter with a gain of about 1.5 is given. Lattice filters are symmetrical bridge circuits which are mostly used to synthesize complex zeros. To reduce the number of reactive elements, lattice filters can be reduced to half lattices using an ideal inverting transformer. A practical example of a delay equalizer using a half-lattice filter is presented. Parallel resonant circuits are used in tuning circuits for oscillators and IF amplifier stages. Two important variations of the parallel resonant circuit are the tapped parallel resonant circuit and the three-winding transformer resonant circuit. Among the infinite networks discussed are the infinite grid, infinite ladder and fractal networks. One of the interesting features of fractal networks is that their phase response above or below a certain frequency can asymptote to a fraction of $90°$ while their magnitude response asymptotes to a slope which is a fraction of $20\,dB/dec$. Another interesting result that can be derived from the calculus of fractal networks is that the voltage and current relationships of an inductor, resistor and a capacitor can be combined into a single equation.

Problems

7.1 RC and RL ladder networks. Consider the ladder network shown in Fig. 7.34 in which *all* the impedance elements, Z_n, are either composed of capacitors and resistors or inductors and resistors. The transfer function of such a ladder network cannot have a magnitude larger than unity. To show this, recall that the phase angle of a passive RC impedance branch, no matter how complex, is always in the range $(0, -90°)$ so that:

$$Z(j\omega) = X_r(\omega) + jX_i(\omega); \ X_r(\omega) > 0 \text{ and } X_i(\omega) < 0 \tag{7.162}$$

(For a passive RL circuit, the range of the phase angle is $(0, 90°)$ so that $X_r(\omega) > 0$ and $X_i(\omega) > 0$). In Fig. 7.34, the voltage at node 1 is related to the input by voltage division:

$$v_1 = v_{in}\frac{1}{1 + \dfrac{Z_1}{Z_{N1}}} \tag{7.163}$$

in which Z_{N1} is the impedance connected from node 1 to return. Use the result in

Eq. (7.162) to show that the magnitude of v_1 is always less than the magnitude of v_{in}. Proceeding in the same manner all the way to the output node, we prove the desired result.

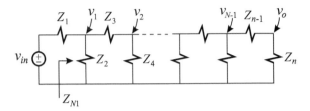

Figure 7.34

7.2 Fourth-order passive *RC* filter with gain. The circuit in Fig. 7.35 produces more gain than the second-order circuit in Fig. 7.1*a*. Show that the transfer function of this circuit is given by:

$$A(s) = \frac{v_o(s)}{v_{in}(s)} = \frac{1 + a_1 s + a_2 s^2 + a_3 s^3}{1 + a_1 s + a_2 s^2 + a_3 s^3 + a_4 s^4} \tag{7.164}$$

in which:

$$
\begin{aligned}
a_1 &= R_1 C_1 + (R_1 + R_2)C_2 + (R_2 + R_3)C_3 + (R_3 + R_4)C_4 \\
a_2 &= R_1 C_1 R_2 C_2 + R_1 C_1 (R_2 + R_3)C_3 + R_1 C_1 (R_3 + R_4)C_4 \\
&\quad + (R_1 + R_2)C_2 (R_3 + R_2 \| R_1)C_3 + (R_1 + R_2)C_2 (R_3 + R_4)C_4 \\
&\quad + (R_2 + R_3)C_3 (R_4 + R_3 \| R_2)C_4 \\
a_3 &= R_1 C_1 R_2 C_2 R_3 C_3 + R_1 C_1 R_2 C_2 (R_3 + R_4)C_4 \\
&\quad + R_1 C_1 (R_2 + R_3)C_3 (R_4 + R_3 \| R_2)C_4 \\
&\quad + (R_1 + R_2)C_2 (R_3 + R_2 \| R_1)C_3 (R_4 + R_3 \| R_2 \| R_1)C_4 \\
a_4 &= R_1 C_1 R_2 C_2 R_3 C_3 R_4 C_4
\end{aligned}
\tag{7.165a–d}
$$

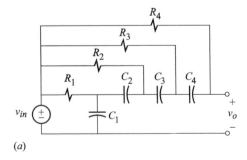

(*a*)

Figure 7.35

Let the circuit elements satisfy the following conditions:

$$R_1C_1 = R_2C_2 = R_3C_3 = R_4C_4 \left.\begin{matrix} \\ \\ \end{matrix}\right\}$$
$$R_4 > R_3 > R_2 > R_1$$

(7.166a, b)

With these conditions satisfied, show that the transfer function reduces to:

$$A(s) = \frac{1 + 4s\tau + 6s^2\tau^2 + 4s^3\tau^3}{(1 + s\tau)^4}$$

(7.167)

The magnitude response of this transfer function is shown in Fig. 7.35b for $n = 4$ and $\tau = 1\,\mu s$. for which the maximum gain is seen to be $A_{max} = 1.38$ at $f_{max} \approx 244\,\text{kHz}$. Also, shown in Fig. 7.35b is the response for a sixth-order version of the network for which the maximum gain is seen to be $A_{max} = 1.51$ at $f_{max} \approx 363\,\text{kHz}$.

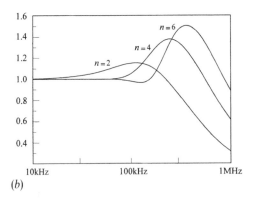

(b)

Figure 7.35 (cont.)

It can be shown that for low values of n, say $2 < n \leq 8$, the maximum and the frequency at which it occurs are approximately given by:

$$A_{max} \approx \frac{(n-1)^{\frac{n-1}{2}}}{n^{\frac{n}{2}-1}} \left.\begin{matrix} \\ \\ \\ \\ \end{matrix}\right\}$$
$$f_{max} \approx \frac{1}{2\pi\tau}\sqrt{n-1}$$

(7.168a, b)

To show these results, recognize that the nth-order transfer function is given by:

$$A(s) = \frac{1 + ns\tau + \binom{n}{2}s^2\tau^2 + \cdots + ns^{n-1}\tau^{n-1}}{(1 + s\tau)^n}$$

(7.169)

For small values of n, the transfer function can be approximated at frequencies in the vicinity of the maximum and beyond by:

$$A(s) \approx \frac{ns^{n-1}\tau^{n-1}}{(1+s\tau)^n} \tag{7.170}$$

Note that it is impractical to build very high-order filters because of the wide spread of the resistance and capacitance values as required by Eq. (7.166).

7.3 Null response analysis. Another way of determining the numerator in Eqs. (7.26) and (7.27) is to look for conditions in the transform network which result in a null in the response as explained in Chapter 2. This is shown in Fig. 7.36 whence you can determine:

$$i(s)N(s) = 0 \tag{7.171}$$

where $N(s)$ is given by Eq. (7.26). Recognize that Eq. (7.171) is satisfied either when $i(s) = 0$ or $N(s) = 0$. Hence, we examine Fig. 7.36 to see if there is yet another condition which makes $i(s) = 0$. In fact we can easily see that for $s = 0$, C_3 acts as an open circuit and prevents the flow of $i(s)$. Hence, the zeros at the origin and at ω'_o in Eq. (7.28a) can be determined using this alternative technique.

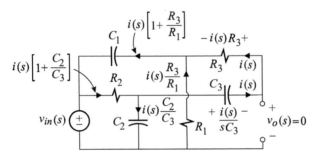

Figure 7.36

7.4 Lattice-equivalent filters. Use any two-port parameter representation to show that the circuits in Figs. 7.8a and b are equivalent.

7.5 Identical time constants. Show that the impedance of n parallel RC branches, shown in Fig. 7.37, with identical time constants, τ, is the same as that of a single RC branch in which:

$$\left.\begin{array}{l} R_e = R_1 \parallel R_2 \parallel \cdots \parallel R_n \\ C_e = \dfrac{\tau}{R_e} \end{array}\right\} \tag{7.172a, b}$$

Hint: Show that the above is true for two branches and deduce the result for n branches by successive paralleling one at a time.

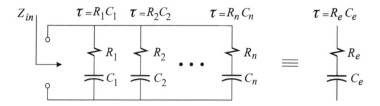

Figure 7.37

7.6 Equivalent circuit model of a three-winding transformer. To show the equivalence between the two models of the three-winding transformer shown in Figs. 7.24 and 7.25a, write the equations of the terminal voltages and currents for each model. For the model in Fig. 7.24, we have:

$$
\left.\begin{aligned}
v_1 &= L_1 \frac{di_1}{dt} + M_{12} \frac{di_2}{dt} + M_{13} \frac{di_3}{dt} \\[2mm]
v_2 &= M_{21} \frac{di_1}{dt} + L_2 \frac{di_2}{dt} + M_{23} \frac{di_3}{dt} \\[2mm]
v_3 &= M_{31} \frac{di_1}{dt} + M_{32} \frac{di_2}{dt} + L_3 \frac{di_3}{dt}
\end{aligned}\right\}
\qquad (7.173a\text{–}c)
$$

in which $M_{ij} = M_{ji}$. Next, show that the equations above are consistent with the equations of the terminal voltages and currents of the model in Fig. 7.25a.

7.7 Comparison of exact and approximate transfer functions. The equivalent circuit model of the three-winding transformer is simulated using OrCAD as shown in Fig. 7.38a in which TX2 is an ideal transformer whose primary inductance and turns ratio are equal to M_1 and n_2, respectively. The element values used in reference to the model, given in Fig. 7.24, are:

$L_1 = 2\,\mu\text{H}, L_2 = 100\,\mu\text{H}, L_3 = 0.01\,\mu\text{H}$

$k_{12} = 0.75, k_{13} = 0.65, k_{23} = 0.5$

$R_2 = 400\,\text{k}\Omega, R_3 = 10\,\Omega$

The frequency response of the transimpedance functions Z_{12} and Z_{13} discussed in the text are shown in Fig. 7.38b.

Using the results derived in the text, verify the frequency response for Z_{12} and Z_{13} shown in Fig. 7.38b.

7.8 Resistance inside infinite grid. Show that the resistance between any two nodes within a finite neighborhood inside the infinite grid in Fig. 7.27 is $R/2$.

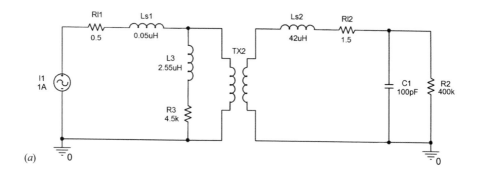

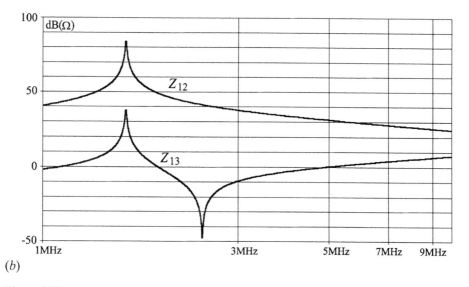

(b)

Figure 7.38

Hint: Follow the same argument for the resistance between two adjacent nodes.

7.9 Resistance of infinite ladder. Since the network is infinite, the input resistance seen by the source is the same as the impedance looking into the network immediately past the first stage. Hence, show that:

$$R_{in} = R(1 + \sqrt{3}) \tag{7.174}$$

7.10 Capacitive fractal network. The fractal model[6] in Fig. 7.39a, in which $a > 2$, is proposed to model the constant phase angle behavior observed in the reactance of a rough metal–electrolyte interface at very low frequencies. Show that the model in Fig. 7.39a can be reduced to the one in Fig. 7.39b and that the impedance at low frequencies is given by:

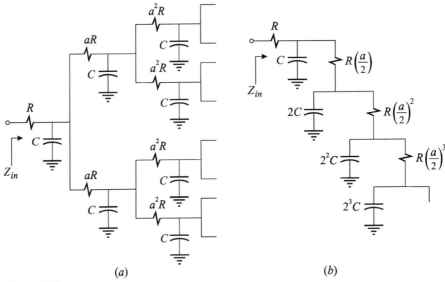

Figure 7.39

$$Z_{in}(j\omega) = R\left(\frac{\omega_o}{j\omega}\right)^{\eta_f} \tag{7.175}$$

in which:

$$\eta_f = 1 - \frac{\ln 2}{\ln a} \tag{7.176}$$

Note that the phase angle of $Z_{in}(j\omega)$ is $-90\eta_f$.

REFERENCES

1. J. Holbrook, *Laplace Transform for Electronic Engineers*, Pergamon Press, Oxford, 1966.
2. G. C. Temes and S. K. Mitra, *Modern Filter Theory and Design*, John Wiley, New York, 1973.
3. K. K. Clark and D. T. Hess, *Communication Circuits: Analysis and Design*, Don Mills, Reading, 1971. Reprinted by Addison-Wesley, 1978.
4. A. G. Ganz, "A Simple, Exact Equivalent Circuit for the Three-Winding Transformer", *IRE Transactions on Component Parts*, CP-9, No. 4, Dec. 1962, pp. 212–213.
5. Shi-Ping Hsu, "Problems in Analysis and Design of Switching Regulators", PhD Dissertation, California Institute of Technology, September 1979.
6. S. H. Liu, "Fractal Model for the ac Response of a Rough Interface", *Physical Review Letters*, 1985, No. 55, pp. 529–532.
7. V. Vorpérian, "A Fractal Model of Anomalous Losses in Ferromagnetic Materials", *Proceedings of the 1992 Power Electronics Specialists Conference*, PESC 92, June 1992, pp. 1277–1283.

8 PWM switching dc-to-dc converters
Introducing the PWM switch

8.1 Introduction

An electronic circuit which transforms the dc level of a voltage or current source in a controllable manner without dissipating power is called an *ideal* dc-to-dc converter or simply a converter. This is shown in Fig. 8.1 in which α is a control parameter on which the output voltage or current depends. When feedback is used as shown in Fig. 8.2, either the output voltage or output current can be regulated against variations in the input voltage or load current. A vast majority of applications, such as logic and sensitive instrumentation circuits, require a converter whose output voltage, rather than the output current, is tightly regulated. A converter may also have more than one control parameter so that α can be taken as a control vector.

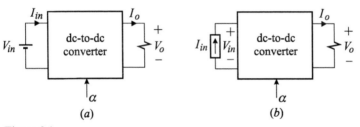

Figure 8.1

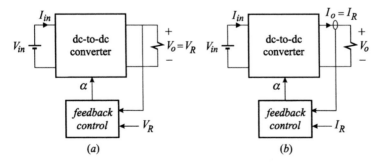

Figure 8.2

In the following sections we shall determine the basic characteristics of dc-to-dc converters, determine their equivalent circuits using a model of the PWM switch, and analyze their dynamics using the techniques developed in this book.

8.2   Basic characteristics of dc-to-dc converters

The voltage conversion ratio of a converter is defined as the ratio of the dc output voltage to the dc input voltage:

$$M_V = \frac{V_o}{V_{in}} \tag{8.1}$$

In a similar manner, the current conversion ratio is defined as the ratio of the dc output current to the dc input current:

$$M_I = \frac{I_o}{I_{in}} \tag{8.2}$$

Since in an ideal converter there can be no power dissipation, the input and output powers must be equal:

$$P_{in} = P_{out} \Rightarrow V_{in}I_{in} = V_oI_o \Rightarrow M_VM_I = 1 \tag{8.3}$$

No practical converter can satisfy the condition in Eq. (8.3) because of losses associated with nonideal components. In designing a converter, every effort is made to keep these losses to a minimum in order to maximize the efficiency, which is defined as the ratio of the output power to the input power:

$$\eta = \frac{P_o}{P_{in}} = M_VM_I \tag{8.4}$$

One may wonder what kind of ideal components go into an ideal converter. Resistors, which may be arranged as a voltage divider to achieve down conversion, are precluded because they dissipate power and lack controllability. A bipolar transistor connected between a voltage source and a load can achieve controllability by adjusting V_{CE}, but such a circuit, known as a series regulator (see Problem 8.1), dissipates power by an amount given by:

$$P_d = V_{CE}I_o = (V_{in} - V_o)I_o \tag{8.5}$$

Hence, the series regulator and its dual, the shunt regulator (see Problem 8.2), are not classified as converters. In fact, these circuits are nothing more than linear dc amplifiers.

The two mechanisms required for nondissipative power conversion are *switching* and *filtering*. Switching is achieved by a *minimum* of two switches while filtering is achieved by inductors and capacitors arranged in an effective low-pass filter configuration. The purpose of the filters is to attenuate the pulsating currents, generated by the switches, down to a specified level at the input and output ports of the converter. Since ideal switches, inductors and capacitors do not dissipate any power, the ideal conversion efficiency is 100%. In reality, all practical inductors and capacitors dissipate a very small percentage of the peak energy stored in them owing to dielectric, conductive and magnetic losses. Also, all practical semiconductor switches have finite conductive and switching losses. The switching frequency is one of the most critical parameters in the design of a switching converter. Increasing the switching frequency results in smaller inductors and capacitors but higher switching losses and lower efficiency. The tradeoff between size, cost, weight and efficiency has never been an exact science and is usually driven by market requirements. For example, a commercially available 48 V to 5 V, 100-W converter may have an efficiency of 80% at full load with linear dimensions $3.0 \times 1.5 \times 0.38''$, whereas a similar custom-designed converter may in comparison have an efficiency of 93% and twice the linear dimensions.

We consider now the input and output port characteristics of an ideal converter with and without feedback. In what follows we shall assume that a converter is made of *linear* inductors and capacitors. For a converter without feedback, the output voltage is *linearly* related to the input voltage by the conversion ratio M_V:

$$\text{No feedback} \Rightarrow V_o = M_V V_{in}; \ M_V \neq M_V(V_{in}) \tag{8.6}$$

The reason for this is that the switching action in any dc-to-dc converter, as we shall see, produces a periodic sequence of linear networks. The conversion ratio is obtained by piecing together the solutions of the individual switched linear networks which are linear functions of the input voltage. Hence, the piecewise composite solution is also a linear function of the input voltage. Following the same argument we conclude that M_I is also independent of the input voltage. Hence, the ideal voltage converter in Fig. 8.1a can be modeled by an ideal transformer with turns ratio M_V as shown in Fig. 8.3. In general M_V is a function of the control parameter, α, the output voltage, V_o, and the output current, I_o:

$$M_V = M_V(\alpha, V_o, I_o) \tag{8.7}$$

Also, a converter can in general have several modes of operation so that a unique M_V may not be sufficient for modeling purposes. In the very important class of converters discussed in this chapter, M_V is only a function of α.

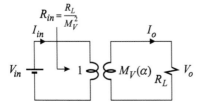

Figure 8.3

The input and output characteristics of an ideal converter can now be ascertained using the equivalent circuit in Fig. 8.3. For an unregulated converter feeding a resistive load R_L, the incremental input resistance seen by the source according to Fig. 8.3 is:

$$R_{in} = \frac{R_L}{M_V^2} \tag{8.8}$$

For an ideal converter in which the output voltage is regulated by the feedback arrangement shown in Fig. 8.4, the output voltage is equal to the reference voltage:

$$V_o = V_R \tag{8.9}$$

Hence, when the load is fixed, the input power of a regulated converter is fixed and independent of the source voltage. It follows that any increase in the source voltage is accompanied by a decrease in the source current, which implies that the incremental input resistance of a regulated converter is negative. Proceeding as follows, we obtain:

$$\frac{\partial P_{in}}{\partial V_{in}} = 0 \Rightarrow V_{in} \frac{\partial I_{in}}{\partial V_{in}} + I_{in} = 0 \Rightarrow \frac{\partial I_{in}}{\partial V_{in}} = -\frac{I_{in}}{V_{in}} \tag{8.10}$$

Substituting Eqs. (8.1), (8.2) and (8.3) in Eq. (8.10) we obtain several useful expressions for the incremental input resistance:

$$
\left.
\begin{aligned}
R_{in} &= -\frac{V_{in}^2}{P_{in}} \\[2mm]
&= -\frac{(V_o/I_o)}{M_V^2} \\[2mm]
&= -\frac{R_L}{M_V^2}
\end{aligned}
\right\} \tag{8.11a–c}
$$

In Eq. (8.11c) we have made use of the fact that $R_L = V_o/I_o$ whenever the load actually consists of a resistor as shown in Fig. 8.4.

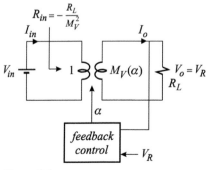

Figure 8.4

The incremental output resistance of regulated and unregulated ideal converters can be determined in a similar manner (see Problem 8.3).

Example 8.1   An ideal battery charging circuit is shown in Fig. 8.5. This circuit is somewhat tricky because it has two voltage sources connected in parallel with zero resistance between them. If it were not for the control circuit, it would not have been clear which of the two voltage sources was charging or being charged and the charging current would have been undetermined. In this ideal case, the feedback circuit adjusts the control parameter α, and hence the conversion ratio M_V, by monitoring the charging current I_o to ensure $I_o = I_{ch}$. The value of α determined by the control circuit is given by:

$$\alpha = M_V^{-1}\left(\frac{V_B}{V_{in}}\right) \tag{8.12}$$

If α were to deviate by the slightest amount from the value in Eq. (8.12), the charging current would become infinite, causing an infinite error signal, which would be instantaneously corrected by the negative feedback circuit. The current drawn from the source, when α is set exactly to the value in Eq. (8.12), is determined from the current conversion ratio:

$$I_{in} = \frac{I_o}{M_I} = I_{ch}M_V = I_{ch}\frac{V_B}{V_{in}} \tag{8.13}$$

where we have made use of the fact that $M_I M_V = 1$ for an ideal converter.

The incremental input impedance is determined next following the procedure in Eq. (8.10):

$$\frac{\partial I_{in}}{\partial V_{in}} = -\frac{I_{in}}{V_{in}} = -\frac{I_{ch}V_B}{V_{in}^2} \tag{8.14}$$

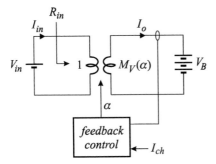

Figure 8.5

It follows that:

$$R_{in} = -\frac{V_{in}^2}{I_{ch}V_B} \tag{8.15}$$

The same result could have been obtained directly from Eq. (8.11a).   ☐

8.3   The buck converter

Consider the task of converting an unregulated dc voltage source, V_g, to a regulated dc voltage source $V_o < V_g$ which supplies power to a load R_L. One simple and efficient way of doing this is to chop the source and generate a *unipolar* voltage pulse train with an amplitude V_g, a fixed period T_s, and a variable width, or on-time T_{on}, as shown in Fig. 8.6. Such a waveform is known as a pulse-width-modulated (PWM) waveform. It is relatively easy to see that the dc component of the PWM waveform in Fig. 8.6 can be controlled by varying the pulse width, T_{on}. To make use of this dc component, the high-frequency components must be filtered out by a nondissipative low-pass filter. A simple converter which can chop and filter as described above is the buck converter shown in Fig. 8.7.

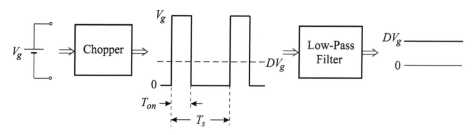

Figure 8.6

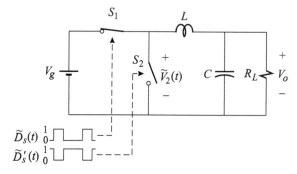

Figure 8.7

The chopper section of the buck converter consists of two switches S_1 and S_2 which are driven by the complementary *switching functions* $\tilde{D}_s(t)$ and $\tilde{D}_s'(t) = 1 - \tilde{D}_s(t)$, respectively. The switching function is a unit pulse train defined:

$$\tilde{D}_s(t) = \begin{cases} 1 & nT_s < t < nT_s + T_{on} \\ 0 & nT_s + T_{on} < t < (n+1)T_s \end{cases} \tag{8.16}$$

Looking back into the chopper circuit from the terminals of S_2, we see an effective voltage source $\tilde{V}_2(t)$ given by:

$$\tilde{V}_2(t) = V_g \tilde{D}_s(t) \tag{8.17}$$

The dc component of $\tilde{V}_2(t)$ is given by:

$$\frac{1}{T_s} \int_0^{T_s} \tilde{V}_2(t)dt = \frac{1}{T_s} \int_0^{T_s} V_g \tilde{D}_s(t)dt = DV_g \tag{8.18}$$

in which D is defined as the *duty-ratio function*, or duty cycle, and is given by:

$$D = \frac{1}{T_s} \int_0^{T_s} \tilde{D}_s(t)dt = \frac{T_{on}}{T_s} \tag{8.19}$$

The LC filter following the chopper is the simplest, lossless, low-pass filter that can extract the dc component of $\tilde{V}_2(t)$ and generate a dc output voltage V_o given by:

$$V_o = DV_g \tag{8.20}$$

It is clear from Eq. (8.20) that the dc output voltage can be regulated by varying the duty cycle, D, so that the duty cycle serves as the control parameter, i.e. $\alpha = D$. It follows from Eq. (8.20) that the voltage conversion ratio, M_V, of the buck converter is given by:

$$M_V = \frac{V_o}{V_g} = D \tag{8.21}$$

In order to determine the steady-state voltages and currents of the buck converter, we need to study the response of the low-pass LC filter to $\tilde{V}_2(t)$. The following transfer function can be easily verified (see Problem 8.4):

$$H(s) = \frac{v_o(s)}{v_2(s)} = \frac{1}{1 + \dfrac{s}{\omega_o Q} + \dfrac{s^2}{\omega_o^2}} \tag{8.22}$$

in which:

$$\left. \begin{array}{l} \omega_o = \dfrac{1}{\sqrt{LC}} \\[4mm] Q = \dfrac{R}{\sqrt{L/C}} \end{array} \right\} \tag{8.23a, b}$$

The magnitude and phase response of $H(s)$ is shown in Fig. 8.8. In order to study the interaction of $\tilde{V}_2(t)$ with $H(s)$, we write $\tilde{V}_2(t)$ as:

$$\tilde{V}_2(t) = DV_g + \tilde{V}'(t) \tag{8.24}$$

in which $\tilde{V}'(t)$ is a periodic waveform which contains only the fundamental and harmonics of $\tilde{V}_2(t)$. From the magnitude response of $H(s)$, it can be seen that the dc component, DV_g, of $\tilde{V}_2(t)$ will pass through and become the dc component of the output voltage. It can also be seen from the same figure that if the corner frequency of the filter, f_o, is chosen well below the switching frequency of the converter, $F_s = 1/T_s$, then all the harmonics of $\tilde{V}_2(t)$ at nF_s will fall on the $-40\,\mathrm{dB/dec}$ asymptote so that the ac components of $\tilde{V}_2(t)$ will be *attenuated* and *integrated twice*. Therefore, the actual output voltage consists of the dc component $V_o = DV_g$ and a small periodic waveform, $\tilde{V}_r(t)$, known as the output ripple voltage:

$$\tilde{V}_o(t) = V_o + \tilde{V}_r(t); \; |\tilde{V}_r(t)| \ll V_o \tag{8.25}$$

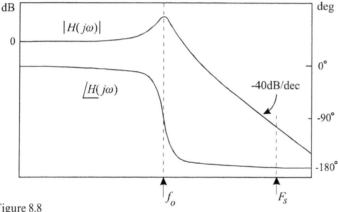

Figure 8.8

Since the output ripple voltage is given by the high-frequency response of $H(s)$ to $V'(t)$, we can approximate $H(s)$ as:

$$H(s) \approx \frac{\omega_o^2}{s^2} \tag{8.26}$$

Equation (8.26) corresponds to a double integration so that we have:

$$\tilde{V}_r(t) \approx \omega_o^2 \iint \tilde{V}'(t)dt \tag{8.27}$$

in which $\tilde{V}'(t)$, according to Eqs. (8.16), (8.17) and (8.24), is given by:

$$\tilde{V}'(t) = \begin{cases} V_g - V_o & nT_s < t < nT_s + T_{on} \\ -V_o & nT_s + T_{on} < t < (n+1)T_s \end{cases} \tag{8.28}$$

Since $\tilde{V}'(t)$ is a constant in each subinterval, the double integration in Eq. (8.27) yields a parabolic segment in each subinterval as shown in Fig. 8.9. Note that the ripple voltage, $\tilde{V}_r(t)$, and $\tilde{V}'(t)$ are inverted with respect to each other because all the frequency components of $\tilde{V}'(t)$ are phase shifted by $180°$ owing to the phase response of $H(s)$ at high frequencies. Each parabolic segment of the voltage ripple is given by:

$$\tilde{V}_{r1}(t) = -V_{p1} + \omega_o^2(V_g - V_o)\frac{t^2}{2}$$

$$\left.\begin{array}{c}\\\\\\\\\end{array}\right\} \qquad (8.29a, b)$$

$$\tilde{V}_{r2}(t) = V_{p2} - \omega_o^2 V_o \frac{t^2}{2}$$

in which the time origin of each segment is taken at its peak. The peak-to-peak output voltage ripple follows from Eqs. (8.29a, b) which, when normalized with respect to the output voltage (see Problem 8.5), is given by:

$$\delta V_r = \frac{V_{r_{p-p}}}{V_o} = \frac{\pi^2}{2}(1 - D)\left(\frac{f_o}{F_s}\right)^2 \qquad (8.30a)$$

For a regulating converter, the worse case normalized ripple, δ_{vr}, occurs at D_{min} or M_{min} so that we have:

$$\delta_{vr} = \frac{\pi^2}{2}(1 - M_{min})\left(\frac{f_o}{F_s}\right)^2 \qquad (8.30b)$$

Hence, the LC filter is designed with a resonant frequency f_o given by:

$$f_o = \frac{F_s}{\pi}\sqrt{\frac{2\delta_{vr}}{1 - M_{min}}} \qquad (8.31)$$

Note that f_o is expressed in terms of the *design specifications* (δ_{vr}, M_{min}) and a *design parameter* F_s which the designer chooses.

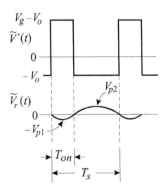

Figure 8.9

Example 8.2   A 100-kHz buck converter is to be used in the design of a regulated converter which operates from an unregulated bus voltage of $28 \pm 4\,\text{V}$ and delivers an output of 12 V at 2.3 A. The maximum output ripple voltage is specified to be 150 mV. Determine the range of the duty ratio and f_o of the output filter.

According to the given variation in the bus voltage the specified range of the duty cycle and the conversion is:

$$M_{min} = D_{min} = \frac{V_o}{V_{in} + \Delta V_{in}} = \frac{12}{28 + 4} = 0.375$$

$$M_{max} = D_{max} = \frac{V_o}{V_{in} - \Delta V_{in}} = \frac{12}{28 - 4} = 0.5$$

(8.32a, b)

According to Eq. (8.31), the worse case, or the lowest corner frequency, requirement is dictated by M_{min} so that we have:

$$\frac{f_o}{F_s} = \frac{1}{\pi}\sqrt{\frac{2\delta_{vr}}{1 - M_{min}}} = \frac{1}{\pi}\sqrt{\frac{2(0.15/12)}{1 - 0.375}} = 0.064$$

(8.33)

Hence, the corner frequency of the filter must be selected less than or equal to 6.4 kHz to have an output ripple voltage of 150 mV or less.   □

Determination of the corner frequency of the filter in the example above does not uniquely determine the inductor and the capacitor. We shall now determine the current in the inductor and show that its ripple component provides another design equation for the selection of L and C. The current in the inductor consists of a dc and a ripple component. The dc component of the inductor current is the same as the dc component of the output current because the capacitor does not carry any dc current. Hence, the inductor current can be written as:

$$\tilde{I}_L(t) = I_o + \tilde{I}_r(t)$$

(8.34)

in which $I_o = V_o/R$.

The ripple current in the inductor is determined by integrating the voltage across it, which is given by:

$$\tilde{V}_L(t) = \tilde{V}_2(t) - \tilde{V}_o(t)$$

$$= V_o + \tilde{V}'(t) - V_o - \tilde{V}_r(t)$$

$$\approx \tilde{V}'(t)$$

(8.35)

In the last step we have ignored the output ripple voltage because it is much smaller in comparison with $\tilde{V}'(t)$. Since $\tilde{V}'(t)$ has a constant value in each subinterval, the ripple current consists of two linear segments as shown in Fig. 8.10. The first segment of the inductor current is given by:

$$\tilde{I}_{r1}(t) = \frac{1}{L}\int_0^t (V_g - V_o)dt = I_{p1} + \frac{V_g - V_o}{L}t$$

(8.36)

in which the time origin has been taken at I_{p1}. For the second segment we have:

$$\tilde{I}_{r2}(t) = \frac{1}{L}\int_0^t (-V_o)dt = I_{p2} - \frac{V_o}{L}t \qquad (8.37)$$

in which the time origin has been taken at I_{p2}. Since the ripple current is linear, the average value of the inductor current is midway between I_{p1} and I_{p2} and the peak-to-peak ripple current can be determined from either segment. Hence, letting $t = D'T_s$ in Eq. (8.37), we obtain:

$$I_{p1} = I_{p2} - \frac{V_oD'T_s}{L} \qquad (8.38)$$

It follows that the peak-to-peak ripple current is given by:

$$I_{r_{p-p}} = I_{p2} - I_{p1} = \frac{V_oD'T_s}{L} \qquad (8.39)$$

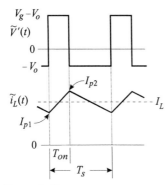

Figure 8.10

It can be seen from Eq. (8.39) that for a buck converter with a regulated output voltage, the maximum ripple current occurs when D' is a maximum, which occurs when the input voltage is a maximum. When the ripple current is normalized with respect to the average inductor current, $I_L = I_o$, we obtain:

$$\delta I_r = \frac{I_{r_{p-p}}}{I_L} = \frac{V_oD'T_s}{I_oL} = \frac{RD'T_s}{L} \qquad (8.40)$$

The amount of ripple current determines the peak current in the inductor and the switches S_1 and S_2. This peak is given by:

$$I_p = I_o + \frac{I_o\delta I_r}{2} \qquad (8.41)$$

The maximum value of the peak current in Eq. (8.41) occurs at the specified maximum load current so that we have:

$$I_{P_{max}} = I_{o_{max}}(1 + \delta_{ir})$$ (8.42)

in which δ_{ir} is the worst case value of $\delta I_r/2$ and is given by:

$$\delta_{ir} = \frac{V_o(1 - M_{min})T_s}{2LI_{o_{max}}}$$ (8.43)

It follows that, for a design choice of δ_{ir}, L is determined uniquely according to:

$$L = \frac{1}{F_s\delta_{ir}} \frac{V_o(1 - M_{min})}{2I_{o_{max}}}$$ (8.44a)

The value of C follows from the resonant frequency given in Eq. (8.31):

$$C = \frac{\delta_{ir}}{F_s} \frac{I_{o_{max}}}{4V_o\delta_{vr}}$$ (8.44b)

Note that Eqs. (8.44a, b), just like Eq. (8.31), express L and C in terms of the design specifications (V_o, M_{min}, $I_{o_{max}}$, δ_{vr}) and the design parameters (F_s, δ_{ir}).

The maximum peak current is an important design consideration and is expressed in terms of δ_{ir}. For example, it can be seen that designing with a larger value of δ_{ir} results in a smaller inductor, a larger capacitor and a larger peak current. There are many design tradeoff considerations which we will not discuss here and which result in practical values of δ_{ir} in the range 0.1–1.0.

Although not necessary, the design equations of the filter elements can be expressed in terms of the resonant frequency f_o given by Eq. (8.31) and the Q-factor, which we shall determine next. Hence, after performing the necessary substitutions, we can rewrite Eq. (8.43) as:

$$\delta_{ir} = \pi(1 - M_{min})Q_o\frac{f_o}{F_s}$$ (8.45)

in which Q_o is given by:

$$\left.\begin{array}{l} Q_o = \dfrac{(V_o/I_{o_{max}})}{\omega_o L} \\[3mm] = \omega_o C(V_o/I_{o_{max}}) \end{array}\right\}$$ (8.46a, b)

If the load is a simple resistor R, then $V_o/I_o = R$ and Q_o is the same as the Q-factor of the LC filter. If the load, on the other hand, is a current source, another regulated converter, or a battery as in Example 8.1, then Q_o and the actual Q-factor are different. Performing the necessary substitutions for Q_o, we obtain an alternate set of design equations for the LC filter:

$$
\left.
\begin{aligned}
f_o &= \frac{F_s}{\pi}\sqrt{\frac{2\delta_{vr}}{1 - M_{min}}} \\[2ex]
Q_o &= \frac{\delta_{ir}}{\sqrt{2\delta_{vr}(1 - M_{min})}}
\end{aligned}
\right\}
\tag{8.47a, b}
$$

Example 8.3 If the maximum peak current in the buck converter of Example 8.2 is to be 3 A, determine the values of L and C.

The value of δ_{ir}, given by Eq. (8.42), is computed first:

$$
\delta_{ir} = \frac{I_{p_{max}}}{I_{o_{max}}} - 1 = \frac{3}{2.3} - 1 = 0.3
\tag{8.48}
$$

Next we compute Q_o given by Eq. (8.47):

$$
Q_o = \frac{0.3}{\sqrt{2(1 - 0.375)(0.15/12)}} = 2.4
\tag{8.49}
$$

The values of L and C are obtained from Eqs. (8.46a, b):

$$
\left.
\begin{aligned}
L &= \frac{(12/2.3)}{(2\pi 6400)2.4} = 54\,\mu\text{H} \\[2ex]
C &= \frac{2.4}{(2\pi 6400)(12/2.3)} = 11.14\,\mu\text{F}
\end{aligned}
\right\}
\tag{8.50a, b}
$$

These are exact values, which may or may not be available. Typically, the tolerances on power inductors and large capacitors can be of the order 5–10%. Hence, when the actual components are chosen, the nearest available values are selected such that their worst case values are greater or equal to those determined above. ☐

The currents and voltages of the switches S_1 and S_2 are examined next. These are shown in Fig. 8.11 and can be verified easily. Switch S_1 is called the *active* switch and it carries the inductor current during DT_s. Switch S_2 is called the *passive* switch and it carries the inductive current during $D'T_s$. Two practical realizations of the switches are shown in Figs. 8.12a and b. The difference between these two realizations is in their modes of operation at low-output currents as shown in Fig. 8.13. When a MOSFET is turned on, it can conduct in both directions whereas a diode can conduct only in one direction. Hence, in Fig. 8.12a, when Q_1 is turned off, the current in the inductor turns on the diode to initiate the $D'T_s$ subinterval. The diode will conduct as long as the inductor current is positive, which is expressed quantitatively:

$$
\frac{I_{r_{p-p}}}{2} \leq I_o
\tag{8.51}
$$

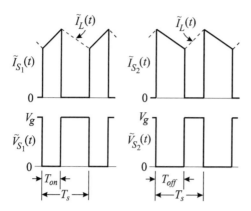

Figure 8.11

If this condition is not satisfied, then the diode will stop conducting before the end of $D'T_s$ and a third interval of operation will occur during which both switches will be in the off state and the output capacitor will discharge into the load. This is called the discontinuous conduction mode (DCM), which we shall not discuss here. In DCM, the conversion ratio is no longer the same as in the continuous conduction mode (CCM) and depends on the output current and the switching frequency in addition to the duty-cycle D. The current waveforms in DCM are shown in Fig. 8.13a.

When a MOSFET is used instead of a diode, as shown in Fig. 8.12b, the converter is capable of operating in CCM down to zero load current, because when a MOSFET is turned on it can conduct in both directions allowing the current in the inductor to reverse direction. The currents in both MOSFETs at low-output currents, when the condition in Eq. (8.51) is not satisfied, are shown in Fig. 8.13b.

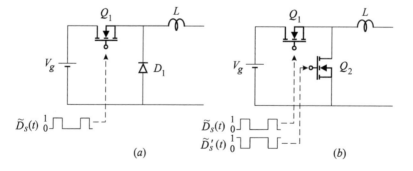

Figure 8.12

For the buck converter discussed above, the input current is pulsating and is the same as the current in S_1. The input current of an ideal converter, however, must be smooth and only follow the variations in the output current as required by the equivalent circuit model of an ideal converter shown in Fig. 8.3. Hence, another filter must be added on the input side to reduce the input ripple current down to an acceptable level. It is interesting to point out that the end-user of a converter usually does not care how much ripple current the converter generates on the input power bus. But, high frequency ripple currents on power lines, called conduction emissions, create serious interference problems for other users. Hence, for commercial applications, conducted emissions are specified and enforced by regulatory agencies such as the FCC in the USA, the CSA in Canada and the VDE in Germany. (For military and aerospace applications, conducted emissions must

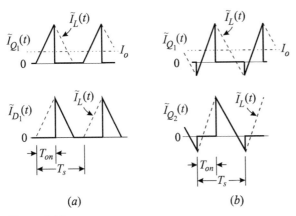

Figure 8.13

comply with US Military Standards 461.) The simplest type of input filter, with series or shunt damping, is shown in Figs. 8.14a and b in which $R_{id}L_{id}$ and $R_{id}C_{id}$ are the series and shunt damping networks, respectively. The purpose of the damping networks is to prevent oscillations or ringing, and these will be discussed shortly.

The input ripple current is calculated by determining the transfer function $H_r(s)$ shown in Fig. 8.15, in which the current source $I_p(s)$ is the pulsating current drawn by the converter. In this figure, we have assumed that the ripple voltage across C_i is small in comparison with the dc voltage across it so that the shape of $\tilde{I}_p(t)$ is essentially the same with or without the input filter. In determining this transfer function, the effect of the damping network can be ignored so that we have from Fig. 8.15:

$$\frac{I_{in}(s)}{I_p(s)} = \frac{1}{1 + s^2/\omega_{oi}^2} \tag{8.52}$$

in which:

$$\omega_{oi} = \frac{1}{\sqrt{L_i C_i}} \tag{8.53}$$

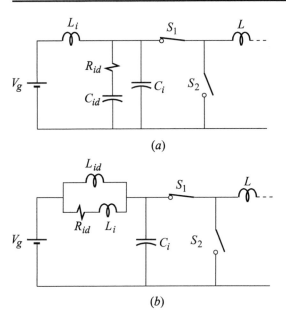

(a)

(b)

Figure 8.14

According to Eq. (8.52), the dc component of $\tilde{I}_p(t)$ will pass through the input circuit, while its ac component, $\tilde{I}'_p(t)$, will be attenuated, if its frequency spectrum falls above ω_{oi}. Hence, as in the design of the output filter, we choose the corner frequency of the input filter to be much lower than the switching frequency so that $\tilde{I}'_p(t)$ will be attenuated and integrated twice. Hence, the input ripple current is given by:

$$\frac{I_{in_r}(s)}{I'_p(s)} \approx \frac{\omega_{oi}^2}{s^2} \Rightarrow \tilde{I}_{in_r}(t) = \omega_{oi}^2 \int \int \tilde{I}'_p(t) dt \tag{8.54}$$

The exact calculation of the peak-to-peak ripple from the double integral in Eq. (8.54) is somewhat tedious because of the cubic term that arises from the current

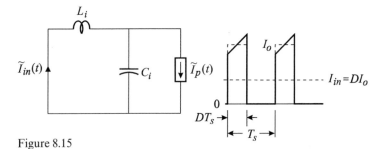

Figure 8.15

slope during DT_s. This calculation can be simplified if the trapezoidal pulse is replaced with a rectangular pulse of the same area and duration. The height of this equivalent rectangular pulse, of course, is equal to the output current. This simplification is justified because the maximum input ripple current occurs at maximum load current; which, in most designs, corresponds to an $\tilde{I}_p(t)$ that is close to rectangular. The resulting expression of the peak-to-peak input ripple current is given (see Problem 8.6) by:

$$I_{ir_{p-p}} \approx I_o \frac{\pi^2 DD'}{2}\left(\frac{f_{oi}}{F_s}\right)^2 \tag{8.55}$$

It is clear from Eq. (8.55) that the worst case value of the input ripple current occurs at maximum load. The maximum value of the term DD', however, depends on the specified range of $D_{min} \leq D \leq D_{max}$. Hence, we have from Eq. (8.55):

$$I_{ir_{p-p}} \approx I_{o_{max}} \frac{\pi^2 (\max[DD'])}{2}\left(\frac{f_{oi}}{F_s}\right)^2 \tag{8.56}$$

in which

$$\max[DD'] = \begin{cases} D_{max}(1 - D_{max}); & D_{max} \leq 1/2 \\ 1/4; & D_{max} \geq 1/2 D_{min} \leq 1/2 \\ D_{min}(1 - D_{min}); & D_{min} \geq 1/2 \end{cases} \tag{8.57}$$

With these design equations we can only determine the resonant frequency of the input filter but not C_i and L_i individually. We could derive another design equation in terms of the voltage ripple on the input capacitor, C_i, so that C_i and L_i could be determined uniquely, but the input voltage ripple is hardly a design consideration. Hence, typically, once f_{oi} is determined, L_i and C_i are usually determined such that their combined volume is as small as possible.

Example 8.4 Determine the resonant frequency of the input filter for the buck converter in Example 8.3 so that $I_{ir_{p-p}} = 10 \, \text{mA}$. Since $D_{max} = 0.5$, we have from Eqs. (8.56) and (8.57);

$$0.01 \approx 2.3 \frac{\pi^2}{2}\frac{1}{4}\left(\frac{f_{oi}}{F_s}\right)^2 \Rightarrow \frac{f_{oi}}{F_s} \approx 0.059 \tag{8.58}$$

Hence, the input filter is designated with $f_{oi} = 5.9 \, \text{kHz}$. □

Since the input impedance of a regulating converter is negative, the addition of an LC input filter can easily result in an instability because of the negative damping effect. Hence, it is always a good idea to damp the input filter of a regulating converter by adding either a series or shunt damping branch as shown in Figs. 8.14a and b. The design of the damping network can be optimized if the frequency response of the input impedance of the converter is known. We shall be brief and present a much simpler technique that assumes that the input impedance is real and negative. In Fig. 8.14a, if C_{id} is chosen to be much larger than C_i, then at resonance the capacitive reactance of C_{id} will be much higher than that of C_i so that the damping resistor, R_{id}, will appear effectively in parallel with the negative resistance of the input converter resulting in a Q given by:

$$Q_{oi} \approx \frac{R_{id} \parallel R_{in}}{\sqrt{L_i/C_i}} \qquad (8.59)$$

in which:

$$R_{in} = -\frac{V_o}{I_o}\frac{1}{M^2} \qquad (8.60)$$

Typically, R_{id} is chosen such that $Q_{oi} \approx 1$. Note that Q_{oi} cannot be chosen to be very small simply because as R_{id} gets smaller, then C_{id} begins to appear effectively in parallel with C_i, which in turn gives rise to a new, lower, undamped resonance formed by $C_i + C_{id}$ and L_i (see Problem 8.7).

Example 8.5   For the buck converter discussed in the previous examples, we select the input filter inductor to be 86 µH. With the resonant frequency determined to be $f_{oi} = 5.9$ kHz, the value of the input filter capacitor is given by:

$$C_i = \frac{1}{(2\pi 5900)^2 86 \times 10^{-6}} = 8.46 \,\mu\text{F} \qquad (8.61)$$

The smallest value of the negative input resistance of the converter occurs at M_{max}, so that we have:

$$R_{in} = -\frac{12 \text{ V}}{2.3 \text{ A}}\frac{1}{(0.5)^2} = -20.9 \,\Omega \qquad (8.62)$$

The value of the damping resistor for the shunt branch is determined from Eq. (8.59) in which we select $Q_{oi} = 1$. This yields:

$$1 = \frac{R_{id} \parallel (-20.9)}{\sqrt{86/8.46}} \Rightarrow R_{id} = 2.8 \,\Omega \qquad (8.63)$$

The value of C_{id} is chosen to be about ten times that of C_i:

$$C_{id} \approx 10 C_i = 84 \,\mu\text{F} \qquad (8.64)$$

An OrCAD/Pspice simulation of the actual filter and its frequency response is shown in Figs. 8.16a and b. The response is seen to be properly damped so that any transient disturbance in the input voltage will not generate any ringing in the input filter.

A simulation of the actual input ripple current is shown in Figs 8.17a and b in

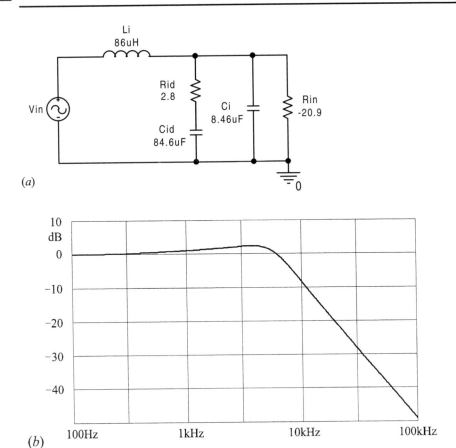

(a)

(b)

Figure 8.16

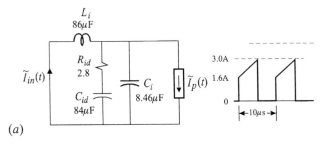

(a)

Figure 8.17

which we see that the peak-to-peak ripple current is 10.11 mA, which is very close to the specified design value of 10 mA.

In this example we have shown that the approximate design equations for the damping network and for the determination of the input ripple current yield very accurate results when compared with actual simulations.   □

For the series damping branch in Fig. 8.14b, if L_{id} is chosen much larger than L_i,

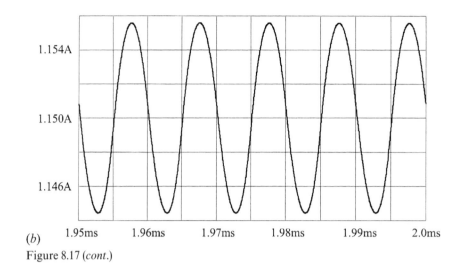

(b) 1.95ms   1.96ms   1.97ms   1.98ms   1.99ms   2.0ms

Figure 8.17 (*cont.*)

then, at resonance, R_{id} will effectively appear in series with L_i and will counteract the negative input resistance of the converter. In this case (see Problem 8.8) the overall Q is given by:

$$Q_{oi} \approx \frac{\sqrt{L_i/C_i}}{R_{id}} \left\| \frac{R_{in}}{\sqrt{L_i/C_i}} \right. \tag{8.65}$$

As in the case of the shunt damping, R_{id} is chosen such that $Q_{oi} \approx 1$. Observe that in this case too it is not possible to design with a much lower Q_{oi} by making R_{id} very large because that would effectively create a new, lower, undamped resonance formed by L_{id} and C_i.

Example 8.6   In this example we shall design a series damping branch for the input filter of the buck converter in Example 8.5. In this case, we chose L_{id} about ten times larger than L_i so that we have:

$$L_{id} \approx 10L_i = 860 \,\mu\text{H} \tag{8.66}$$

The value of the damping resistor is determined from Eq. (8.65):

$$1 \approx \frac{\sqrt{86/8.46}}{R_{id}} \left\| \frac{-20.9}{\sqrt{86/8.46}} \right. \Rightarrow R_{id} = 3.67 \,\Omega \tag{8.67}$$

An OrCAD/Pspice simulation of this circuit and its frequency response are shown in Figs. 8.18*a* and *b*. In comparison to the response of the shunt damping network, the peaking seems to be a little higher because this transfer function has a zero at about 617 Hz and a pole at about 824 Hz, whereas the shunt network has

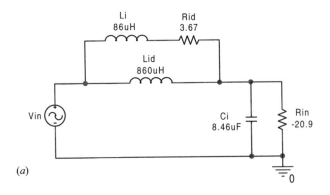

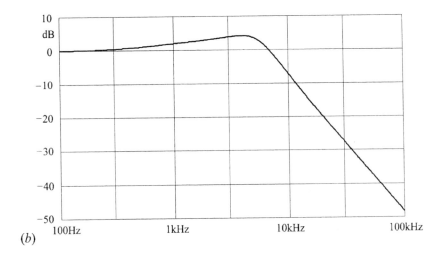

(a)

(b)

Figure 8.18

essentially a pole-zero cancellation at about 676 Hz (see Problems 8.7 and 8.8).

A simulation run of the input ripple current yields a peak-to-peak ripple current of 11.14 mA, which is about 10% higher than the parallel-damped filter in the previous example. The reason for this slight increase is that, at the switching frequency, the pulsating current is divided between C_i and $L_{id} \parallel L_i$ instead of C_i and L_i. (Here we have ignored the resistance of R_{id} with respect to the reactance of L_i at the switching frequency.) Since $L_{id} = 10L_i$, the reactance of their parallel combination is about 10% smaller than the reactance of L_i, which is what accounts for the 10% reduction in the ripple attenuation. Since L_{id} carries the dc input current, it may not be as economical as the shunt damping element for high input currents. The size and cost of an inductor increases with its inductance and its current handling capacity.   □

8.4   The boost converter

By rearranging the filter elements around the switches in the buck converter we obtain different converters with different conversion ratio characteristics. All of these converters operate on the same basic principle of commutating the inductor current between the input and output circuits. When the elements of the buck converter are rearranged as shown in Fig. 8.19, we obtain the boost converter. The voltage and current waveforms of this converter are shown in Fig. 8.20. In this

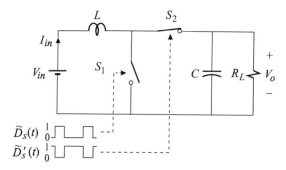

Figure 8.19

circuit when S_1 is turned on during T_{on}, while S_2 is turned off, the inductor charges linearly from I_{p1} to I_{p2} and the output filter capacitor discharges into the load with a time constant much longer than T_{on}. When S_1 is turned off and S_2 is turned on during T_{off}, the current in the inductor discharges linearly from I_{p2} to I_{p1} into R_L and C. Since the capacitive reactance of C at the switching frequency is much lower than the load resistance, the ac component of the pulsating current in S_2 is almost entirely absorbed by C while the dc component is absorbed by the load R_L. From the current waveforms in Fig. 8.20 we obtain the current conversion ratio:

$$I_o = D'I_{in} \Rightarrow M_I = \frac{I_o}{I_{in}} = D' \tag{8.68}$$

The voltage conversion ratio for the ideal boost converter follows immediately and is given by:

$$M_V = \frac{1}{D'} > 1 \tag{8.69}$$

Hence, in contrast to the buck converter, the output voltage of the boost converter is always larger than its input voltage.

During the on-time, the inductor current charges up linearly with a slope V_{in}/L so that the peak-to-peak ripple current shown in Fig. 8.21 is given by:

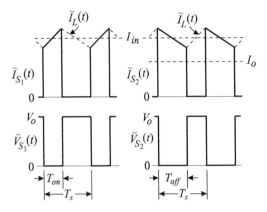

Figure 8.20

$$I_{r_{p-p}} = \frac{V_{in}}{L} DT_s = V_o \frac{DD'}{F_s L} \tag{8.70}$$

When $I_{r_{p-p}}$ is normalized with respect to the average inductor current, $I_L = I_{in}$, we obtain:

$$\delta I_r = \frac{I_{r_{p-p}}}{I_L} = \frac{V_o}{I_{in}} \frac{DD'}{F_s L} = \frac{V_o}{I_o} \frac{DD'^2}{F_s L} \tag{8.71}$$

The peak value of the inductor current is given by:

$$I_p = I_{in} + \frac{I_{r_{p-p}}}{2} \tag{8.72}$$

Substituting Eqs. (8.68) and (8.70) in Eq. (8.72) we obtain:

$$I_p = \frac{V_o}{2LF_s} DD' + \frac{I_o}{D'} \tag{8.73}$$

Hence, for a regulating converter in which V_o is a constant, I_p depends on I_o and D. In order to design the inductor, we need to know the operating conditions for which I_p is a maximum. Clearly, one of the operating conditions, according to Eq.

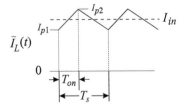

Figure 8.21

(8.73), is that the load current be at its maximum value. To determine the value of D which causes I_p to be a maximum, we set its derivative with respect to D to zero and obtain:

$$\frac{\partial I_p}{\partial D} = 0 \Rightarrow (1 - 2D) + \frac{2LF_s}{V_o}\frac{I_{o_{max}}}{D'^2} = 0 \tag{8.74}$$

For typical designs, the average inductor current at *maximum* load is always larger than half the ripple current through it, so that according to Eqs. (8.72) and (8.73) we have:

$$\frac{I_{o_{max}}}{D'} > \frac{V_o}{2LF_s}DD' \Rightarrow \frac{I_{o_{max}}}{D'^2}\frac{2LF_s}{V_o} > D \tag{8.75}$$

It follows that, for typical designs, there is no value of D which satisfies Eq. (8.74), and the maximum value of I_p occurs at D_{max}. Hence, the worst case value of the design parameter for the inductor, according to Eq. (8.71), is given by:

$$\delta_{ir} \equiv \left(\frac{\delta I_r}{2}\right)_{w.c.} = \frac{V_o}{2LF_sI_{o_{max}}}D_{max}(1 - D_{max})^2 \tag{8.76}$$

Practical values of δ_{ir} range from 0.1 to 1.0. Expressing the duty cycle in terms of the conversion ratio in Eq. (8.76), we obtain the design equation for the inductor:

$$L = \frac{V_o}{2I_{o_{max}}}\frac{M_{max}-1}{M_{max}^3}\frac{1}{\delta_{ir}F_s} \tag{8.77}$$

The output voltage ripple can be easily determined from the on-time when the capacitor is discharging into the load as shown in Fig. 8.22. Since the output time constant is much shorter than the on-time, we can assume the capacitor discharges linearly and write:

$$C\frac{\Delta v}{\Delta t} = C\frac{V_{r_{p-p}}}{DT_s} = I_o \tag{8.78}$$

The normalized output voltage ripple follows:

$$\frac{V_{r_{p-p}}}{V_o} = \frac{I_o}{V_o}\frac{DT_s}{C} \tag{8.79}$$

Clearly, the capacitor now must be chosen in such a way as to meet the specified output ripple voltage under worst case conditions which correspond to maximum load current and maximum duty cycle. Hence we have:

$$\delta_{vr} \equiv \left(\frac{V_{r_{p-p}}}{V_o}\right)_{w.c.} = \frac{I_{o_{max}}}{V_o}\frac{D_{max}}{F_sC} \tag{8.80}$$

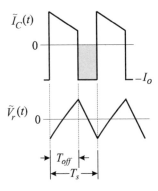

Figure 8.22

Hence the design equation for the capacitor is given by:

$$C = \frac{I_{o_{max}}}{V_o} \frac{M_{max} - 1}{M_{max}} \frac{1}{\delta_{vr} F_s} \tag{8.81}$$

Equation (8.79) is not very accurate at very light load currents as the current in the switches (assuming MOSFETs are used for both S_1 and S_2) begins to reverse. At zero load current Eq. (8.79) suggests that the output ripple voltage is zero, which certainly is not accurate (see Problem 8.9).

Example 8.7   An OrCAD/Pspice simulation of a practical boost converter is shown in Fig. 8.23. The converter operates at 100 kHz and has a duty cycle of $D = 7/12$. The inductor current and the output ripple voltage are shown in Fig. 8.24 at $I_o = 0\,\text{A}$, 1 A. The ideal output voltage is given by:

$$V_o = \frac{5}{1 - 7/12} = 12\,\text{V} \tag{8.82}$$

Since the MOSFETs have a finite resistance when turned on, the output voltage is slightly less than 12 V at $I_o = 1\,\text{A}$.

According to Eq. (8.79), the output ripple voltage at full load is given by:

$$V_{r_{p-p}} = 1 \frac{(7/12)(10 \times 10^{-6})}{50 \times 10^{-6}} = 117\,\text{mV} \tag{8.83}$$

which is in exact agreement with the ripple in Fig. 8.24a. At zero load current, we need to use the result derived in Problem 8.9:

$$V_{r_{p-p}} = V_o \frac{D'^2 D T_s^2}{8LC}$$

$$= 12 \frac{(1 - 7/12)^2(7/12)(10 \times 10^{-6})^2}{8(10 \times 10^{-6})(50 \times 10^{-6})} \tag{8.84}$$

$$= 30\,\text{mV}$$

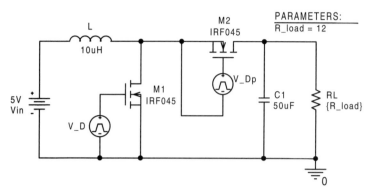

Figure 8.23

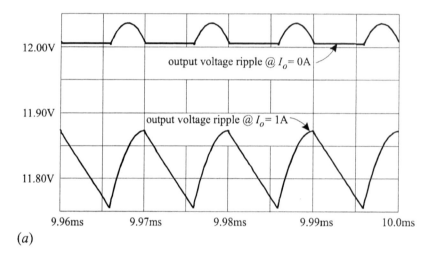

(a)

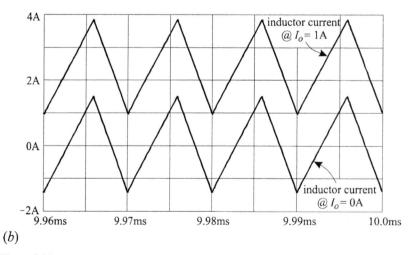

(b)

Figure 8.24

which is in exact agreement with the ripple in Fig. 8.24a.   □

Example 8.8   If a feedback loop is added to the boost converter in the previous example to maintain the output voltage at 12 V as the input voltage varies from 5 V to 10 V, then the maximum value of the peak current would occur at D_{max} and $I_{o_{max}} = 1$ A (assuming the *specified* maximum load current is 1 A). The maximum and minimum values of D are given by:

$$
\left.
\begin{aligned}
D_{max} &= 1 - \frac{V_{g_{min}}}{V_o} = 1 - \frac{5}{12} = 0.5833 \\[2mm]
D_{min} &= 1 - \frac{V_{g_{max}}}{V_o} = 1 - \frac{10}{12} = 0.1666
\end{aligned}
\right\}
\qquad (8.85a, b)
$$

According to Eq. (8.73), the maximum peak current is given by:

$$
\begin{aligned}
I_{p_{max}} &= \frac{V_o}{2LF_s} D_{max}(1 - D_{max}) + \frac{I_{o_{max}}}{1 - D_{max}} \\[2mm]
&= \frac{12(7/12)(1 - 7/12)}{2(10 \times 10^{-6})(100 \times 10^3)} + \frac{1}{1 - 7/12}
\end{aligned}
\qquad (8.86)
$$

This is in agreement with the waveforms shown in Fig. 8.25 in which the inductor current is shown at D_{max} and D_{min}.

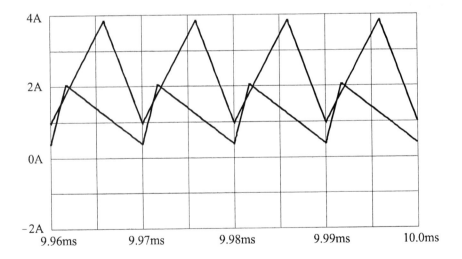

Figure 8.25

If this were an unconventional design, then the maximum specified load current would have been fairly low, say 60 mA. Also, if the converter had to operate from an input voltage range 3.5 V $< V_{in} <$ 10 V, then the range of D would have been given by:

$$0.1666 < D < 0.708 \tag{8.87}$$

In this case, Eq. (8.74) would yield the value of D at which the maximum value of I_p would occur:

$$(1 - 2D) + \frac{2LF_s}{V_o} \frac{I_{o_{max}}}{(1 - D)^2} = 0 \tag{8.88}$$

A numerical solution of Eq. (8.88) yields $D = 0.483$. The peak inductor current at $D = 0.483$ is equal to 1.614 A, whereas at $D_{max} = 0.708$ it is equal to 1.445 A, as can be seen in Fig. 8.26 in which the inductor currents at D_{min}, D_{max} and $D = 0.483$ are shown.

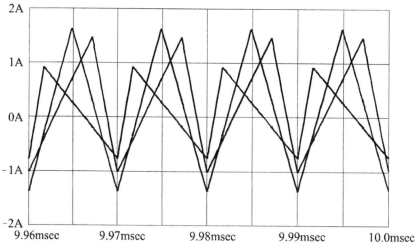

Figure 8.26

The purpose of this exercise was to demonstrate the use of Eq. (8.74) even though the difference in the peak currents between the two operating points is not significant.   □

8.5   The buck-boost converter

The converter shown in Fig. 8.27 is called the buck-boost because it is capable of either stepping up or down the input voltage. When S_1 is closed and S_2 is opened during the on-time, the inductor charges up linearly from I_{p1} to I_{p2} with a slope V_{in}/L, while the output capacitor discharges into the load with a slope I_o/C. When S_1 is opened and S_2 is closed during T_{off}, the inductor discharges into the output circuit with a slope V_o/L.

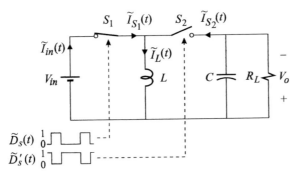

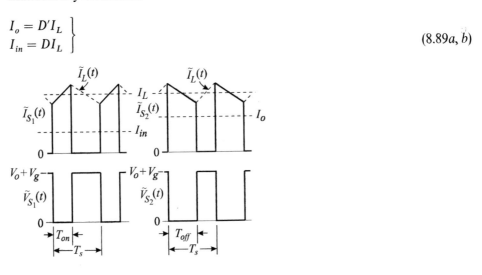

Figure 8.27

The voltages and currents of the switches are shown in Fig. 8.28 whence we can immediately determine:

$$\left.\begin{array}{l} I_o = D'I_L \\ I_{in} = DI_L \end{array}\right\} \qquad (8.89a, b)$$

Figure 8.28

Hence the current conversion ratio is given by:

$$M_I = \frac{I_o}{I_{in}} = \frac{D'}{D} \qquad (8.90)$$

The ideal voltage conversion ratio follows immediately:

$$M_V = \frac{1}{M_I} = \frac{D}{D'} \qquad (8.91)$$

The output ripple voltage is determined in the same manner as the output ripple voltage of the boost converter and is given by Eq. (8.79). Substituting Eq. (8.91) in (8.79), we obtain:

$$\frac{V_{r_{p-p}}}{V_o} = \frac{I_o}{V_o F_s C} \frac{1}{1+M_V} \frac{M_V}{1+M_V} \tag{8.92}$$

As in the case of the boost converter, this expression is not accurate at very light load currents (see Problem 8.10).

The output filter capacitor must be chosen to meet the output ripple voltage specification, δ_{vr}, under worst case conditions which, for a converter with a regulated output voltage, occurs when the load current and the conversion ratio are at a maximum. Hence, we have from Eq. (8.92):

$$\delta_{vr} = \left(\frac{V_{r_{p-p}}}{V_o}\right)_{w.c.} = \frac{I_{o_{max}}}{V_o} \frac{1}{F_s C} \frac{M_{V_{max}}}{1+M_{V_{max}}} \tag{8.93}$$

The peak-to-peak ripple current in the inductor is determined from the on-time and is given by:

$$\left.\begin{aligned} I_{r_{p-p}} &= V_{in} \frac{DT_s}{L} \\[2ex] &= \frac{V_o}{F_s L} \frac{1}{M_V + 1} \end{aligned}\right\} \tag{8.94a, b}$$

The average inductor current can be seen to be given by the sum of the average input and output currents so that when Eq. (8.94) is normalized with respect to the average inductor current, we obtain:

$$\delta I_r = \frac{I_{r_{p-p}}}{I_L} = \frac{V_o}{I_o} \frac{1}{F_s L} \frac{1}{(M_V + 1)^2} \tag{8.95}$$

The peak value of the inductor current is given by:

$$\left.\begin{aligned} I_p &= I_L + \frac{I_{r_{p-p}}}{2} \\[2ex] &= I_o(1 + M_V) + \frac{V_o}{2F_s L} \frac{1}{M_V + 1} \end{aligned}\right\} \tag{8.96a, b}$$

Clearly, one of the operating conditions which maximizes the peak current is the maximum load current. The other condition is the duty cycle, which can be determined by setting the derivative of I_p with respect to D to zero. This yields:

$$\frac{\partial I_p}{\partial D} = 0 \Rightarrow 1 - D = \sqrt{\frac{2F_s L}{V_o/I_{o_{max}}}} \tag{8.97}$$

For most typical designs, no value of D satisfies this condition so that the maximum value of the peak occurs at $M_{V_{max}}$. Hence we have:

$$I_{p_{max}} = I_{o_{max}}(1 + M_{V_{max}}) + \frac{V_o}{2F_sL} \frac{1}{M_{V_{max}} + 1} \tag{8.98}$$

It follows that the worst case design parameter for the inductor is given by:

$$\delta_{ir} = \left(\frac{\delta I_r}{2}\right)_{w.c.} = \frac{V_o}{I_{o_{max}}} \frac{1}{2F_sL} \frac{1}{(M_{V_{max}} + 1)^2} \tag{8.99}$$

Typical values of δ_{ir} range from 0.1 to 1.0. Hence, the design equations for the values of L and C of the buck-boost converter are given by:

$$L = \frac{V_o}{I_{o_{max}}} \frac{1}{2F_s\delta_{ir}} \frac{1}{(M_{V_{max}} + 1)^2}$$

$$C = \frac{I_{o_{max}}}{V_o} \frac{1}{F_s\delta_{vr}} \frac{M_{V_{max}}}{1 + M_{V_{max}}} \tag{8.100a, b}$$

The input current of the buck-boost converter in Fig. 8.27 is pulsating and must be filtered as shown in Fig. 8.29. The design procedure of the input filter and its damping elements is the same as that of the buck converter.

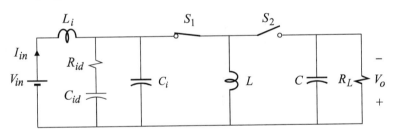

Figure 8.29

Example 8.9   The buck-boost converter in Fig. 8.30 generates a -12-V output from a $+5$-V source. The load current varies from 0 to 1 A. The output ripple voltage and the inductor current at three different load currents are shown in Figs. 8.31a and b below.

The peak-to-peak ripple voltage according to Eq. (8.92) is given by:

$$V_{r_{p-p}} = I_o \frac{1}{F_sC} \frac{M_V}{1 + M_V}$$

$$= I_o \frac{1}{(100 \ 10^3)(50 \ 10^{-6})} \frac{12/5}{1 + 12/5} \tag{8.101}$$

$$= 0.141I_o$$

At $I_o = 1$ A, the peak-to-peak ripple voltage is computed to be 141 mV, which is very close to the observed value of 143 mV in Fig. 8.31a. At $I_o = 0.5$ A, we compute

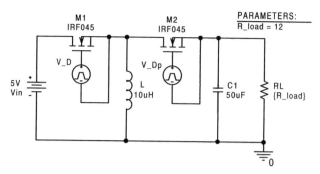

Figure 8.30

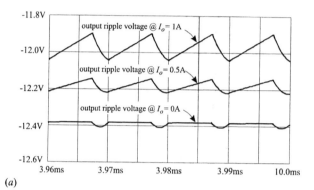

(a)

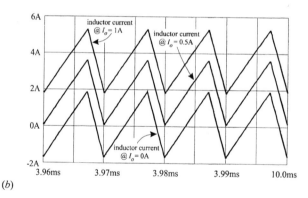

(b)

Figure 8.31

$V_{r_{p-p}} = 70\,\text{mV}$, while the observed value is $76\,\text{mV}$. At no load current, we can no longer use Eq. (8.92) and we must use Eq. (8.278b), see Problem 8.10, instead. This yields:

$$V_{r_{p-p}} = \frac{V_o}{8LC}\left(\frac{T_s}{1 + M_V}\right)^2$$

$$= \frac{12}{8(10 \, 10^{-6})(50 \, 10^{-6})} \left[\frac{10 \, 10^{-6}}{1 + 12/5} \right]^2 \tag{8.102}$$

$$= 26 \, \text{mV}$$

which is in close agreement with the 25 mV seen in Fig. 8.31a.

Note that the output voltage in Fig. 8.31a changes slightly as a function of the load current because the converter is not regulated and the MOSFETs have a resistance of 15 mΩ when turned on with an 8-V gate-drive signal.   □

8.6   The Cuk converter

This converter, named after its inventor Slobodan Cuk[1-3] (pronounced *chook*), is shown in Fig. 8.32. The input loop, consisting of V_{in}, L_1 and S_1, looks like a boost converter while its output loop, consisting of S_2, L_2 and V_o, looks like a buck converter. The switching voltage and current waveforms are shown in Fig. 8.33. Since the dc current in the capacitor C_c must be zero, we have:

$$I_{in}D' = I_oD \tag{8.103}$$

It follows that the current conversion ratio is given by:

$$M_I = \frac{D'}{D} \tag{8.104}$$

The ideal voltage conversion ratio follows:

$$M_V = \frac{D}{D'} \tag{8.105}$$

The output voltage, as in the case of the buck-boost converter, is inverted with respect to the input. An isolated output can be easily obtained by introducing a transformer between S_1 and S_2 (see Problem 8.11).

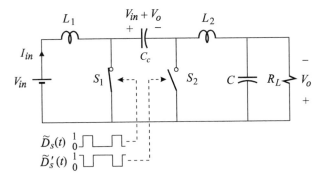

Figure 8.32

Since the dc voltage across each of L_1 and L_2 is zero, as we go around the outer loop containing V_{in}, V_o, L_1, L_2 and C_c, we find that the dc voltage across C_c is equal to $V_{in} + V_o$. It follows that the voltage waveforms across L_1 and L_2, excluding the ripple component of the voltages across C and C_c, are essentially identical, as shown in Fig. 8.33. The peak-to-peak ripple current in L_1 and L_2 can be determined from the on-time, DT_s:

$$I_{ir_{p-p}} = V_{in}\frac{DT_s}{L_i}; i = 1, 2$$

$$= \frac{V_o}{F_sL_i}\frac{1}{M_V + 1}$$

$$(8.106a, b)$$

An important feature of this converter is that L_1 and L_2 can be coupled[2] as shown in Fig. 8.34 and the ripple current in either inductor can essentially be reduced to zero by adjusting the coupling coefficient. This can be shown by writing the equations for $v_1(t)$ and $v_2(t)$ in Fig. 8.34:

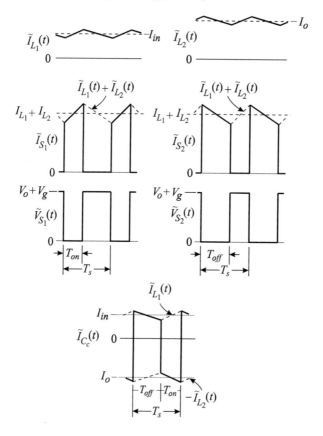

Figure 8.33

$$v_1(t) = L_1 \frac{di_1}{dt} + M \frac{di_2}{dt}$$

$$v_2(t) = L_2 \frac{di_2}{dt} + M \frac{di_1}{dt}$$

$$(8.107a, b)$$

in which M is the mutual inductance between the windings. In the Cuk converter, if we ignore the ripple voltages on C_c and C, then v_1 and v_2 are identical so that we have from Eq. (8.107):

$$(L_1 - M)\frac{di_1}{dt} - (L_2 - M)\frac{di_2}{dt} = 0 \qquad (8.108)$$

We can deduce from this equation that:

$$L_1 - M = 0 \Rightarrow \frac{di_2}{dt} = 0$$

$$L_2 - M = 0 \Rightarrow \frac{di_1}{dt} = 0$$

$$(8.109a, b)$$

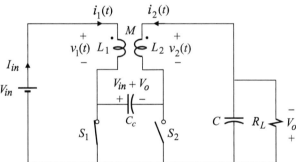

Figure 8.34

The disappearance of the derivative of the current implies that the current ripple is reduced to zero. Hence, if the mutual inductance is set according to Eq. (8.109a), then the ripple current in L_2 is reduced to zero. This can be expressed in terms of the coupling coefficient:

$$L_1 - M = 0 \Rightarrow L_1 - k\sqrt{L_1 L_2} = 0 \Rightarrow k = \sqrt{\frac{L_1}{L_2}} < 1 \qquad (8.110)$$

This result states that if we want to null the ripple current in L_2, we must make L_1 smaller than L_2 so that we can adjust k to be equal to $\sqrt{L_1/L_2}$. The coupling coefficient can be set by adjusting the air gap between the two cores. Similarly, we see from Eq. (8.109) that the ripple current in L_1 can be reduced to zero if we set

$k = \sqrt{L_2/L_1}$, in which case L_2 has to be made smaller than L_1.

In an isolated Cuk converter,[3] it is possible to reduce the ripple currents in both inductors to zero by coupling them to the isolation transformer. In reality, the ripple current is not exactly reduced to zero but is highly reduced. The reason for this is that the voltage across the inductors are not exactly identical because of the ripple voltage across the capacitors.

8.7   The PWM switch and its invariant terminal characteristics

The active and passive switches in the four converters discussed earlier can be lumped together in a single-pole–double-throw switch called the PWM switch,[4] as shown in Fig. 8.35. The terminals designations a, p and c refer to *active, passive* and *common*, respectively. The common terminal is designated as such simply because it is common to both switches. All of the four converters discussed earlier are redrawn in Fig. 8.36 with the PWM switch identified as a three-terminal switching device. The important thing to see in this figure is that all the elements outside the PWM switch are linear passive elements which provide filtering, whereas the PWM switch is the only nonlinear element which performs the dc-to-dc conversion process. We shall capitalize on this point and show that the conversion process in the PWM switch is independent of the particular converter in which it occurs and can be described by a set of invariant equations. This invariance will lead to a very simple model of the PWM switch which can be used towards the determination of the dynamics of any (two switched-state) PWM converter.

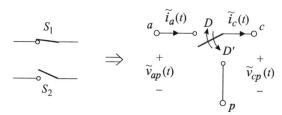

Figure 8.35

We begin with a qualitative study of the invariance of the terminal voltages and currents of the PWM switch. The active terminal current is the current in the active switch S_1 which has the invariant shape shown in Fig. 8.37a. By invariance we simply mean that one cannot tell which converter the active switch is in by looking at its current waveform. The common terminal current is the same as the *total switched inductive* current in the converter and has the invariant shape shown in Fig. 8.37b. For example in the Cuk converter, the current in the common terminal is the sum of the currents in the two switched inductors L_1 and L_2. The inductor in the *input* filter of a buck converter is an example of an inductor which is

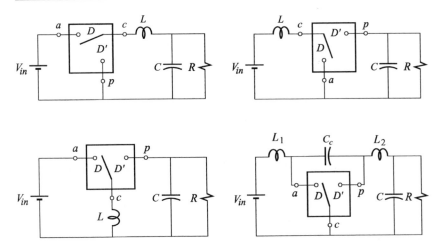

Figure 8.36

not switched. Hence, in the buck converter with an input filter, the common terminal current is the same as the current in the output filter inductor only and not the sum of the currents in the input and output filter inductors. Next we examine the port voltages of the PWM switch. Ignoring ripple voltages, the voltage across port *ap* in all the converters is a dc voltage while the voltage across port *cp* is a pulsating voltage, as shown in Figs. 8.37c and d. We can now write the following set of invariant equations for the terminal currents and port voltages of the PWM switch:

$$\tilde{i}_a(t) = \tilde{d}_s(t)\tilde{i}_c(t) \left.\begin{array}{l} \\ \\ \end{array}\right\} \qquad (8.111a, b)$$
$$\tilde{v}_{cp}(t) = \tilde{d}_s(t)\tilde{v}_{ap}(t)$$

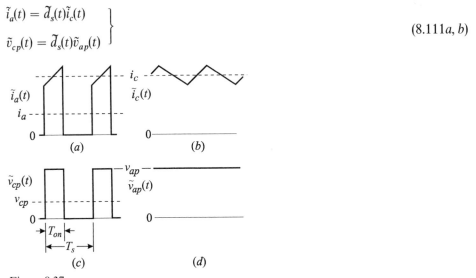

Figure 8.37

in which $\tilde{d}_s(t)$ is the switching function defined in Eq. (8.16). These equations describe the entire switching action responsible for the dc-to-dc conversion process in a converter. By taking the time average of Eqs. (8.111a,b) we obtain the following invariant average equations for the PWM switch:

$$i_a = di_c \left.\begin{matrix} \\ \\ \end{matrix}\right\}$$
$$v_{cp} = dv_{ap}$$

(8.112a, b)

These are the invariant equations that describe the dc-to-dc conversion function of any two switched-state PWM converter in continuous conduction mode.

If we allow for small-signal variations in the duty-ratio function and other voltages and currents in the converter about a steady-state operating point, then the propagation of these variations through the PWM switch can be determined by taking the differentials in Eqs. (8.112a, b):

$$\hat{i}_a = D\hat{i}_c + I_c\hat{d} \left.\begin{matrix} \\ \\ \end{matrix}\right\}$$
$$\hat{v}_{cp} = D\hat{v}_{ap} + V_{ap}\hat{d}$$

(8.113a, b)

in which D, I_c and V_{ap} are the steady-state dc operating points of the PWM switch which satisfy Eqs. (8.112a, b), i.e. $I_a = DI_c$ and $V_{cp} = DV_{ap}$.

8.8   Average large-signal and small-signal equivalent circuit models of the PWM switch

An equivalent circuit model[4] for Eqs. (8.112a, b) can most easily be constructed using dependent sources, as shown in Fig. 8.38. This is a nonlinear average model because it involves products of the time functions $d(t)i_c(t)$ and $d(t)v_{ap}(t)$. This is also a large-signal model because it places no restriction on the magnitude of variations in the time functions. For small-signal variations, we can use the linearized small-signal equations in (8.113a, b) to construct the equivalent circuit model in Fig. 8.39a or b. In Fig. 8.39b the dependent sources $D\hat{i}_c$ and $D\hat{v}_{cp}$ have been replaced with a $1:D$ transformer and the control source $\hat{d}V_{ap}$ has been moved from the common terminal side to the active terminal side. Note that the small-signal sources are evaluated at the dc operating point. Under steady-state conditions, the large- and small-signal models reduce to the same transformer model shown in Fig. 8.40.

The model of the PWM switch is used very much in the same way as the

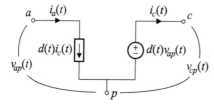

Figure 8.38

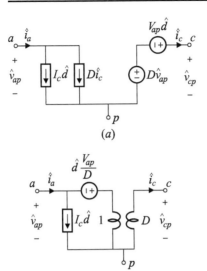

(a)

(b)

Figure 8.39

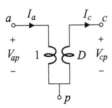

Figure 8.40

equivalent circuit model of a transistor or a vacuum tube. First, the device is replaced point-by-point with its equivalent circuit model and a dc analysis is carried out to determine the operating point (D, I_c, V_{ap}). As usual, in a dc analysis all reactive elements and small-signal sources vanish. Second, the small-signal analysis is carried out using the small-signal model of the PWM switch evaluated at the dc operating point. As usual, in an ac analysis, all dc sources vanish. In a small-signal analysis, one of the most commonly determined transfer functions is the control-to-output transfer function:

$$H_d(s) = \frac{\hat{v}_o(s)}{\hat{d}(s)} \tag{8.114}$$

This transfer function is necessary for the design of a stable feedback loop for a converter whose output voltage is regulated. Other transfer functions of interest are the line-to-output transfer function and the input and output impedances. The line-to-output transfer function relates variations in the output voltage to variations in the input voltage:

$$H_v(s) = \frac{\hat{v}_o(s)}{\hat{v}_{in}(s)} \tag{8.115}$$

Ideally, for a regulating converter $H_v(s)$ must be zero because the input voltage is simply a disturbance in the closed-loop system which the feedback circuit must attenuate and prevent from reaching the output.

Example 8.10 For the buck-boost converter in Fig. 8.41, determine:

(*i*) The voltage conversion ratio.
(*ii*) The control-to-output transfer function.
(*iii*) The line-to-output transfer function.
(*iv*) The input admittance.
(*v*) The output impedance.

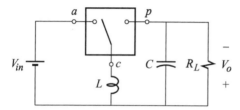

Figure 8.41

The complete dc and small-signal equivalent circuit model of the buck-boost converter is obtained by replacing the PWM switch with its equivalent circuit model as shown in Fig. 8.42. To determine the conversion ratio, we perform a dc analysis by setting all the small-signal sources to zero, shorting the inductors and opening all the capacitors, as shown in Fig. 8.43.

In Fig. 8.43, we can see that the voltage across port ap is simply V_o/D so that writing KVL around the outer loop we obtain:

$$V_{in} = \frac{V_o}{D} - V_o = V_o \frac{1-D}{D} = V_o \frac{D'}{D} \tag{8.116}$$

The voltage conversion ratio follows immediately:

$$M_V = \frac{V_o}{V_{in}} = \frac{D}{D'} \tag{8.117}$$

Hence, we see how the invariant conversion function $1:D$ of the PWM switch can produce the conversion ratio D/D' of the buck-boost converter by a simple rotation of the PWM switch. Before going to the small-signal analysis, we should determine the dc operating point of the PWM switch. The quiescent common-terminal current is given by:

$$I_c = I_{in} + I_o = I_o\left(1 + \frac{1}{M_I}\right) = I_o\left(1 + \frac{D}{D'}\right) = \frac{I_o}{D'} \tag{8.118}$$

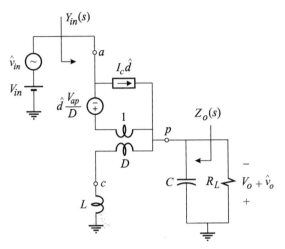

Figure 8.42

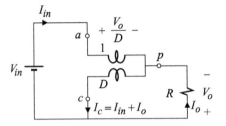

Figure 8.43

This establishes I_c in terms of the output current of the converter, which can be expressed in several different convenient forms using the results of the dc analysis:

$$I_c = \frac{I_o}{D'} = \frac{V_o/R}{D'} = \frac{V_{in}}{R}\frac{D}{D'^2} \tag{8.119}$$

The quiescent port voltage V_{ap} is readily seen to be V_o/D, which can be expressed in terms of the input voltage:

$$V_{ap} = \frac{V_o}{D} = \frac{V_{in}}{D'} \tag{8.120}$$

All of the small-signal transfer functions have the same denominator which can be determined by setting all the independent excitations in Fig. 8.42 to zero as shown in Fig. 8.44a. The 2-EET can now be applied by taking out the inductor and the capacitor as shown in the reference circuit in Fig. 8.44b. The following port resistances required for the 2-EET are determined in reference to Fig. 8.45:

$$\left.\begin{array}{l} R^{(1)} = D'^2 R \\[2mm] R^{(2)} = 0 \\[2mm] R^{(2)}_{(1)} = R \end{array}\right\} \tag{8.121a–c}$$

Hence, the denominator is given by:

$$D(s) = 1 + \frac{sL}{R^{(1)}} + sCR^{(2)} + \frac{sL}{R^{(1)}} sCR^{(2)}_{(1)}$$

$$= 1 + s\frac{L}{D'^2 R} + s^2\frac{LC}{D'^2} \tag{8.122}$$

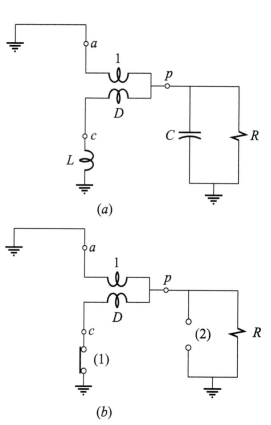

(a)

(b)

Figure 8.44

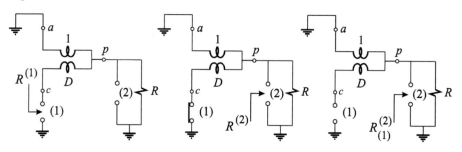

Figure 8.45

The line-to-output transfer function and the input admittance are determined by retaining only the small-signal source $\hat{v}_{in}$ in the complete equivalent circuit model in Fig. 8.42 and examining the transform circuit for nulls in the response of the particular transfer function. For the line-to-output transfer function we see in Fig. 8.46 that there are no conditions in the transform circuit which result in a null in the response $\hat{v}_o(s)$ so that the line-to-output transfer function is given by:

$$H_l(s) \equiv \frac{\hat{v}_o(s)}{\hat{v}_{in}(s)} = M_V \frac{1}{D(s)} = \frac{D}{D'} \frac{1}{1 + s\dfrac{L}{D'^2 R} + s^2 \dfrac{LC}{D'^2}} \tag{8.123}$$

in which M_V is clearly the low-frequency asymptote of $H_l(s)$.

The input admittance function is given by:

$$G_{in}(s) \equiv \frac{\hat{i}_{in}(s)}{\hat{v}_{in}(s)} = G_{in0} \frac{N_{in}(s)}{D(s)} \tag{8.124}$$

in which G_{in0} is the low-frequency asymptote of the input conductance discussed earlier and shown in Fig. 8.3. Hence we have:

$$G_{in0} = \frac{M_V^2}{R} = \left(\frac{D}{D'}\right)^2 \frac{1}{R} \tag{8.125}$$

The numerator $N_{in}(s)$ corresponds to conditions of the transform circuit that results in a null in the response $\hat{i}_{in}(s)$. Referring to Fig. 8.46, we see that a null in either $\hat{i}_c(s)$ or $\hat{i}_p(s)$ would result in a null in $\hat{i}_{in}(s)$ simply because when one of the terminal currents of the $1:D$ transformer vanishes, the other two vanish as well. The impedance encountered by $\hat{i}_p(s)$ is:

$$\frac{R}{1 + sCR} \tag{8.126}$$

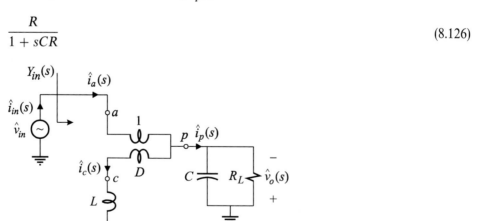

Figure 8.46

The pole at $s_p = -1/RC$ corresponds to an "open-circuit" in the transform domain, which results in a null in $\hat{i}_p(s)$ and, hence, in $\hat{i}_{in}(s)$. It follows that $s_p = -1/RC$ is a zero of $G_{in}(s)$ and that $N_{in}(s) = 1 + sCR$. Hence, the input admittance is given by:

$$G_{in}(s) = \left(\frac{D}{D'}\right)^2 \frac{1}{R} \frac{1 + sCR}{1 + s\dfrac{L}{D'^2 R} + s^2\dfrac{LC}{D'^2}} \tag{8.127}$$

The impedance encountered by $\hat{i}_c(s)$ is sL, which has no poles and hence does not contribute a zero to $\hat{i}_{in}(s)$.

The output impedance is determined from the equivalent circuit shown in Fig. 8.47, whence we see that a zero of the impedance in the common terminal branch would cause the terminals a and c to be at the same potential, which in turn would require that the voltages on both sides of the $1:D$ transformer be the same, which could only happen if $\hat{v}_o(s) = 0$. The impedance connected to terminal c is sL, which has a zero at the origin, so that the output impedance must have a zero at the origin too. Hence we have:

$$Z_o(s) \equiv \frac{\hat{v}_o(s)}{\hat{i}_T(s)} = \frac{sLk}{D(s)} \tag{8.128}$$

in which k is a constant, which can be determined by examining the circuit. It can be seen from Fig. 8.47 that at high frequencies we have:

$$\lim_{s \to \infty} Z_o(s) \to \frac{1}{sC} \tag{8.129}$$

Substituting this result in Eq. (8.128), we obtain $k = 1/D'^2$, and the output impedance can be written as:

$$Z_o(s) = \frac{sL/D'^2}{1 + s\dfrac{L}{D'^2 R} + s^2\dfrac{LC}{D'^2}} \tag{8.130}$$

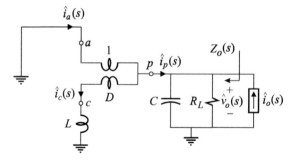

Figure 8.47

Finally, we determine the control-to-control transfer function by retaining the control sources as shown in Fig. 8.48. This transfer function can be written as:

$$H_d(s) \equiv \frac{\hat{v}_o(s)}{\hat{d}(s)} = H_{do} \frac{N_d(s)}{D(s)} \tag{8.131}$$

The low-frequency asymptote, H_{do}, is simply the derivative of the output voltage with respect to the duty ratio and is given by:

$$H_{do} \equiv \frac{dV_o}{dD} = V_g \frac{D' - (-D)}{D'^2} = \frac{V_g}{D'^2} \tag{8.132}$$

The numerator, $N_d(s)$, is determined by examining the transform circuit for a null in the output voltage. This is shown in Fig. 8.49 in which we see that the only way to have a null in $\hat{v}_o(s)$ is to have a null in $\hat{i}_p(s)$. With $\hat{v}_o(s) = \hat{i}_p(s) = 0$, from Fig. 8.49 we have:

$$\left. \begin{array}{l} \hat{i}_a(s) = \hat{i}_c(s) \\[6pt] \hat{v}_{ap}(s) = 0 \end{array} \right\} \tag{8.133a, b}$$

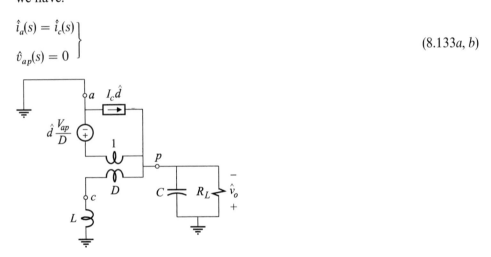

Figure 8.48

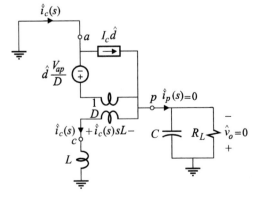

Figure 8.49

Hence:

$$D\hat{i}_c + I_c\hat{d} = \hat{i}_c$$

$$\frac{\hat{i}_c(s)sL}{D} = \frac{V_{ap}}{D}\hat{d}$$

$$\left.\right\} \qquad (8.134a, b)$$

Solving these we obtain:

$$1 - s\frac{L}{D'}\frac{I_c}{V_{ap}} = 0 \Rightarrow N_d(s) = 1 - s\frac{L}{D'}\frac{I_c}{V_{ap}} \qquad (8.135)$$

Substituting for I_c and V_{ap} from the operating point determined earlier in Eqs. (8.119) and (8.120), we obtain:

$$N_d(s) = 1 - s\frac{L}{D'}\frac{I_o}{D'}\frac{D}{V_o}$$

$$= 1 - s\frac{L}{R}\frac{D}{D'^2} \qquad (8.136a, b)$$

Hence, the control-to-output transfer function has a RHP zero and is given by:

$$H_d(s) = \frac{V_g}{D'^2}\frac{1 - s\dfrac{L}{R}\dfrac{D}{D'^2}}{1 + s\dfrac{L}{D'^2R} + s^2\dfrac{LC}{D'^2}} \qquad (8.137)$$

The presence of the RHP zero in $H_d(s)$ above implies that the output voltage momentarily dips before rising when the duty cycle is increased abruptly (step function). The physical explanation for this is that an abrupt increase in the duty cycle d causes an abrupt decrease in d', which in turn causes an initial decrease in the average terminal current $i_p(t)$ because the average inductor current cannot reach its higher steady-state value instantaneously. It is this momentary decrease in $i_p(t)$ that causes the initial dip in the output voltage. In fact, we can use this physical explanation to determine the RHP zero as follows. Let us assume that the RHP zero in Eq. (8.137) is ω_a so that its high-frequency response is given by:

$$H_d(s) \approx \frac{V_g}{D'^2}\frac{-s/\omega_a}{s^2\dfrac{LC}{D'^2}} = -\frac{V_g}{s\omega_a LC} \qquad (8.138)$$

The initial response to a step increase, Δ_d, in the duty cycle is then given by:

$$\hat{v}_o(s) \approx -\hat{d}(s)\frac{V_g}{s\omega_a LC} = -\frac{\Delta_d}{s}\frac{V_g}{s\omega_a LC} \qquad (8.139)$$

Taking the inverse Laplace transform, we have:

$$\hat{v}_o(t) \approx -\frac{\Delta_d V_g}{\omega_a LC} t \qquad (8.140)$$

Next, we look at the circuit to determine the initial response to a sudden increase in the duty cycle by an amount Δ_d. We can see from Fig. 8.50 that the passive terminal current $\tilde{i}_p(t)$ for the first few cycles experiences a *step* reduction in its *average* value by an amount:

$$\delta i_p = -\Delta_d \frac{I_o}{D'} \qquad (8.141)$$

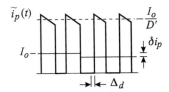

Figure 8.50

This sudden reduction in the average current is absorbed by the output capacitor so that the change in the output voltage is given by:

$$C\frac{dv_o}{dt} = -\Delta_d \frac{I_o}{D'} \Rightarrow \hat{v}_o(t) = -\frac{\Delta_d}{C}\frac{I_o}{D'} t \qquad (8.142)$$

Comparing Eqs. (8.140) and (8.142), we determine the RHP zero:

$$\omega_a = \frac{D' V_g}{I_o L} = \frac{D'^2 V_o}{D I_o L} = \frac{D'^2 R}{DL} \qquad (8.143)$$

which is the same as the RHP in Eq. (8.137). $\qquad\qquad\square$

8.9   The PWM switch in other converter topologies

The three different types of conversion ratios we have seen so far, D, $1/D'$ and D/D', were generated by the three possible orientations of the PWM switch. It is possible to generate other types of conversion ratios by arranging the switches differently, using tapped inductors or inverting transformers, and using more switches. In all such converters, the PWM switch can be identified after a few simple circuit manipulations. In this section, we shall consider four such converters which cover most known types of basic topological variations.

Any of the basic converters discussed earlier can be modified by tapping the inductor and connecting one of the two switches to the tap point. The two variations of the tapped buck converter are shown in Figs. 8.51a, b.

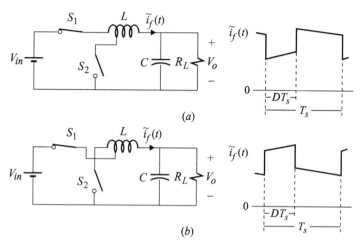

Figure 8.51

The PWM switch cannot be identified directly in these converters because the common terminal, through which the inductive current flows, is lost. Since S_1 and V_g are in series, one may interchange their positions and establish a common point between S_1 and S_2 and *incorrectly* identify the PWM switch. The mistake here is that the current through this common point is $\tilde{i}_f(t)$, which is *not* a purely inductive current and has the shape shown in Figs. 8.51a,b. Immediately we recognize that the tapped inductor in these converters is actually acting as an auto-transformer so that the current i_f is actually a winding current rather than a magnetizing or inductive current. In order to recover the inductive current, we replace the tapped inductor with an untapped inductor L and an ideal transformer with the same turn-ratio as the tapped inductor as shown in Fig. 8.52a. The energy stored in L is the inductive energy of the converter and the current in L is the inductive current in the converter. The PWM switch can now be identified by moving S_2 to the opposite side of the ideal transformer as shown in Fig. 8.52b. Now, the common terminal between S_1 and S_2' clearly carries a purely inductive current. The dc and small-signal characteristics can be determined by replacing the PWM switch with its equivalent circuit model as explained in the following example.

Example 8.11   The voltage conversion ratio of the tapped buck converter is determined by replacing the PWM switch with its dc model as shown in Fig. 8.53. By applying KVL around the loop shown, we obtain:

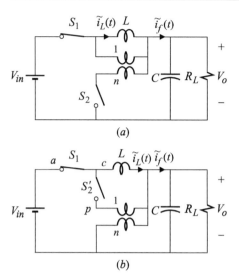

(a)

(b)

Figure 8.52

$$V_{in} = \frac{V_o}{nD} - \frac{V_o}{n} + V_o \tag{8.144}$$

from which the voltage conversion ratio is obtained:

$$M_V = \frac{1}{1 + \dfrac{D'}{nD}} \tag{8.145}$$

The operating point of the PWM switch is given by:

$$\left.\begin{aligned} V_{ap} &= \frac{V_o}{nD} = \frac{V_{in}}{nD + D'} \\[2mm] I_c &= \frac{I_{in}}{D} = \frac{M_V I_o}{D} \end{aligned}\right\} \tag{8.146a, b}$$

The advantage of tapping the inductor can be immediately seen from Eq. (8.145) in which we see that by letting $n < 1$ we can obtain large step-down conversion

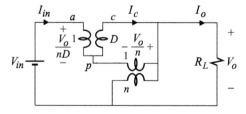

Figure 8.53

ratios without making the duty cycle too small. For example, we can convert 120 to 5 V by making $D = 0.25$ and $n = 3/23$.

The control-to-output transfer is obtained by replacing the PWM switch with its small-signal equivalent circuit model as shown in Fig. 8.54, in which the input dc voltage is set to signal ground. This transfer function is given (see Problem 8.12) by:

$$H_d(s) = H_{d0} \frac{1 + \dfrac{s}{\omega_z}}{1 + \dfrac{s}{\omega_o Q} + \dfrac{s^2}{\omega_o^2}}$$ (8.147)

in which:

$$\left.\begin{aligned}
H_{d0} &= \frac{V_g}{(nD + D')^2} \\[2ex]
\omega_z &= \frac{DR}{M_V^2 L}\frac{1}{n-1} \\[2ex]
\omega_o &= \frac{D}{M_V}\frac{1}{\sqrt{LC}} \\[2ex]
Q &= \frac{R}{\omega_o L}\left(\frac{D}{M_V}\right)^2
\end{aligned}\right\}$$ (8.148)

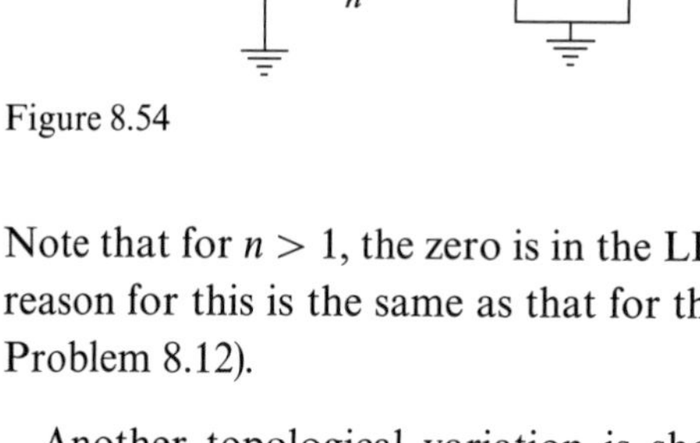

Figure 8.54

Note that for $n > 1$, the zero is in the LHP, while for $n < 1$, zero is the RHP. The reason for this is the same as that for the buck-boost converter given earlier (see Problem 8.12). □

Another topological variation is shown in Fig. 8.55, in which an inverting transformer is used. This converter, sometimes referred to as the Watkins–Johnson[5] converter, is analyzed in the same way as the tapped buck converter. This is shown in Fig. 8.56, in which the inverting transformer is replaced with its equivalent circuit model and S_2 is moved to the N_1-side of the transformer in order to identify the PWM switch.

The complete equivalent circuit is the same as that of the tapped buck converter except for the $1:n$ transformer, which in this case is inverting. It follows that the voltage conversion ratio is the same as that of the tapped buck converter given in Eq. (8.145) with n replaced by $-n$:

$$M_V = \frac{1}{1 - \dfrac{D'}{nD}} \tag{8.149}$$

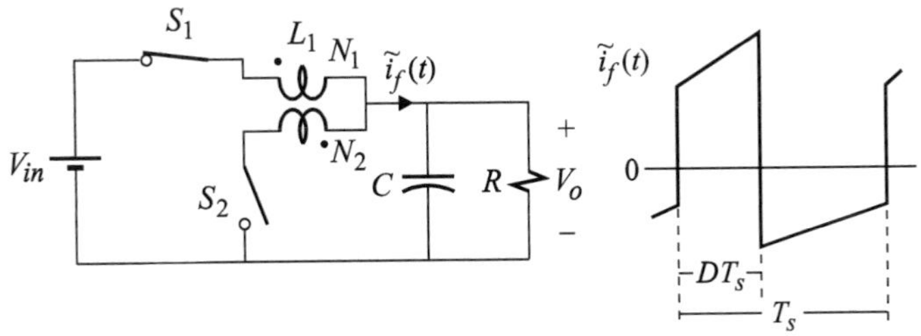

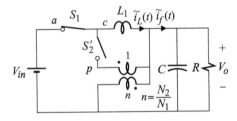

Figure 8.55

Figure 8.56

A plot of M_V for $n = 1$ is shown in Fig. 8.57, in which we notice that not only the conversion ratio becomes singular at $D = 0.5$ but also the polarity of the output voltage changes. It is relatively easy to show that M_V becomes zero, and not infinite, at $D = 0.5$, if the slightest parasitic resistance is included in the circuit (see Problem 8.13). By interchanging the source and load, another variation of this converter can be obtained, as shown in Fig. 8.58, whose conversion ratio monotonically increases from negative to positive values passing through zero at $D = 0.5$ (see Problem 8.14).

By rearranging the switches in a fourth-order converter, such as the Cuk converter, one can obtain new converters such as the one shown in Fig. 8.59a. In such a converter, the PWM switch cannot be identified immediately, but a simple circuit transformation can do the trick. The idea, of course, is to move one of the switches without altering the operation of the circuit. This is shown in Fig. 8.59b in

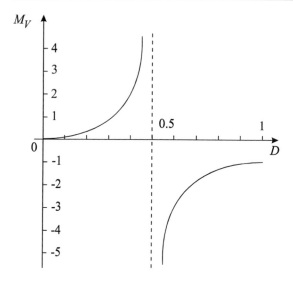

Figure 8.57

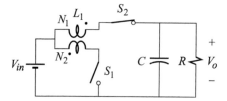

Figure 8.58

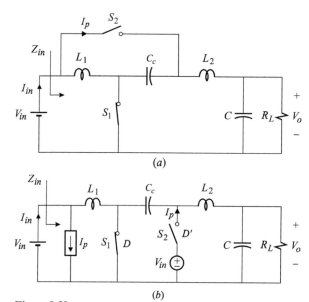

Figure 8.59

which S_2 is lifted from the input side and brought to its new position while maintaining the same potential difference across it with the help of the dependent source V_{in}. To preserve the input current, the dependent current source I_p is introduced. The PWM switch is identified in Fig. 8.59c simply by interchanging the positions of S_2 and v_{in}. The voltage conversion ratio is determined by replacing the PWM switch with its dc model as shown in Fig. 8.60 in which, going around the outer loop, we have:

$$V_{in} + \frac{V_o - V_{in}}{D} + V_{in} = V_o \tag{8.150}$$

It follows that the conversion ratio is given by:

$$M_V = \frac{1 - 2D}{D'} \tag{8.151}$$

A plot of M_V is shown in Fig. 8.61, which shows that the polarity of the output voltage changes as the duty cycle is increased beyond 0.5.

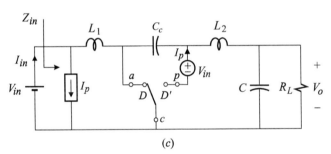

(c)

Figure 8.59 (cont.)

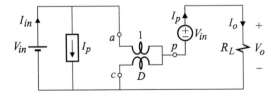

Figure 8.60

It is possible to cascade any two basic converters to obtain an ideal conversion ratio which has a quadratic dependence on the duty cycle. Furthermore, it is possible to rearrange the switches in a cascaded converter in such a way to drive only a *single* active switch instead of two. The formal synthesis procedure of such converters has been reported by Maksimovich and Cuk[6] and one such converter is shown in Fig. 8.62. In this converter, D_1 turns on in synchronism with the main active switch S_1 during T_{on}, while D_2 and D_3 are turned off. During T_{off}, D_1 turns

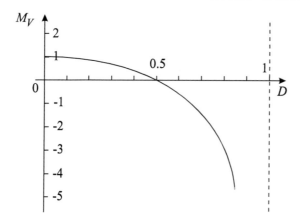

Figure 8.61

off in synchronism with S_1, while D_2 and D_3 are turned on together. For purposes of analysis in continuous conduction mode, the nature of these switches is immaterial. Our aim is to analyze this converter using the model of the PWM switch, which clearly cannot be identified directly. Since the PWM switch is applicable to all converters that are switched between two states, all we need to know are the two states of the converter during T_{on} and T_{off}. These are shown in Fig. 8.63a. In

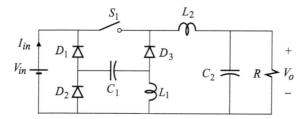

Figure 8.62

Fig. 8.63b, the on-state is redrawn by separating the L_1 and L_2 loops using the dependent sources V_{in} and I_{a1}. The purpose of I_{a1} in shunt with V_{in} is to preserve the input current I_{in}. The off-state is redrawn simply by interchanging the positions of L_1 and C_1. We can now see that the on- and off-states in Fig. 8.63b correspond to the operation of two buck converters connected in cascode as shown in Fig. 8.63c. The PWM switch in each converter is clearly identified and the complete dc and small-signal equivalent circuit model is shown in Fig. 8.64.

Example 8.12   To determine the voltage conversion ratio of the converter in Fig. 8.62 we set all the small-signal sources to zero, short the inductors, and open the capacitors in the equivalent circuit in Fig. 8.64. This yields the circuit in Fig. 8.65, whence we have for the first PWM switch:

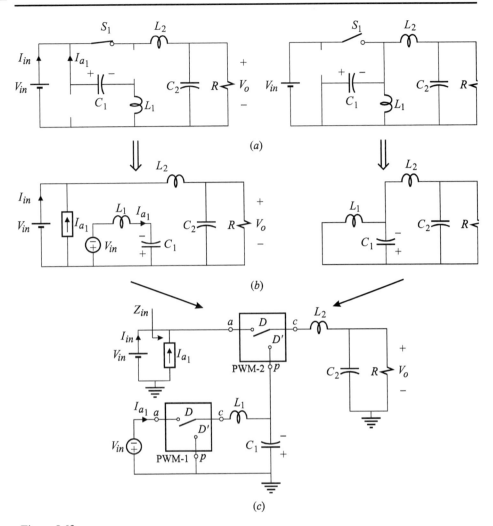

Figure 8.63

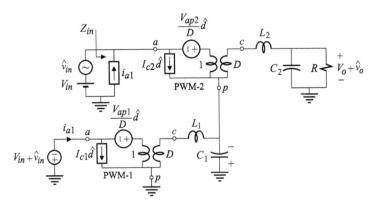

Figure 8.64

$$V_{cp1} = -DV_{in} \tag{8.152}$$

For the second PWM switch we have:

$$V_{ap2} = V_{a2} - V_{p2} = V_{in} - (-DV_{in}) = V_{in}(1 + D) \tag{8.153}$$

The output voltage is given by:

$$V_o = V_{cp2} + V_{cp1} = DV_{ap2} + V_{cp1} \tag{8.154}$$

Substituting Eqs. (8.152) and (8.153) in (8.154), we obtain:

$$V_o = DV_{in}(1 + D) - DV_{in} = D^2 V_{in} \tag{8.155}$$

Hence, the voltage conversion ratio is given by:

$$M_V = D^2 \tag{8.156}$$

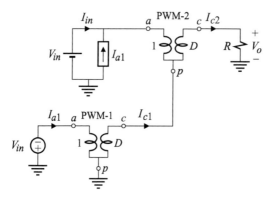

Figure 8.65

The dc operating point of each PWM switch is determined before performing the small-signal analysis. For the first and second PWM switches we have:

$$\left.\begin{aligned}
V_{ap1} &= V_{in} \\
I_{c1} &= I_{c2} - DI_{c2} = D'I_{c2} \\
V_{ap2} &= V_{in}(1 + D) \\
I_{c2} &= \frac{V_o}{R} = \frac{D^2 V_{in}}{R}
\end{aligned}\right\} \tag{8.157a–d}$$

The control-to-output transfer function has the following form:

$$H_d(s) = \frac{\hat{v}_o(s)}{\hat{d}(s)} = H_{d0} \frac{N_d(s)}{D(s)} \tag{8.158}$$

in which H_{d0} is the low-frequency asymptote and is given by:

$$H_{d0} = \frac{dV_o}{dD} = 2DV_{in} \tag{8.159}$$

The denominator is obtained by the application of the 4-EET to the reference circuit in Fig. 8.66, whence we have by inspection:

$$\left.\begin{array}{l} R^{(1)} = R^{(2)} = 0 \\ R^{(3)} = R \end{array}\right\} \tag{8.160a, b}$$

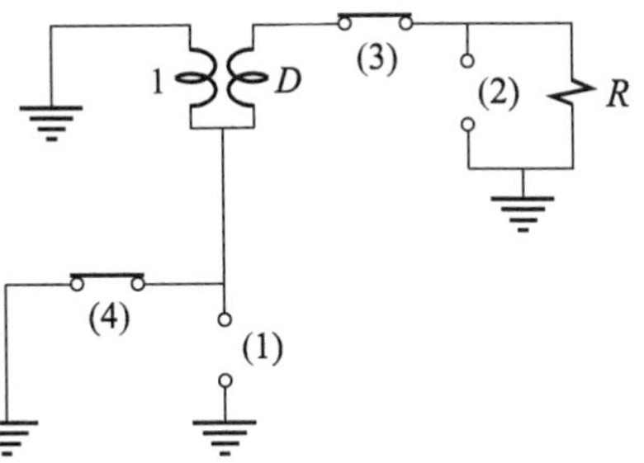

Figure 8.66

From Fig. 8.67 we have:

$$R^{(4)} = \frac{R}{D'^2} \tag{8.160c}$$

Since L_1 is connected across port (4) and L_2 is connected across port (3), we shall renumber them after their port numbers to make the writing of the terms in the 4-EET easier. Hence, we shall temporarily reassign L_1 and L_2:

$$L_1 \to L_4; L_2 \to L_3$$

The fourth-order denominator is given by:

$$D(s) = 1 + a_1 s + a_2 s^2 + a_3 s^3 + a_4 s^4 \tag{8.161}$$

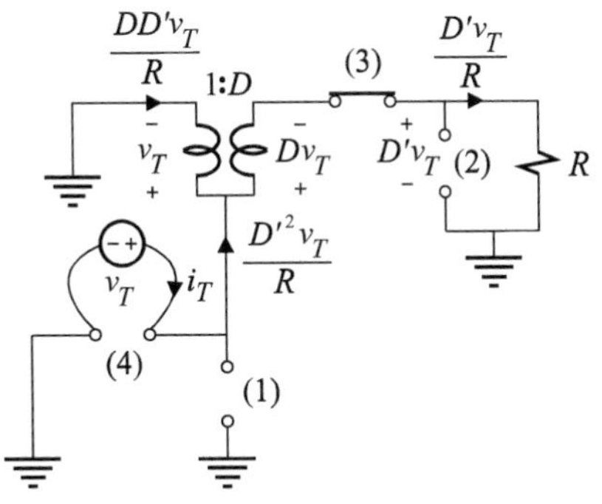

Figure 8.67

Substituting for $R^{(i)}$ in the coefficient a_1, we obtain:

$$a_1 = R^{(1)}C_1 + R^{(2)}C_2 + \frac{L_3}{R^{(3)}} + \frac{L_4}{R^{(4)}}$$

$$= \frac{L_3 + D'^2 L_4}{R} \tag{8.162}$$

Since the reference circuit is almost purely reactive, we should expect indeterminate forms in the higher-order coefficients. We shall remove these indeterminate forms by changing the order of the ports. The coefficient a_2 is given by:

$$a_2 = R^{(1)}C_1 R_{(1)}^{(2)}C_2 + R^{(1)}C_1 \frac{L_3}{R_{(1)}^{(3)}} + R^{(1)}C_1 \frac{L_4}{R_{(1)}^{(4)}}$$

$$+ R^{(2)}C_2 \frac{L_3}{R_{(2)}^{(3)}} + R^{(2)}C_2 \frac{L_4}{R_{(2)}^{(4)}} + \frac{L_3}{R^{(3)}}\frac{L_4}{R_{(3)}^{(4)}} \tag{8.163}$$

By inspection of Fig. 8.66, we have:

$$\left. \begin{array}{l} R_{(1)}^{(2)} = R_{(1)}^{(4)} = R_{(2)}^{(3)} = R_{(2)}^{(4)} = 0 \\[2mm] R_{(3)}^{(4)} = \infty \\[2mm] R_{(1)}^{(3)} = R \end{array} \right\} \tag{8.164a–c}$$

It follows that the first two terms and the last term in the expression of a_2 are all zero and the remaining terms in the middle have an indeterminacy of the form 0/0, which can easily be removed by changing the order in which the ports are taken:

$$a_2 = \frac{L_4}{R^{(4)}} R_{(4)}^{(1)}C_1 + \frac{L_3}{R^{(3)}} R_{(3)}^{(2)}C_2 + \frac{L_4}{R^{(4)}} R_{(4)}^{(2)}C_2 \tag{8.165}$$

Once again by inspecting Fig. 8.66, we have:

$$\left. \begin{array}{l} R_{(4)}^{(1)} = R^{(4)} \\[2mm] R_{(3)}^{(2)} = R \\[2mm] R_{(4)}^{(2)} = R \end{array} \right\} \tag{8.166a–c}$$

Substituting these in Eq. (8.165), we obtain:

$$a_2 = L_4 C_1 + (L_3 + D'^2 L_4)C_2 \tag{8.167}$$

The coefficient a_3 is given by:

$$a_3 = R^{(1)}C_1 R_{(1)}^{(2)}C_2 \frac{L_3}{R_{(1,2)}^{(3)}} + R^{(1)}C_1 R_{(1)}^{(2)}C_2 \frac{L_4}{R_{(1,2)}^{(4)}}$$

$$+ R^{(1)}C_1 \frac{L_3}{R_{(1)}^{(3)}} \frac{L_4}{R_{(1,3)}^{(4)}} + R^{(2)}C_1 \frac{L_3}{R_{(2)}^{(3)}} \frac{L_4}{R_{(2,3)}^{(4)}} \tag{8.168}$$

in which we can determine the following by inspection:

$$R_{(1,2)}^{(3)} = R_{(1,2)}^{(4)} = R_{(1,3)}^{(4)} = 0$$

$$R_{(2,3)}^{(4)} = \infty \tag{8.169}$$

The indeterminacy in the first term of a_3 is removed by changing the order of the ports from 1, 2, 3 to 1, 3, 2. Using the fact that $R_{(3,1)}^{(2)} = 0$, we obtain:

$$R^{(1)}C_1 R_{(1)}^{(2)} C_2 \frac{L_3}{R_{(1,2)}^{(3)}} = R^{(1)}C_1 \frac{L_3}{R_{(1)}^{(3)}} R_{(3,1)}^{(2)} C_2 = 0 \tag{8.170}$$

The indeterminacy in the second term of a_3 is removed by changing the order of the ports from 1, 2, 4 to 4, 1, 2. Using the fact that $R_{(4,1)}^{(2)} = 0$:

$$R^{(1)}C_1 R_{(1)}^{(2)} C_2 \frac{L_4}{R_{(1,2)}^{(4)}} = \frac{L_4}{R^{(4)}} R_{(4)}^{(1)} C_1 R_{(4,1)}^{(2)} C_2 = 0 \tag{8.171}$$

In the third term of a_3, the indeterminacy is removed by changing the order from 1, 3, 4 to 4, 1, 3. Using the fact that $R_{(4,1)}^{(3)} = R$, we obtain:

$$R^{(1)}C_1 \frac{L_3}{R_{(1)}^{(3)}} \frac{L_4}{R_{(1,3)}^{(4)}} = \frac{L_4}{R^{(4)}} R_{(4)}^{(1)} C_1 \frac{L_3}{R_{(4,1)}^{(3)}} = L_4 C_1 \frac{L_3}{R} \tag{8.172}$$

In the last term of a_3, the indeterminacy is removed by changing the order from 2, 3, 4 to 3, 2, 4. Using the fact that $R_{(3,2)}^{(4)} = \infty$:

$$R^{(2)}C_1 \frac{L_3}{R_{(2)}^{(3)}} \frac{L_4}{R_{(2,3)}^{(4)}} = \frac{L_3}{R^{(3)}} R_{(3)}^{(2)} C_1 \frac{L_4}{R_{(2,3)}^{(4)}} = 0 \tag{8.173}$$

Hence, a_3 is given by:

$$a_3 = L_4 C_1 \frac{L_3}{R} \tag{8.174}$$

Finally, we obtain a_4 by continuing the order in a_3:

$$a_4 = \frac{L_4}{R^{(4)}} R_{(4)}^{(1)} C_1 \frac{L_3}{R_{(4,1)}^{(3)}} R_{(4,1,3)}^{(2)} C_2 = L_4 C_1 \frac{L_3}{R} R C_2 \tag{8.175}$$

Substituting for a_i in Eq. (8.161) and reverting the notation to L_1 and L_2 from L_4 and L_3, we obtain the denominator:

$$D(s) = 1 + s \frac{L_2 + D'^2 L_1}{R} + s^2 [L_2 C_2 + (C_1 + D'^2 C_2)L_1]$$

$$+ s^3 L_1 C_1 \frac{L_2}{R} + s^4 L_1 L_2 C_1 C_2 \tag{8.176}$$

This denominator can be factored into the product of two quadratics:

$$D(s) = \left(1 + \frac{s}{\omega_{o1} Q_{o1}} + \frac{s^2}{\omega_{o1}^2}\right)\left(1 + \frac{s}{\omega_{o2} Q_{o2}} + \frac{s^2}{\omega_{o2}^2}\right) \tag{8.177}$$

whose frequencies and damping factors are approximately given by:

$$\omega_{o1} \approx \frac{1}{\sqrt{L_2 C_2 + (C_1 + D'^2 C_2) L_1}}$$

$$Q_{o1} \approx \frac{R}{\omega_{o1}(L_2 + D'^2 L_1)}$$

$$\omega_{o2} \approx \sqrt{\frac{1}{L_1 C_1} + \frac{1}{L_2 \left(C_2 \| \dfrac{C_1}{D'^2}\right)}} \tag{8.178}$$

$$Q_{o2} \approx \frac{R C_2}{\omega_{o2} C_1 L_1}\left[\frac{\omega_{o2}^2}{\omega_{o1}^2} - 1\right]$$

The approximations for ω_{o1} and ω_{o2} are fairly good when they are separated by a factor of 2.5 or better. The second Q-factor, Q_{o2}, generally is very high and the accuracy of its approximation above is only moderate. In a real converter, Q_{o2} is a strong function of parasitic resistances, so that the accuracy of Eq. (8.178d) is not very relevant.

The numerator is determined by studying the nulls of $\hat{v}_o(s)$ as shown in Fig. 8.68 in which we see that a null in $\hat{v}_o(s)$ requires a null in $\hat{i}_{L2}(s)$ which in turn requires $\hat{i}_{p2}(s) = I_{c2}\hat{d}(s)$. It follows that the transform voltage across C_1 is given by:

$$\hat{v}_{C_1}(s) = V_{ap1}\hat{d}\frac{1/sC_1}{sL_1 + 1/sC_1} + I_{c2}\hat{d}[sL_1 \| (1/sC_1)]$$

$$= \hat{d}\left(V_{ap1}\frac{1}{1 + s^2 L_1 C_1} + I_{c2}\frac{sL_1}{1 + s^2 L_1 C_1}\right) \tag{8.179}$$

This transform voltage must be the same as the transform voltage across the D-side of the second PWM switch so that we have:

$$\hat{v}_{C_1}(s) = -\left[\frac{V_{ap2}}{D}\hat{d} - \hat{v}_{C_1}(s)\right]D \tag{8.180}$$

Equations (8.179) and (8.180) yield:

$$D'V_{ap1}\frac{1}{1+s^2L_1C_1} + D'I_{c2}\frac{sL_1}{1+s^2L_1C_1} + V_{ap2} = 0 \tag{8.181}$$

Multiplying out and substituting for V_{ap1}, V_{ap2} and I_{c2} in Eq. (8.157), we obtain the numerator:

$$N_d(s) = 1 + sL_1\frac{DD'}{2R} + s^2L_1C_1\frac{1+D}{2D} \tag{8.182}$$

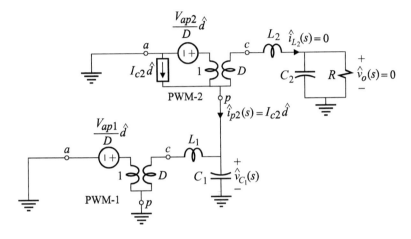

Figure 8.68

We can rewrite this in standard quadratic form:

$$N_d(s) = 1 + \frac{s}{\omega_{oz}Q_{oz}} + \frac{s^2}{\omega_{oz}^2} \tag{8.183}$$

in which:

$$\left. \begin{aligned} \omega_{oz} &= \frac{1}{\sqrt{L_1C_1}}\sqrt{\frac{2}{1+1/D}} \\[2mm] Q_{oz} &= \frac{R}{\sqrt{L_1/C_1}}\frac{1}{DD'}\sqrt{\frac{1+D}{2D}} \end{aligned} \right\} \tag{8.184a, b}$$

To summarize, the control-to-output transfer function is given by:

$$H_d(s) = H_{d0}\frac{1 + \dfrac{s}{\omega_{oz}Q_{oz}} + \dfrac{s^2}{\omega_{oz}^2}}{\left(1 + \dfrac{s}{\omega_{o1}Q_{o1}} + \dfrac{s^2}{\omega_{o1}^2}\right)\left(1 + \dfrac{s}{\omega_{o2}Q_{o2}} + \dfrac{s^2}{\omega_{o2}^2}\right)} \tag{8.185}$$

A plot of $H_d(s)$ using numerical values is discussed in Problem 8.15.   □

8.10   The effect of parasitic elements on the model of the PWM switch

The four important parasitic elements in a converter that affect the average model of the PWM switch are:

(a) the high-frequency resistance across port a–p
(b) the on-resistance of the switches
(c) the storage-time modulation in bipolar transistors
(d) the general saturation characteristics of the switches.

These parasitic elements affect only the relationship between the average port voltages of the PWM switch but not its average terminal currents. In this section, we shall formulate their effect in an *invariant* way and model them by a resistor in series with the common terminal of the PWM switch. The first of these parasitic elements is the most subtle because it concerns the general problem of modeling the nonlinear part of a system (in our case the PWM switch) by separating it from its linear part. Naturally, the accuracy of such a modeling technique depends on how carefully the quantitative interaction between the linear and nonlinear parts is accounted for when separating both parts. In the absence of parasitic elements and in continuous conduction mode there is no quantitative interaction between the linear elements of the filter and the average voltages and currents of the PWM switch, so that the switch can easily be separated from the rest of the circuit and modeled accurately. In the presence of certain parasitic elements, a small amount of interaction between these elements and the average voltages and currents of the PWM switch exists which warrants a refinement of the ideal models in Figs. 8.38 and 8.39.

(a) The high-frequency resistance across port a–p   The pulsating active terminal current, $\tilde{i}_a(t)$, in a converter is absorbed entirely by those elements which lie between terminals a and p of the PWM switch. All other elements lying between terminals c and p absorb the continuous inductive current in the common terminal. Hence, if $\tilde{i}_a(t)$ encounters a resistance r_e it will give rise to a pulsating voltage ripple, $r_e\tilde{i}_a(t)$, which will be superimposed on top of the smooth voltage across port a–p. To determine r_e, one simply shorts the capacitive and input voltage ports, opens the inductive ports and determines the resistance between terminals a and p. This is best illustrated by examples. For the buck converter with an input filter shown in Fig. 8.69a, r_e is equal to the parasitic equivalent series resistance (ESR) of the input filter capacitor, i.e. $r_e = r_{C_i}$. For the buck-boost converter with an input filter shown in Fig. 8.69b, $r_e = r_{C_i} + R \parallel r_c$. For the converter in Fig. 8.59 and for the Cuk converter in Fig. 8.32, $r_e = r_{C_c}$. For other converters, r_e is determined in a similar manner (see Problem 8.16).

In all cases, we see that the high-frequency parasitic resistance across port a–p is due to the ESR of the filter capacitors connected effectively across that port. Since the pulsating current is absorbed by the filter capacitors, the ripple voltage it generates is balanced about the average value of the port voltage V_{ap} as shown in Fig. 8.70. There is of course another ripple component in the port voltage $\tilde{v}_{ap}(t)$ (not shown in Fig. 8.70) which is essentially a triangular one and due to the capacitance, as explained earlier and shown in Fig. 8.22. The main difference between these two ripple components is that the average value of the capacitive ripple component in Fig. 8.22 is zero in *each* subinterval, T_{on} and T_{off}, whereas the average value of the ripple due to the ESR in Fig. 8.70 is zero only over the entire switching interval, T_s, and not over each subinterval. Hence, the average value of $\tilde{v}_{ap}(t)$ during T_{on} is not equal to the average value of $\tilde{v}_{ap}(t)$ over T_s, which is simply the average value $\tilde{v}_{ap}(t)$. According to Fig. 8.70, the average value of $\tilde{v}_{ap}(t)$ during T_{on} is given by:

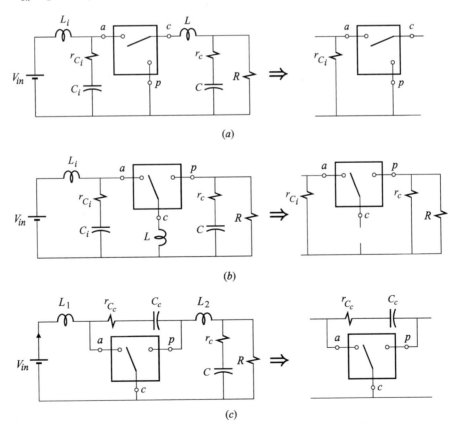

(a)

(b)

(c)

Figure 8.69

$$\langle \tilde{v}_{ap}(t) \rangle_{T_{on}} = v_{ap} - d'i_c r_e \qquad (8.186)$$

All we have to do now is fix the relationship between the average values of the voltages across ports a–p and c–p:

$$v_{cp} = d\langle \tilde{v}_{ap}(t) \rangle_{T_{on}}$$

$$= dv_{ap} - dd'i_c r_e \qquad (8.187)$$

The relationship between the average terminal currents i_a and i_c remains unaffected and is still given by $i_a = di_c$. The interaction term, $dd'i_c r_e$, can now be easily modeled by a parasitic resistance $dd'r_e$ in series with the common terminal in the large-signal average model of the PWM switch, as shown in Fig. 8.71. The small-signal model is obtained by perturbing Eq. (8.187):

$$\hat{v}_{cp} = D\hat{v}_{ap} + V_D\hat{d} - \hat{i}_c r_e DD' \qquad (8.188)$$

in which:

$$V_D = V_{ap} + I_c r_e(D - D') \approx V_{ap} \qquad (8.189)$$

The relationship between $\hat{i}_a$ and $\hat{i}_c$ remains unaffected and is still given by Eq. (8.116a). The dc and small-signal model corresponding to Eqs. (8.188) and (8.116) now follows and is shown in Fig. 8.72.

(a)

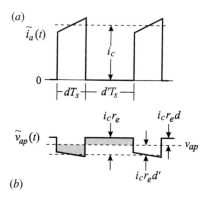

(b)

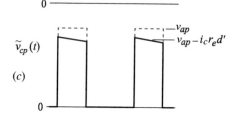

(c)

Figure 8.70

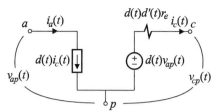

Figure 8.71

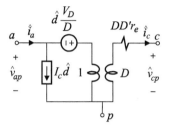

Figure 8.72

(b) The on-resistance of the switches   When MOSFETS are used to realize the active and passive switches, the effect of their finite on-resistance can easily be incorporated in the model of the PWM switch. Let r_t and r_d be the resistances of the active and passive switches of the PWM switch, respectively, as shown in Fig. 8.73. Intuitively, one can see that since the common terminal spends $D\%$ of its time in series with r_t and $D'\%$ of its time in series with r_d, the effective resistance appearing in series with the common terminal must be $Dr_t + D'r_d$. We can verify this by re-examining the relationship between v_{cp} and v_{ap}. We can see from Fig. 8.73 that during T_{on} the voltage across port c–p is less than v_{ap} by a factor of $i_c r_t$, while during T_{off} it is less than zero by a factor of $i_c r_d$. It follows that the relationship between v_{cp} and v_{ap} is:

$$\left. \begin{aligned} v_{cp} &= d(v_{ap} - i_c r_e d' - i_c r_t) - d'i_c r_d \\ &= dv_{ap} - i_c(r_e dd' + dr_t + d'r_d) \end{aligned} \right\} \tag{8.190a, b}$$

This result is consistent with our expectation, so that the effect of r_t and r_d can be modeled by adding another parasitic resistor, $dr_t + d'r_d$, in series with $dd'r_e$ in the

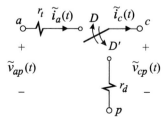

Figure 8.73

average large-signal model of the PWM switch in Fig. 8.71. The small-signal model is determined by perturbing Eq. (8.190) and obtaining:

$$\hat{v}_{cp} = D\hat{v}_{ap} + V_D\hat{d} - \hat{i}_c r_c \tag{8.191}$$

in which:

$$\left.\begin{array}{l} V_D = V_{ap} + I_c[r_e(D - D') + r_d - r_t] \\ r_c = DD'r_e + Dr_t + D'r_d \end{array}\right\} \tag{8.192}$$

Since the relationship between $\hat{i}_a$ and $\hat{i}_c$ is still given by Eq. (8.116), the small-signal model corresponding to Eqs. (8.191) and (8.116a) looks the same as in Fig. 8.72 in which V_D is given by Eq. (8.191b) and $Dr_t + D'r_d$ is added to $DD'r_e$.

(c) The storage-time modulation of BJTs   When a bipolar junction transistor is used to realize the active switch, the modulation in the duty ratio at the collector generally differs from the modulation in the duty ratio applied to the base of the BJT. It has been shown that:[7]

$$\hat{d} = \hat{d}_B - \frac{\hat{i}_c}{I_{me}} \tag{8.193}$$

in which:

$$\left.\begin{array}{l} \hat{d} \equiv \text{modulation in the duty ratio at the collector of the BJT} \\ \hat{d}_B \equiv \text{modulation in the duty ratio applied to the base of the BJT} \\ I_{me} \equiv \text{modulation parameter} \end{array}\right\} \tag{8.194a-c}$$

The modulation parameter has a negative value for constant base drive and a positive value for proportional base drive. More sophisticated types of base drive can render the value of I_{me} infinite so that $\hat{d} = \hat{d}_B$. Since storage-time modulation modifies the duty-ratio, the relationships between the port voltages and the terminal currents are both affected. The effect of storage-time modulation on the port voltages is determined by substituting Eq. (8.193) in (8.191):

$$\hat{v}_{cp} = D\hat{v}_{ap} + V_D\hat{d}_B - \hat{i}_c(r_{me} + r_c) \tag{8.195}$$

in which:

$$r_{me} = \frac{V_D}{I_{me}} \equiv \text{Storage-time modulation resistance} \tag{8.196}$$

The effect of storage-time modulation on the average terminal currents is determined by substituting Eq. (8.193) in (8.113a):

$$\hat{i}_a = \left(D - \frac{I_c}{I_{me}}\right)\hat{i}_c + \hat{d}_B I_c \tag{8.197}$$

Typically $D \gg I_c/I_{me}$, so that Eq. (8.197) reduces back to its original form:

$$\hat{i}_a \approx D\hat{i}_c + \hat{d}_B I_c \tag{8.198}$$

Equations (8.197) and (8.198) correspond to the small-signal model shown in Fig. 8.74 in which r_{me} is an ac resistance which, unlike r_c, vanishes under dc conditions.

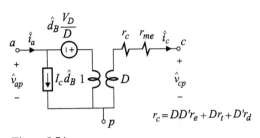

$$r_c = DD'r_e + Dr_t + D'r_d$$

Figure 8.74

Storage-time modulation can have a pronounced effect on the high-frequency response of certain converters (see Problem 8.17).

(d) Simple and general saturation models for the switches   When a BJT and a diode are used for the active and passive switches, respectively, their simple saturation models can be easily included in the model of the PWM switch. If we assume that a transistor or a diode in its on-state sustains a fixed voltage across its terminals, then we can write the relationship between v_{cp} and v_{ap}:

$$v_{cp} = d(v_{ap} - V_{CE_{sat}}) - d'V_{AK} \tag{8.199}$$

in which $V_{CE_{sat}}$ is the saturation voltage of the transistor and V_{AK} is the diode voltage. The relationship between the average terminal currents remains unaffected and is still given by $i_a = di_c$. Equation (8.199) can be rewritten as:

$$v_{cp} = dv_{ap} - (dV_{CE_{sat}} + d'V_{AK}) \tag{8.200}$$

which, along with $i_a = di_c$, can be incorporated in the large-signal average model of the PWM switch, as shown in Fig. 8.75a. It can easily be shown that the small-signal model is still given by Fig. 8.72, in which V_D is given by:

$$V_D = V_{ap} - V_{CE_{sat}} + V_{AK} \approx V_{ap} \tag{8.201}$$

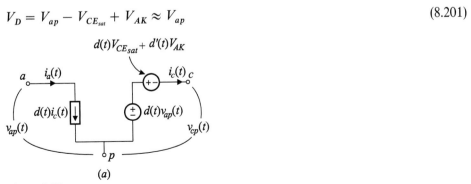

(a)

Figure 8.75

Instead of assuming a fixed on-voltage or an on-resistance when a switch is closed, we can assume a general relationship $V_S(i_c, \ldots)$ so that:

$$v_{cp} = d[v_{ap} - i_c r_e - V_{S_1}(i_t, \ldots)] - d' V_{S_2}(i_c, \ldots)$$

$$i_a = d i_c$$

$$d = d(v_{cont}, i_c, \ldots)$$

$$(8.202a\text{–}c)$$

In these equations, $V_{S_1}(i_c, \ldots)$ and $V_{S_2}(i_c, \ldots)$ are the on-voltages of the active and passive switches, respectively, which depend on the current i_c and possibly upon other parameters which depend on drive mechanisms. The duty ratio is shown to be a function of the control voltage, v_{cont}, and possibly the current i_c and other parameters which also depend on drive mechanisms. An equivalent circuit model corresponding to Eqs. (8.202a–c) is shown in Fig. 8.75b.

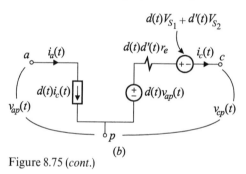

Figure 8.75 (cont.)

8.11 Feedback control of dc-to-dc converters

The output voltage or current of a dc-to-dc converter can be regulated using one of the feedback schemes shown earlier in Fig. 8.2. With output voltage feedback, an optional current feedback loop can also be added to form a two-loop, or two-state, feedback system. There are several ways of implementing the current feedback loop, all of which require special modeling techniques. We shall discuss only one popular method of current feedback, known as peak current-mode control, and show two different ways of modeling it. We shall use a buck converter to illustrate the use of single-loop and two-loop feedback control circuits.

Regardless of the type of feedback used, the most common way of converting the control signal to a PWM waveform is to compare it to a sawtooth waveform. This is shown in Fig. 8.76, in which a pulse with a fixed repetition rate, T_s, is initiated at the beginning of the ramp and is terminated when the ramp voltage exceeds the control signal. In this figure we have ignored the small ripple component in control voltage. By similar triangles we have, from Fig. 8.76:

$$\frac{v_c - V_1}{V_P} = \frac{t_{on}}{T_s} = d \tag{8.203}$$

It follows that:

$$\hat{d} = \frac{\hat{v}_c}{V_P} = K_M \hat{v}_c \tag{8.204}$$

in which:

$$K_M = \frac{1}{V_P} \equiv \text{modulator gain} \tag{8.205}$$

Typical values of V_P range from 1 to 2 V.

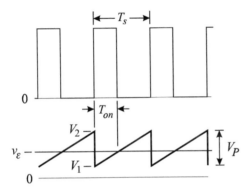

Figure 8.76

8.11.1 Single-loop voltage feedback control

The buck converter in Fig. 8.77 operates from a fixed bus voltage of 5 V and delivers 0–10 A at 2.5 V. The converter is designed to operate at 100 kHz and uses power MOSFET switches. The feedback compensation is designed to yield a stable loop which has a very large gain ($\sim 10^6$) at dc and a crossover frequency of 17 kHz. The large loop gain at dc ensures excellent regulation against quasi-static load and line variations (even though the line in this application is fixed at 5 V). The crossover frequency at 17 kHz ensures the smallest deviation in the output voltage when the load current changes abruptly from 0 to 10 A, or vice versa. An approximate expression of the overshoot (or undershoot) in the output voltage is given by:

$$\Delta V_o \approx \frac{\Delta I_o}{2\pi f_c C}; r_C < \frac{1}{2\pi f_c C} \tag{8.206}$$

in which f_c is the crossover frequency, ΔI_o is the step load current and C is the output capacitor. This expression is valid when the change in the load current takes place over one or more switching cycles. For faster changes in the load current, other factors need to be considered which will not be discussed here. Another restriction on the validity of Eq. (8.206) is that the value of the ESR has to be less than the reactance of the output filter capacitor at the crossover frequency.

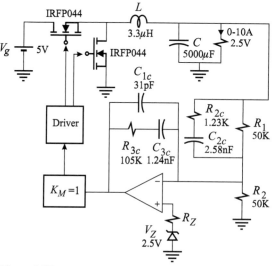

Figure 8.77

The purpose of the relatively large value of the output capacitor is to ensure that the output voltage ripple stays within 0.3% of the output voltage (7.5 mV) under worst case conditions. Such a large capacitance is typically obtained by paralleling several capacitors made of solid tantalum which have significant equivalent series resistance (ESR). The capacitor in this converter has an ESR of $2\,\text{m}\Omega$ and is obtained by paralleling twenty 250-μF solid tantalum capacitors each with an ESR of $40\,\text{m}\Omega$. When the capacitive reactance at the switching frequency is much smaller than the ESR, the output ripple voltage, instead of being piecewise parabolic as discussed earlier, is triangular and is given by the product of the ripple current in the inductor and the value of the ESR. In this converter, the reactance of C at 100 kHz is $3.2 \times 10^{-4}\,\Omega$, which is much less than $2 \times 10^{-3}\,\Omega$.

The control-to-output transfer function is determined from the equivalent circuit diagram shown in Fig. 8.78, where:

$$r_c = Dr_t + D'r_d = r_{DS_{on}}$$

$$V_{ap} = V_g$$

$$(8.207a, b)$$

In Eq. (8.207b), $r_{DS_{on}}$ is the on-resistance of the MOSFETs and has a typical value of 28 mΩ at a gate-drive voltage of 10 V. The transfer function relating the output voltage to the control voltage is given by:

$$G_c(s) = \frac{\hat{v}_o(s)}{\hat{v}_c(s)} = \frac{\hat{d}}{\hat{v}_c}\frac{\hat{v}_o}{\hat{d}} = K_M G_d(s) \tag{8.208}$$

in which $G_d(s)$ is given by:

$$G_d(s) = V_g \frac{1 + s/\omega_{z1}}{1 + s/(\omega_o Q) + (s/\omega_o)^2} \tag{8.209}$$

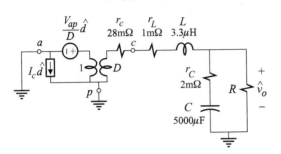

Figure 8.78

Using the numerical values shown, we compute the following values for the zero, the resonant frequency and the characteristic impedance:

$$\omega_{z1} = \frac{1}{r_c C} = (2\pi)15.9 \times 10^3 \text{ rad/s}$$

$$\omega_o = \frac{1}{\sqrt{LC}} = (2\pi)1.24 \times 10^3 \text{ rad/s} \tag{8.210a–c}$$

$$Z_o = \sqrt{\frac{L}{C}} = 25.7 \text{ m}\Omega$$

The Q-factor is computed for the cases of resistive and current loading:

$$Q = \frac{1}{\dfrac{Z_o}{r_c + r_L + R} + \dfrac{r_C + (r_c + r_L)\,\|\,R}{Z_o}} = \begin{cases} 0.83; & R \to \infty \\ 0.85; & R = 250\,\text{m}\Omega \end{cases} \tag{8.211}$$

The compensation network has a transfer function $H(s)$ of the form:

$$H(s) = K\frac{\omega_z}{s}\frac{(1 + s/\omega_z)^2}{[1 + s/(\omega_s/2)]^2} \tag{8.212}$$

An asymptotic construction of $H(s)$ is shown in Fig. 8.79. The pole at the origin provides infinite gain at dc for excellent dc regulation. The double zero at ω_z is placed on top of ω_o of the power stage in Eqs. (8.209) and (8.210) to compensate for the $180°$ phase shift due to ω_o. The double pole at $\omega_s/2$ provides $-20\,\mathrm{dB/dec}$ attenuation for the switching ripple. In terms of the circuit elements in Fig. 8.77, $H(s)$ is given by:

$$H(s) = \frac{1}{sR_1(C_{3c} + C_{1c})} \frac{(1 + sC_{3c}R_{3c})[1 + sC_{2c}(R_1 + R_{2c})]}{(1 + sC_{1c} \| C_{3c}R_{3c})(1 + sC_{2c}R_{2c})} \tag{8.213}$$

Since $\omega_z \ll \omega_s/2$, we can deduce from the expressions of ω_z and $\omega_s/2$ that $C_{1c} \ll C_{3c}$ and $R_{2c} \ll R_1$, so that Eq. (8.213) can be simplified to:

$$H(s) \approx \frac{1}{sR_1C_{3c}} \frac{(1 + sC_{3c}R_{3c})(1 + sC_{2c}R_1)}{(1 + sC_{1c}R_{3c})(1 + sC_{2c}R_{2c})} \tag{8.214}$$

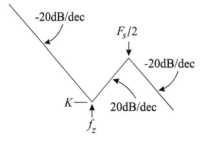

Figure 8.79

The loop gain is given by:

$$T(s) = -G_c(s)H(s) \tag{8.8.215}$$

An asymptotic construction of the magnitude of the loop gain, with ω_z placed on top of ω_o, is shown in Fig. 8.80. The closed-loop model can be simulated using

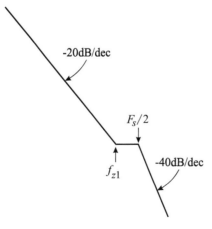

Figure 8.80

OrCAD/Pspice in which the large-signal average model of the PWM switch in continuous conduction mode is available as a subcircuit called VMLSCCM (voltage-mode large-signal continuous conduction mode). The simulation circuit is shown in Fig. 8.81a, and a magnitude and phase plot of $T(j\omega)$ is shown in Fig. 8.81b. The phase margin and the crossover frequency are seen to be 96° and 17.6 kHz, respectively. Note that the simulation program uses the large-signal average model of the PWM switch to determine the dc operating point automatically. Subsequently, the simulation program expands the large-signal model numerically at the dc operating point to determine the small-signal response.

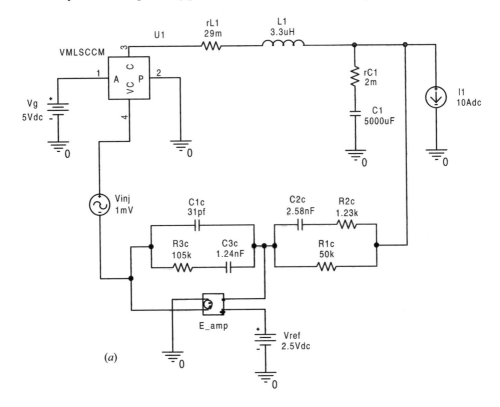

Figure 8.81

To complete the design, a properly damped input filter is added to reduce the input ripple current to about 170 mA, as shown in Fig. 8.82. The effect of the input filter on loop gain is shown in Fig. 8.83, in which the crossover frequency is seen to remain the same as before.

The time-domain response of the converter in Fig. 8.82, to a 0–10-A step change in the load current, is shown in Fig. 8.84. In this figure, the predicted response of the output voltage obtained by the large-signal average model of the PWM switch

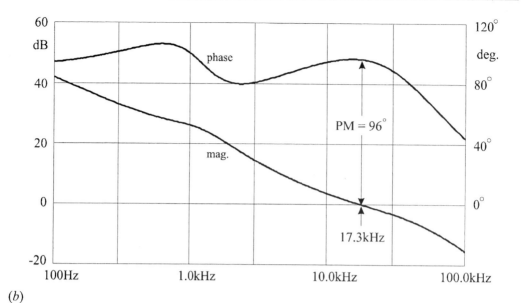

(b)

Figure 8.81 (cont.)

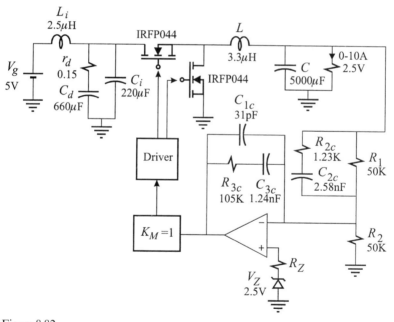

Figure 8.82

and the actual response obtained by cycle-by-cycle simulation of the actual circuit are compared. The agreement between the average and the actual response is quite satisfactory considering the fact that the simulation run time of the average model is much shorter than that of the cycle-by-cycle simulation. If we substitute the

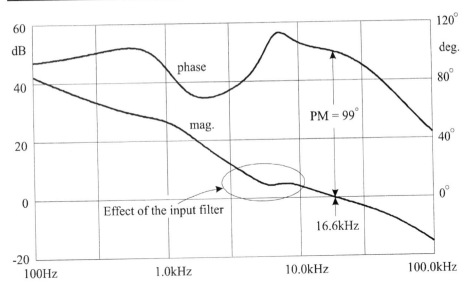

Figure 8.83

value of the crossover frequency from Fig. 8.83 in Eq. (8.206), we obtain the estimated undershoot in the output voltage:

$$\Delta V_o \approx \frac{\Delta I_o}{2\pi f_c C} = \frac{10\text{A}}{2\pi(16.6 \times 10^3)(5000 \times 10^{-6})} = 19\,\text{mV} \tag{8.216}$$

which is about 4 mV away from the average value of 23 mV observed in Fig. 8.84. The reason for this discrepancy is that Eq. (8.206) is at the limit of its validity because the reactance of the output capacitor at the crossover frequency is 1.92 mΩ, which is larger than the value of its ESR.

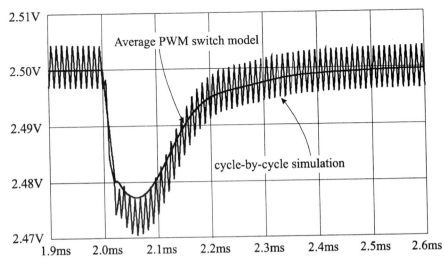

Figure 8.84

8.11.2 Current feedback control

A very popular type of current feedback control, known as peak current-mode control,[8] is shown in Fig. 8.85. In this type of control, the positive ramp of the active switch current during the on-time serves as the sawtooth waveform in the pulse-width-modulating circuit. When the peak current reaches a specified threshold, V_c, the comparator resets a flip–flop terminating the on-time. A clock with a fixed period, T_s, always sets the flip–flop to initiate the on-time. It is not hard to see that it is the common terminal current that is effectively being fedback, since the active and common terminal currents of the PWM switch are coincident during the on-time. Therefore, it should be possible to derive an invariant model for peak current mode control similar to that of the PWM switch. This is shown in Fig. 8.86 in which the PWM switch and the current feedback loop are combined in a new invariant structure called the current-controlled PWM switch[9] (CC-PWM switch). The sawtooth waveform which is added to the current waveform is called the external ramp and its purpose is to provide stability, as we shall see presently.

The steady-state current waveform for $D < 0.5$ is shown in Fig. 8.87a. It is fairly easy to see by a simple geometric construction that a steady-state duty cycle greater than 0.5 cannot be sustained as shown in Fig. 8.87b. Here we see that peak

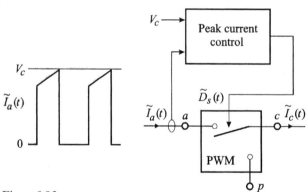

Figure 8.85

current control becomes unstable for $D > 0.5$ where it breaks into subharmonic instability, i.e. the duty cycle increases with a subharmonic periodicity resulting in a periodic waveform at half the switching frequency. In Fig. 8.87a we see how an initial disturbance in the current waveform settles down for $D \leq 0.5$. The addition of an external sawtooth waveform, or ramp, as shown in Fig. 8.88 extends the stable operation of peak current control to $D > 0.5$. It can be seen from these figures that the dynamics of peak current mode control is independent of any particular converter so that it should be possible to derive an invariant model of the CC-PWM switch. Such a model should contribute to the characteristic equation of a converter, a quadratic factor whose frequency is at half the switching

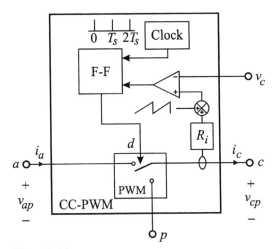

Figure 8.86

frequency and whose damping becomes negative for $D > 0.5$ in the absence of an external ramp.

In order to describe the action of peak current control in invariant terms, all we need to do is describe the slopes of the common terminal current in terms of the terminal quantities of the PWM switch. It is not hard to verify that the slopes of the current signal during the on-time and off-time for *any* converter are given by:

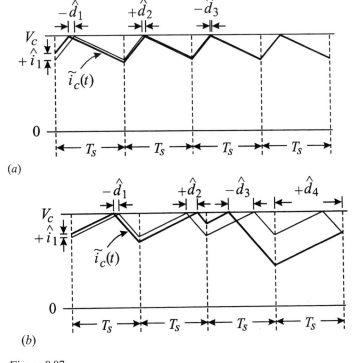

(a)

(b)

Figure 8.87

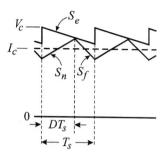

Figure 8.88

$$
\left.\begin{aligned}
S_n &= \frac{V_{ac}}{L} R_i = \frac{V_{ap}D'}{L} R_i \\[2mm]
S_f &= \frac{V_{cp}}{L} R_i
\end{aligned}\right\}
\tag{8.217a, b}
$$

in which R_i is a scaling constant which transforms the current signal into a voltage signal and L is the effective inductance which dictates the slope of the current. In a single inductor converter, L is the same as the switched inductor in that converter. For example, in the buck converter with an input filter, shown earlier in Fig. 8.69a, $L = L$; whereas in the Cuk converter, $L = L_1 \| L_2$. According to Fig. 8.88, the following dc equation holds at the switching instants:

$$
I_c R_i + \frac{S_f D' T_s}{2} + S_e D T_s = V_c
\tag{8.218}
$$

When time variations in the duty cycle are considered, all quantities in Eq. (8.218) must be replaced with their *instantaneous* values resulting in a sampled-data equation. If, for the moment, we ignore the effects of sampling, then we can replace all the quantities in Eq. (8.218) with their average values to determine the effect of the control law on these quantities. Hence, Eq. (8.218) together with the invariant equation of the PWM switch yield:

$$
\left.\begin{aligned}
i_c &= \frac{v_c}{R_i} - S_e \frac{dT_s}{R_i} - \frac{s_f(1-d)T_s}{2R_i} \\[2mm]
i_a &= di_c \\[2mm]
v_{cp} &= dv_{ap}
\end{aligned}\right\}
\tag{8.219a–c}
$$

If we substitute $d = v_{cp}/v_{ap}$ and $s_f = v_{cp}R_i/L$ in Eqs. (8.219a, b), we obtain:

$$i_c = \frac{v_c}{R_i} - \frac{v_{cp}}{v_{ap}} \frac{T_s S_e}{R_i} - v_{cp} \left(1 - \frac{v_{cp}}{v_{ap}}\right) \frac{T_s}{2L}$$

$$i_a = i_c \frac{v_{cp}}{v_{ap}}$$

$$(8.220a, b)$$

These two equations correspond to the large-signal average model of the CC-PWM switch shown in Fig. 8.89. The capacitance C_s in this figure does not follow from Eqs. (8.220a, b) and has been added to predict the subharmonic instability of the current loop, as will be discussed shortly. Equations (8.220a, b) were written assuming positive i_c flowing out of the common terminal. If positive i_c flows into the common terminal, then the direction of the control and external ramp sources should be reversed. Although we can always identify the PWM switch with positive i_c flowing out of the common terminal, we can allow, in general, for i_c to flow in either direction and automatically reverse the control and ramp control sources by multiplying these sources by $i_c/|i_c|$ (with positive i_c defined flowing out of the common terminal). This is particularly useful in creating a user-defined subcircuit for simulation purposes. The dc model of the CC-PWM switch is essentially the same as the large-signal average model without C_s. Generally, it is not possible to solve for the dc operating point in terms of V_c in closed form because the solution is generally given by the roots of a cubic equation, as we shall see in Example 8.13.

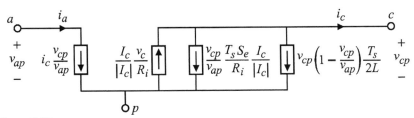

Figure 8.89

The equivalent small-signal circuit model of the CC-PWM switch is obtained by perturbing the large-signal average equations in Eqs. (8.220):

$$\hat{i}_c = \hat{v}_c k_o + \hat{v}_{ap} g_f - g_o \hat{v}_{cp}$$

$$\hat{i}_a = D \hat{i}_c + g_i \hat{v}_{ap} + g_r \hat{v}_{cp}$$

$$(8.221a, b)$$

in which the reader can verify:

$$k_o = \frac{1}{R_i} \frac{I_c}{|I_c|}$$

$$g_o = \frac{T_s}{L}\left(D'\frac{S_e}{S_n} + \frac{1}{2} - D\right)$$

$$g_f = Dg_o - \frac{DD'T_s}{2L}$$

$$g_i = -\frac{I_a}{V_{ap}}$$

$$g_r = \frac{I_c}{V_{ap}}$$

(8.222a–e)

These equations correspond to the circuit model shown in Fig. 8.90. Clearly, there are many different ways of expressing the small-signal parameters in Eqs. (8.222). For example, we have expressed g_o explicitly in terms of S_e/S_n because it is typical to specify the amount of the external ramp by its ratio to the natural current ramp during the on-time.

The capacitance C_s in the model of the CC-PWM switch is determined quite easily by examining the dynamics of peak current control in the vicinity of half the switching frequency, as shown in fig. 8.87b. As mentioned earlier, the dynamics

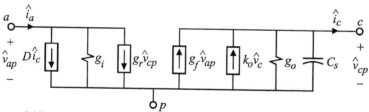

Figure 8.90

shown in Fig. 8.87b is independent of the type of converter and is characteristic of peak current control only. This observation is verified by examining the equivalent circuit of *any* converter with peak current control using the model of CC-PWM switch in Fig. 8.90. Hence, with the control voltage and input held constant and the output voltage essentially at signal ground in the vicinity of half the switching frequency, the equivalent circuit model of *any* converter in peak current control reduces to the parallel resonant circuit shown in Fig. 8.91b. To see this, we use the buck-boost converter in Fig. 8.91a as an example and write:

$$V_c = \text{constant} \Rightarrow \hat{v}_c = 0$$

$$V_g, V_o = \text{constant} \Rightarrow V_{ap} = \text{constant} \Rightarrow \hat{v}_{ap} = 0$$

(8.223)

We could have started with any other converter and arrived at the same circuit in

Fig. 8.91b using the conditions in Eq. (8.223). This is the circuit which contributes to the characteristic equation a quadratic factor at half the switching frequency which becomes unstable for $D > 0.5$ when $S_e = 0$. In fact, the characteristic equation for this parallel resonant circuit is:

$$\Delta(s) = 1 + sLg_o + \frac{s^2}{LC_s}$$

(8.224)

If this circuit were to resonate at half the switching frequency, the value of C_s would have to be such that:

$$\frac{\omega_s}{2} = \frac{1}{\sqrt{LC_s}} \Rightarrow C_s = \frac{4}{L\omega_s^2}$$

(8.225)

According to Eq. (8.222b), the value of g_o in the absence of an external ramp is given by:

$$g_o\big|_{S_e=0} = \frac{T_s}{L}\left(\frac{1}{2} - D\right)$$

(8.226)

which for $D > 0.5$ becomes negative and the parallel resonant circuit oscillates at half the switching frequency. The addition of an external ramp allows g_o to remain

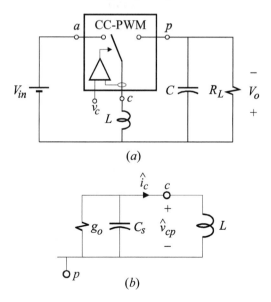

(a)

(b)

Figure 8.91

positive for $D > 0.5$, as can be seen from Eq. (8.222), thereby stabilizing the current loop. Hence, the predictions of the model of the CC-PWM switch in Fig. 8.90 and the behavior of peak current control are consistent. In the following example we shall determine the dynamics of a boost converter with peak current control.

The quadratic factor in Eq. (8.224) is common to the denominator of the small-signal dynamics of PWM converters operating in CCM with peak current control. Performing the necessary substitutions, Eq. (8.224) can be written as:

$$\Delta(s) = 1 + \frac{s}{\omega_n Q_n} + \frac{s^2}{\omega_n^2} \tag{8.227}$$

in which

$$\omega_n = \frac{\omega_s}{2}$$

$$\left. \begin{array}{l} Q_n = \dfrac{1}{\pi\left(D'\dfrac{S_e}{S_n} + \dfrac{1}{2} - D\right)} \\[6mm] \quad = \dfrac{1}{\pi\left[D\left(\dfrac{S_e}{S_f} - 1\right) + \dfrac{1}{2}\right]} \end{array} \right\} \tag{8.228a, b}$$

Example 8.13   A boost converter with peak current control is shown in Figs. 8.92a and b. In Fig. 8.92a the CC-PWM switch is identified with positive i_c flowing into the common terminal and in Fig. 8.92b it is identified with positive i_c flowing out of the common terminal. Using the circuit in Fig. 9.92a, we replace the CC-PWM switch with its dc model as shown in Fig. 8.93, in which the arrows of the ramp and control source have been reversed. According to Fig. 8.93, we have:

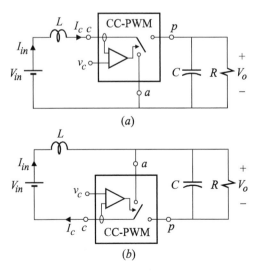

Figure 8.92

$$I_c = \frac{V_c}{R_i} - \frac{V_{cp}}{V_{ap}}\frac{T_s S_e}{R_i} + V_{cp}\left(1 - \frac{V_{cp}}{V_{ap}}\right)\frac{T_s}{2L}$$

$$V_o = \left(I_c - I_c\frac{V_{cp}}{V_{ap}}\right)$$

(8.229a–d)

$$V_{ap} = -V_o$$

$$V_{cp} = V_g - V_o$$

Solving these simultaneously, we obtain the following cubic equation in V_o:

$$V_o^3 + V_o V_g\left(\frac{T_s S_e R}{R_i} + \frac{R T_s V_g}{2L} - \frac{R}{R_i}V_c\right) - \frac{T_s R S_e V_g^2}{R_i} - \frac{R T_s V_g^3}{2L} = 0 \tag{8.230}$$

Clearly it is not possible to obtain an analytical answer for V_o for a given control voltage V_c. Unfortunately, this is the case with peak current control regardless of the type of converter. The best that we can do is get an approximate idea by letting L be very large and ignore S_e. This yields:

$$V_o^3 - V_o V_g V_c\frac{R}{R_i} \approx 0 \Rightarrow V_o \approx \sqrt{V_g V_c\frac{R}{R_i}} \tag{8.231}$$

The control-to-output transfer function is determined by replacing the CC-PWM switch with its small-signal model as shown in Fig. 8.94 and has the form:

$$\frac{\hat{v}_o}{\hat{v}_c} = A_c\frac{N_c(s)}{D(s)} \tag{8.232}$$

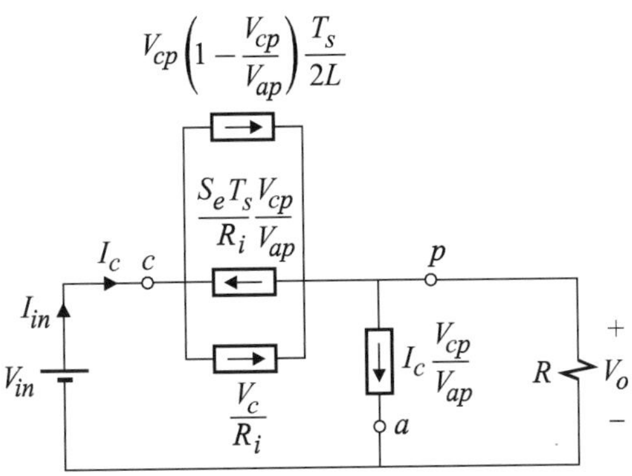

Figure 8.93

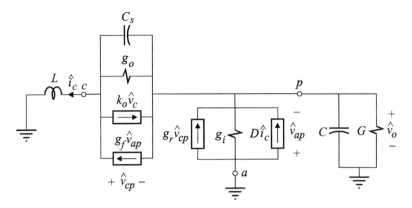

Figure 8.94

At dc, the circuit in Fig. 8.94 reduces to the one in Fig. 8.95, whence the low-frequency asymptote can be obtained:

$$A_c = \frac{k_o D'}{g_i + g_r + G + D'(g_o - g_f)}$$

$$= \frac{R}{R_i} \frac{1}{2M + \frac{RT_s}{LM^2}\left(\frac{1}{2} + \frac{S_e}{S_n}\right)} \qquad (8.233a, b)$$

in which $G = 1/R$. The same result can be obtained by an implicit differentiation of Eq. (8.230) with respect to V_c.

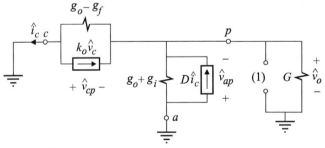

Figure 8.95

The demoninator is obtained by setting the control source $\hat{v}_c$ in Fig. 8.94 to zero. We already know that at high frequencies this circuit reduces to the parallel resonant circuit in Fig. 8.91b, which contributes the quadratic term at half the switching frequency in Eq. (8.224) as discussed earlier. Hence we have:

$$D(s) = (1 + s/\omega_p)[1 + s/(\omega_n Q_n) + s^2/\omega_n^2] \qquad (8.234)$$

in which:

$$\omega_n = \frac{\omega_s}{2}$$

$$Q_n = \frac{1}{\omega_n L g_o} = \frac{1}{\pi \left(D' \dfrac{S_e}{S_n} + \dfrac{1}{2} - D \right)}$$

(8.235a, b)

The dominant pole ω_p is determined by examining the circuit in Fig. 8.94 at low frequencies in which the output filter capacitor is seen to dominate the dynamics. The conductance looking into the capacitive port is determined from Fig. 8.95, which the reader can verify to be:

$$\frac{1}{R^{(1)}} = g_i + g_r + G + D'(g_o - g_f)$$

$$= 2G + \frac{T_s}{LM^3} \left(\frac{1}{2} + \frac{S_e}{S_n} \right)$$

(8.236)

Hence, the dominant pole in Eq. (8.234) is given by:

$$\omega_p = \frac{1}{R^{(1)}C} = \frac{2G + \dfrac{T_s}{LM^3} \left(\dfrac{1}{2} + \dfrac{S_e}{S_n} \right)}{C}$$

(8.237)

The numerator, $N_c(s)$, in Eq. (8.232) is determined by examining the null response of the circuit as shown in Fig. 8.96. A null in the output voltage implies $\hat{v}_{ap} = 0$, which in turn implies that:

$$\hat{v}_{cp} = \hat{i}_c sL$$

$$\hat{i}_c = D\hat{i}_c + g_r\hat{v}_{cp}$$

(8.238a, b)

The simultaneous solution of these equations yields:

$$-D' + g_r sL = 0$$

(8.239)

Substituting $g_r = 1/D'R$ we obtain the numerator:

$$N_c(s) = 1 - s\frac{L}{RD'^2}$$

(8.240)

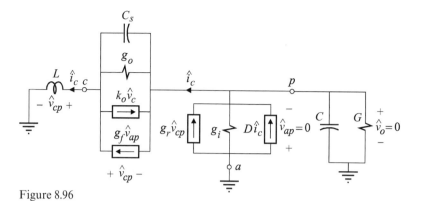

Figure 8.96

This is the same RHP zero in the duty-ratio-to-output transfer function in continuous conduction mode as explained earlier in Fig. 8.50 and in Eqs. (8.141–8.143). This is expected since a step change in the control voltage v_c in peak current control results in a simultaneous step change in the duty cycle.   □

Another model for peak current control, which actually preceded the model of the CC-PWM switch, was developed by R. Ridley[10] and is shown in Fig. 8.97. In this figure, the power stage is modeled using the PWM switch with duty ratio control, while the current loop is modeled using the blocks k_f, k_r, F_m and $H_e(s)$. The blocks k_f and k_r model the effect of the input and output voltages and their expressions for certain converters are:

$$
k_f = \begin{cases} -\dfrac{DT_sR_i}{L}\left(1 - \dfrac{D}{2}\right) & \text{buck, buck-boost} \\[2mm] -\dfrac{T_sR_i}{2L} & \text{boost} \end{cases}
\tag{8.241}
$$

$$
k_r = \begin{cases} \dfrac{T_sR_i}{L} & \text{buck} \\[2mm] \dfrac{D'^2T_sR_i}{2L} & \text{boost, buck-boost} \end{cases}
\tag{8.242}
$$

The modulator gain is modeled by F_m, while the effect of sampling is modeled by $H_e(s)$. These are given by:

$$
\left.\begin{aligned}
F_m &= \frac{1}{(S_n + S_e)T_s} \\[3mm]
H_e(s) &= \frac{sT_s}{e^{sT_s} - 1} \approx 1 - \frac{s}{\omega_n}\frac{\pi}{2} + \frac{s^2}{\omega_n^2}; \ \omega_n = \frac{\omega_s}{2}
\end{aligned}\right\}
\tag{8.243a, b}
$$

The approximation of $H_e(s)$ by a quadratic is valid up to the neighborhood of half the switching frequency.

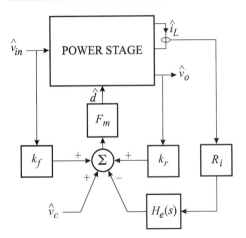

Figure 8.97

Both modeling techniques yield the same results for the control-to-output transfer functions, as shown in the next example.

Example 8.14   The complete small-signal equivalent circuit of the boost converter in peak current control using the model in Fig. 8.97 is shown in Fig. 8.98. The denominator $D(s)$ in Eq. (8.232) is determined by setting all the excitations to zero in Fig. 8.98 and examining the resulting circuit at low and high frequencies shown in Figs. 8.99a, b, respectively. The dominant pole is obtained from Fig. 8.99a and can be shown to be given by Eq. (8.237). The roots of the high-frequency factor are obtained from Fig. 8.99b:

$$[\hat{d}]\frac{V_{ap}}{D}D = [-\hat{i}_L R_i H_e(s)F_m]\frac{V_{ap}}{D}D = sL\hat{i}_L \qquad (8.244)$$

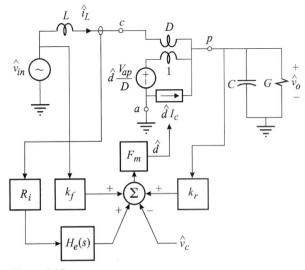

Figure 8.98

The quadratic factor follows by canceling $\hat{i}_L$ in the last two equalities:

$$H_e(s) + \frac{sL}{F_m R_i V_{ap}} = 0 \tag{8.245}$$

Substituting S_n from Eq. (8.217a) in the expression for F_m, we obtain:

$$F_m R_i V_{ap} = \frac{1}{T_s S_n} \frac{R_i V_{ap}}{1 + S_e/S_n} = \frac{L}{T_s D'} \frac{1}{1 + S_e/S_n} \tag{8.246}$$

Substituting Eqs. (8.245) and (8.243) in (8.246), we obtain the same quadratic factor as in Eq. (8.234).

The numerator of the control-to-output transfer function is determined by examining the null response of the circuit as shown in Fig. 8.101, whence we have:

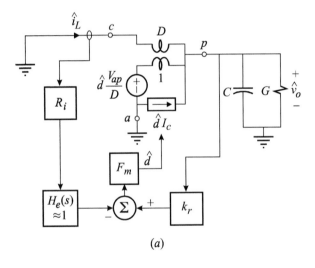

(a)

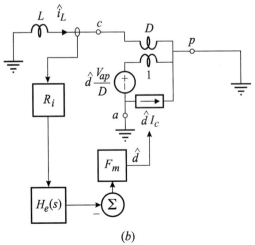

(b)

Figure 8.99

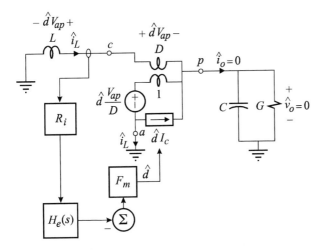

Figure 8.100

$$\left. \begin{array}{l} \left[\hat{d}\dfrac{V_{ap}}{D}\right]D = \hat{i}_L sL \\[3mm] \hat{i}_L = I_c\hat{d} + D\hat{i}_L \end{array} \right\}$$

(8.247a, b)

The simultaneous solution of these equations yields:

$$1 - sL\frac{I_c}{V_{ap}D'} = 0$$

(8.248)

When the dc operating point, $I_c = -I_o/D'$ and $V_{ap} = -V_o$, is substituted in the above, the same RHP zero as in Eq. (8.240) is obtained. ☐

8.11.3 Voltage feedback control with peak current control

In this section we shall demonstrate one of the advantages of adding a peak current feedback loop to a voltage feedback loop. In Eq. (8.215), we saw that the loop gain of the regulated buck converter was directly proportional to the input voltage because of the dc gain G_o of $G_c(s)$ given by Eq. (8.209). Hence, if the input voltage was to vary significantly, the crossover frequency would vary proportionally, which in certain applications could be undesirable. To show how peak-current control renders the crossover frequency of the loop gain insensitive to the input voltage, consider the small-signal equivalent circuit of a simple buck converter without an input filter in peak-current mode control, as shown in Fig. 8.101. The control-to-output transfer function is obtained after setting $\hat{v}_g = 0$ and is given by:

$$\frac{\hat{v}_o}{\hat{v}_c} = G_o\frac{1 + s/\omega_{z1}}{1 + s/\omega_{p1}}\frac{1}{1 + s/(\omega_n Q_n) + (s/\omega_n)^2}$$

(8.249)

in which the quadratic factor is the same as the one discussed earlier in Eqs. (8.234) and (8.235a, b). The zero at ω_{z1} is the usual zero due to r_{C1} of the output filter capacitor and is given by:

$$\omega_{z1} = \frac{1}{r_{C1}C_1} \tag{8.250}$$

If we let $s \to 0$, then the circuit in Fig. 8.101 reduces to the current source $k_o\hat{v}_c$ feeding the parallel combination of the load resistance and g_o. It follows that G_o in Eq. (8.249) is given by:

$$G_o = \frac{k_o}{g_o + G_L} \tag{8.251}$$

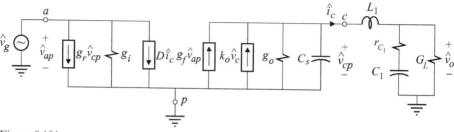

Figure 8.101

in which g_o and k_o are given by Eqs. (8.222a, b).

The low-frequency pole at ω_{p1} is obtained by examining the circuit in Fig. 8.101 at low frequencies with $\hat{v}_c = \hat{v}_g = 0$. At low frequencies, L_1 is ineffective and the dominant time constant is formed by $C_1 + C_s$ parallel with g_o and the load resistance. Hence, the dominant pole is given by:

$$\omega_{p1} = \frac{g_o + G_L}{C_1 + C_s} \approx \frac{g_o + G_L}{C_1} \tag{8.252}$$

The crossover frequency is normally well above ω_p and well below $\omega_n = \omega_s/2$. Hence, in the vicinity of the crossover, G_o and ω_{p1} combine together in the approximate behavior of the control-to-output transfer function:

$$G_c(s) \approx G_o \frac{1 + s/\omega_{z1}}{s/\omega_{p1}} = \frac{k_o}{sC_1}(1 + s/\omega_{z1}) \tag{8.253}$$

in which we have made use of Eqs. (8.251) and (8.252). Since k_o is independent of the input voltage, it follows that the control-to-output transfer function and, hence, the loop gain are both independent of the input voltage near the crossover frequency. It follows that the crossover frequency itself is practically independent of the input voltage. An asymptotic sketch of $G(s)$ is shown in Fig. 8.102.

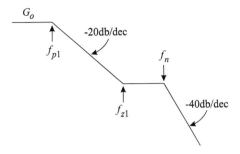

Figure 8.102

The voltage feedback compensation of a dc-to-dc converter with peak-current control is somewhat simpler than that of a converter without peak-current control. The reason of course is that the power stage in current mode has a dominant single pole whereas a power stage without peak-current control and operating in CCM has a dominant complex pole-pair. A typical compensation scheme for converters in peak-current control is shown in Fig. 8.103. The transfer function relating the control voltage to the converter output voltage of the amplifier in Fig. 8.103 is given by:

$$H(s) \equiv \frac{\hat{v}_c(s)}{\hat{v}_o(s)} = H_o \frac{1 + \dfrac{\omega_{zc}}{s}}{1 + \dfrac{s}{\omega_{p2}}} \tag{8.254}$$

in which:

$$H_o = -\frac{R_{2c}}{R_{1c}}$$

$$\omega_{zc} = \frac{1}{R_{2c}C_{2c}} \tag{8.255a–c}$$

$$\omega_{p2} = \frac{1}{R_{2c}C_{2c} \parallel C_{1c}} \approx \frac{1}{R_{2c}C_{1c}}$$

The pole at ω_{p2} is optional and is placed at or above half the switching frequency mainly to attenuate high-frequency switching noise. The zero at ω_{zc} is placed between the dominant pole of the power stage and the desired crossover frequency. It should be placed low enough to provide adequate phase margin but not so low as to cause a sluggish response. A numerical illustration of a practical voltage feedback loop around a power stage with peak-current control is given in the following example.

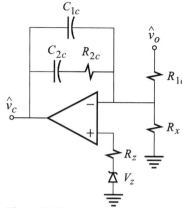

Figure 8.103

Example 8.15 An OrCAD/Pspice simulation of the equivalent circuit of a buck converter with peak-current control and output voltage feedback control is shown in Fig. 8.104. The subcircuit CMLSCCM (current-mode large-signal continuous conduction mode) is a large-signal model of the current-controlled PWM switch in continuous conduction mode. The converter operates at 100 kHz from an input voltage range of 13 to 26 V and has an output voltage of 5 V at 0–10 A. The load is assumed to behave as a current source so that its incremental resistance is infinite, i.e. $G_L = 0$.

The reader can verify that the output ripple voltage is dictated by r_{C1} of the output filter capacitor C_1 instead of C_1 and that it has a maximum value of 25 mV, which occurs when the input voltage is 26 V.

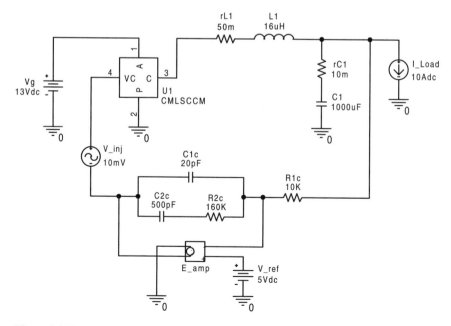

Figure 8.104

The two parameters of peak-current control that must be determined are the amount of the external stabilizing ramp, S_e, and the current-to-voltage conversion resistance R_i, both of which are shown in the model of the CC-PWM switch in Fig. 8.86. The value of R_i is dictated by the maximum value of the current-sense voltage signal which, for most commercially available PWM control chips, is in the range of a few volts. In this design, we shall assume the maximum value of the current-sense signal to be 2.5 V and that the maximum current to be sensed is about 12 A. Hence:

$$R_i = \frac{2.5 \text{ V}}{12 \text{ A}} \approx 0.2 \Omega \tag{8.256}$$

Also, in this design, the amount of external ramp is chosen to yield $Q_n = 0.5$ in the subharmonic quadratic factor in Eq. (8.228). Hence, according to Eq. (8.228):

$$Q_n = \frac{1}{\pi \left[D_{min} \left(\frac{S_e}{S_f} - 1 \right) + \frac{1}{2} \right]} = 0.5 \tag{8.257}$$

The minimum duty cycle, assuming an efficiency of $\eta = 85\%$, is computed to be:

$$D_{min} = \frac{V_o}{\eta V_{in_{max}}} = \frac{5 \text{ V}}{(0.85)26 \text{ V}} = 0.225 \tag{8.258}$$

The slope of the inductor current during the off-time can now be computed:

$$S_f = \frac{V_o R_i}{L} = \frac{5 \text{ V}(0.2 \Omega)}{16 \, \mu\text{H}} = 6.25 \text{ V}/\mu\text{s} \tag{8.259}$$

Substituting Eqs. (8.258) and (8.259) in Eq. (8.257) yields the amount of external ramp:

$$S_e = 10 \text{ V}/\mu\text{s} \tag{8.260}$$

The values of R_i, S_e, L and the switching frequency are the arguments which are supplied to the subcircuit CMLSCCM, which in turn computes the value of C_s according to Eq. (8.225).

The loop gain is given by the product of $H(s)$ and $G_c(s)$:

$$T(s) = T_o \frac{1 + \dfrac{s}{\omega_{z1}} \, 1 + \dfrac{\omega_{zc}}{s}}{1 + \dfrac{s}{\omega_{p1}} \, 1 + \dfrac{s}{\omega_{p2}} \, 1 + \dfrac{s}{\omega_n Q_n} + \dfrac{s^2}{\omega_n^2}} \tag{8.261}$$

in which T_o is given by:

$$T_o = -\frac{k_o}{g_o} \frac{R_{2c}}{R_{1c}} \tag{8.262}$$

The numerical values of k_o and g_o are given by:

$$k_o = \frac{1}{R_i} = 5\,\Omega^{-1} \tag{8.263}$$

$$g_o = \frac{T_s}{L_1}\left(D'\frac{S_e}{S_n} + \frac{1}{2} - D\right) = \begin{cases} 0.416\,\Omega^{-1}; \ V_g = 13\,\text{V} \\ 0.364\,\Omega^{-1}; \ V_g = 26\,\text{V} \end{cases} \tag{8.264}$$

The following values are computed for the numerical determination of the loop gain:

$$T_o = -\frac{k_o}{g_o}\frac{R_{2c}}{R_{1c}} = \begin{cases} 192; \ V_g = 13\,\text{V} \\ 219; \ V_g = 26\,\text{V} \end{cases}$$

$$\omega_{z1} = \frac{1}{r_{C1}C_1} = \frac{1}{(10\,\text{m}\Omega)(1000\,\mu\text{F})} = (2\pi)15.9\,\text{krad/s}$$

$$\omega_{zc} = \frac{1}{R_{2c}C_{2c}} = \frac{1}{(160\,\text{k}\Omega)(500\,\text{pF})} = (2\pi)1989\,\text{rad/s}$$

$$\omega_{p2} \approx \frac{1}{R_{2c}C_{1c}} = \frac{1}{(160\,\text{k}\Omega)(20\,\text{pF})} = (2\pi)50\,\text{krad/s} \qquad (8.265a\text{–}g)$$

$$\omega_n = \frac{\omega_s}{2} = (2\pi)50\,\text{krad/s}$$

$$Q_n = \frac{1}{\pi\left[D\left(\dfrac{S_e}{S_f} - 1\right) + \dfrac{1}{2}\right]} = \begin{cases} 0.5; \ V_g = 26\,\text{V} \\ 0.4; \ V_g = 13\,\text{V} \end{cases}$$

$$\omega_{p1} = \frac{g_o}{C_1} = \begin{cases} (2\pi)66\,\text{rad/s}; \ V_g = 13\,\text{V} \\ (2\pi)58\,\text{rad/s}; \ V_g = 26\,\text{V} \end{cases}$$

Note that the shape of the loop gain past the dominant pole, ω_{p1}, is practically independent of the input voltage for the same reason given earlier in Eq. (8.253).

In Fig. 8.105, a magnitude and phase plot of the loop gain is obtained using an injection signal, as shown in the simulation program in Fig. 8.104. As expected the shape of the loop gain and the crossover frequency are independent of the input voltage.

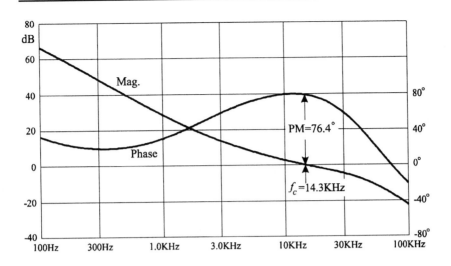

Figure 8.105

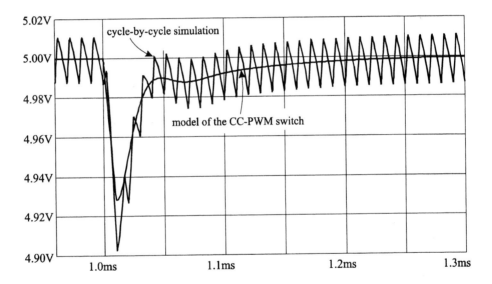

Figure 8.106

The dynamic response of the output voltage to a 0–10 A step change in the output load current is shown in Fig. 8.106. The continuous trace shows the response obtained using the large-signal average model of the CC-PWM switch (CMLSCCM) in Fig. 8.105. Also shown in Fig. 8.106 is the actual response of the converter obtained by cycle-by-cycle simulation. The agreement between the predictions of the cycle-by-cycle simulation and the large-signal average model simulation is generally good. $\square$

8.12 Review

An ideal dc-to-dc converter transforms the dc level of a voltage or a current source in a controllable and nondissipative manner using switches, inductors and capacitors. The inductors and capacitors form effective low-pass filters which extract the dc component of the switching waveforms and provide smooth dc voltages and currents at the input and output ports of the converter. A practical converter dissipates a relatively small amount of power in comparison with its output power because of nonideal components. The ratio of the output voltage to the input voltage is defined as the conversion ratio and is controlled by the duty cycle of the switches. Although, in principle, a myriad number of switches, driven in myriad ways, can be used to generate nonisolated converters *ad nauseam*, a pair of switches, driven in a complementary fashion, is sufficient to generate basic converters which accomplish *all* the necessary functions of dc-to-dc power conversion. The buck, the boost, the buck-boost, the Cuk, the Watkins–Johnson and its bilateral inverse are among the basic converters discussed in this chapter. The buck converter steps down the input voltage and the boost converter steps up the input voltage. The sign of the output voltage in both of these converters follows the sign of the input voltage. The Cuk and the buck-boost converters can either step up or step down the input voltage and the sign of their output voltage is always opposite that of the input voltage. The Watkins–Johnson and its bilateral inverse can either step up or step down the input voltage while the sign of their output voltage can either be the same or opposite that of the input voltage. Other converters, some of which use four switches, are also discussed.

Like amplifiers, dc-to-dc converters are nonlinear circuits whose *exact* solutions can only be determined numerically using circuit simulation programs. Nevertheless, the vast amount of design-oriented analytical techniques available for amplifier circuits are also desirable for dc-to-dc converter circuits. In an amplifier circuit, these techniques are applied once an equivalent circuit model of the amplifier is obtained by replacing the transistor, or the vacuum tube, with its equivalent circuit model. In order to apply these techniques to a PWM converter circuit, we have introduced the concept of the PWM switch and derived an equivalent circuit model for it. The PWM switch, just like the transistor, is a three-terminal nonlinear device which can be replaced with its equivalent circuit model to yield an equivalent circuit model of a PWM converter circuit. The model of the PWM switch does not depend on any particular converter topology, just like the model of a transistor does not depend on any particular amplifier topology. Hence, the analyses of the dynamics of amplifiers and PWM converters are identical. For example, to determine the small-signal characteristics of a PWM converter, first a dc analysis is performed and the dc operating point of the PWM switch is determined. Second, the small-signal parameters in the model of the PWM switch are evaluated at the dc operating point and a small-signal equivalent circuit of the converter is obtained from which the input and output impedance, the line-to-output and the control-to-output transfer functions are determined.

Various feedback control techniques can be used to regulate the output voltage of a converter against variations in the input voltage and the load current. In a single-loop feedback system, only the output voltage is fed back into the PWM control circuit where it is compared with a reference voltage. In a two-loop feedback system, in addition to the output voltage, the current in the active switch is fed back to improve the loop-gain characteristics of the voltage feedback loop. The nature of the sampled data of the active-switch current feedback loop must be correctly accounted for whenever a continuous-time model of the converter is sought. This can be done either by deriving a model for the current feedback loop alone or by deriving a model for the PWM switch and the current feedback loop combined together. The combination of the PWM switch and the current feedback loop is called the current-controlled PWM switch.

Problems

8.1 Linear series voltage regulator. A simplified diagram of a series regulator using a bipolar transistor is shown in Fig. 8.107a. An equivalent circuit diagram is shown in Fig. 8.107b.

(a) Show that the line-to-output transfer function and the output impedance are given by:

$$
\left.
\begin{aligned}
\frac{v_o(s)}{v_g(s)} &= \frac{1}{a_o + \dfrac{r_g}{R_L \alpha}} \frac{1}{1 + sC_L R_L \left\| \dfrac{r_g}{\alpha a_o}\right.} \\[2em]
Z_o(s) &= \frac{R_L \left\| \dfrac{r_g}{\alpha a_o}\right.}{1 + sC_L R_L \left\| \dfrac{r_g}{\alpha a_o}\right.}
\end{aligned}
\right\}
\qquad (8.266a, b)
$$

(b) If a_o is an operational amplifier, then its gain can be approximated by $a(s) \approx \omega_o/s$. Show that the regulator in this case becomes unstable as $R_L \to \infty$. One way to prevent this oscillation is to connect a large capacitance from the emitter to ground in order to shunt r_g. Determine the transfer functions in part

(a) with C connected from emitter to ground and discuss how the circuit becomes stable. (When a linear regulator is placed far from the source, r_g accounts for the resistance of the connecting wires. This is why manufacturers of linear regulator ICs recommend that a large capacitance (100 μF, solid tantalum) be connected immediately to the input terminals of the linear regulator whenever it is placed far from V_g.)

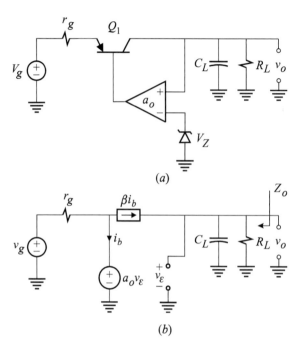

Figure 8.107

8.2 Linear shunt regulator. Show that for the shunt regulator in Fig. 8.108a we have:

$$\frac{v_o(s)}{i_g(s)} = Z_o(s) = Z_{in}(s) = \frac{R_L \| r_g \left\| \dfrac{r_e}{1 + a_o} \right\|}{1 + sC_L R_L \| r_g \left\| \dfrac{r_e}{1 + a_o} \right\|} \tag{8.267}$$

8.3 Incremental output resistance.

(a) Show that the incremental output resistance of an ideal unregulated converter is given by:

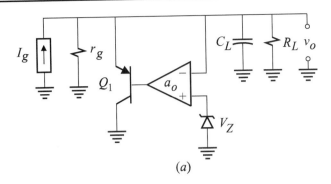

(a)

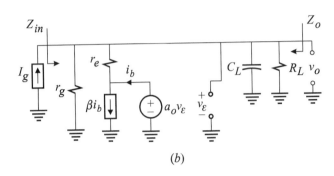

(b)

Figure 8.108

$$r_o = V_g \frac{\partial M_V / \partial I_o}{1 - V_g \partial M_V / \partial V_o} \tag{8.268}$$

Hint: Recall that M_V, in general, can be a function of α, V_o and I_o. Also we have:

$$r_o = \frac{dV_o}{dI_o} = \frac{d(M_V V_g)}{dI_o} = V_g \frac{dM_V}{dI_o} \tag{8.269}$$

(b) Deduce that when M_V is only a function of control parameter, then $r_o = 0$.

(c) Certain converters have certain modes of operation in which they act as gyrators, i.e. they transform the input voltage source to a current source which is *linearly* proportional to the input voltage:

$$I_o = g_T(\alpha) V_g \tag{8.270}$$

in which g_T is some transconductance that depends on the control parameter α. (An example of such a converter is the series resonant converter operating in an even-type discontinuous conduction mode.[11]) It is clear then that the output impedance for such a converter is infinite. Show that r_o indeed is infinite in Eq. (8.268) for such converters.

Hint: Show that M_V can be written as:

$$M_V = g_T R_L = g_T \frac{V_o}{I_o} \tag{8.271}$$

8.4 Transfer function of the _LC_ low-pass filter. Use the 2-EET to determine the transfer function of the _LC_ low-pass filter shown in Fig. 8.109.

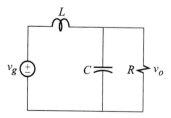

Figure 8.109

8.5 Output ripple voltage in the buck converter.

(a) Derive the expression of the output ripple voltage in the buck converter starting with the expressions of the individual components of the ripple voltage in Eqs. (8.29a, b).

(b) Derive the same expression by recognizing that the peak-to-peak ripple voltage on the capacitor is equal to the amount of charge delivered to the capacitor in the time interval $T_{on}/2 < t < (T_{on} + T_s)/2$ divided by the capacitance. This is shown in Fig. 8.110 in which $\tilde{I}_r(t)$ is the ripple component of the inductor current.

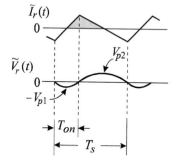

Figure 8.110

8.6 Input ripple current of a buck converter with an input filter. Using simple geometry, derive the expression of the peak-to-peak ripple current of the input filter of the buck converter in Eq. (8.55).

Hint: Recognize that the double integral in Eq. (8.54) is nothing more than the area under the triangle shown in Fig. 8.111, whose height is half the area of the rectangular voltage pulse during T_{on}.

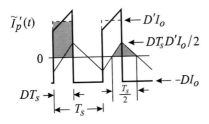

Figure 8.111

8.7 Parallel damping of the input filter.[12] An LC filter with parallel damping is shown in Fig. 8.112a. Use the 3-EET to show that the characteristic equation is given by:

$$D(s) = 1 + s\left(C_d R_d + \frac{L}{R}\right) + s^2\left(LC + LC_d\frac{R_d}{R \parallel R_d}\right)$$

$$+ s^3 LCC_d R_d \tag{8.272}$$

In order to ensure that $R_d C_d$ damps the LC resonance (and does not create a new lower resonance), it should be chosen such that:

$$\frac{1}{R_d C_d} \ll \omega_o = \frac{1}{\sqrt{LC}} \tag{8.273}$$

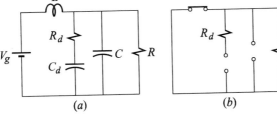

(a)   (b)

Figure 8.112

If $R < 0$, as in the case of loading by a regulating switching converter, we should further require that:

$$R_d < |R| \tag{8.274}$$

so that:

$$R \parallel R_d > 0 \tag{8.275}$$

Hence, provided $|R|$ is not too small, practical values of R_d can be realized. Show that under these considerations $D(s)$ can be factored:

$$D(s) \approx (1 + sC_dR_d)\left(+ s\frac{L}{R_d \| R} + s^2LC \right) \tag{8.276}$$

in which we see that the Q-factor of the quadratic term is given by Q_{oi} in Eq. (8.59).

Hint: Use the reference circuit in Fig. 8.112*b* to derive $D(s)$ in Eq. (8.272).

8.8 Series damping of the input filter.[12] Following the same procedure as in Problem 8.7, derive similar results for the series-damped input filter shown in Fig. 8.113.

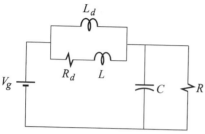

Figure 8.113

8.9 Output ripple voltage of the boost converter under conditions of no load. Show that under conditions of no load ($R_L \to \infty$) the current through S_2 (assuming S_2 is a bi-directional switch) in the boost converter in Fig. 8.19 has the shape shown in Fig. 8.114. Determine the ripple voltage across the capacitor and show that its peak-to-peak value is given by:

$$\left.\begin{aligned}
\frac{V_{r_{p-p}}}{V_o} &= \frac{D'^2 D T_s^2}{8LC} \\[2mm]
&= \frac{T_s^2}{8LC}\frac{M_V - 1}{M_V^3}
\end{aligned}\right\} \tag{8.277a,b}$$

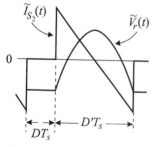

Figure 8.114

8.10 Output ripple voltage of the buck-boost converter under conditions of no load. Show that the peak-to-peak ripple voltage in the buck-boost converter under no-load conditions (assuming bi-directional switches) is given by:

$$\left. \begin{array}{l} \dfrac{V_{r_{p-p}}}{V_o} = \dfrac{(D'T_s)^2}{8LC} \\ \\ = \dfrac{1}{8LC}\left(\dfrac{T_s}{1 + M_V}\right)^2 \end{array} \right\} \tag{8.278a,b}$$

8.11 Isolated Cuk converter. The isolated Cuk converter is shown in Fig. 8.115. Show that the conversion ratio is given by:

$$M_V = n\dfrac{D}{D'} \tag{8.279}$$

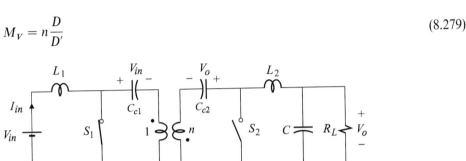

Figure 8.115

The magnetizing inductance of the transformer can have a significant effect on the control-to-output transfer function because it can resonate with C_{c1} and C_{c2}, both of which must be properly damped.[13]

8.12 The control-to-output transfer function of the tapped buck converter.

(a) Determine the transfer function in Eq. (8.147).

(b) Examine the current waveform $\tilde{i}_f(t)$ in Figs. 8.51a, b, and explain why the zero in the control-to-output transfer function is in the right-half plane for $n < 1$ and in the left-half plane for $n > 1$.

(c) Derive the analytical expression of this zero using a similar argument given for the boost converter in Eqs. (8.138–8.143).

8.13 Effect of parasitic elements on the voltage conversion ratio of the Watkins–Johnson converter.

(a) Substitute the dc model of the PWM switch in the converter shown in Fig. 8.56 and include a parasitic resistance, r_e, in series with the active terminal (or at any other place that you like). Determine the new conversion ratio using the EET with r_e as the designated extra element.

(b) Show that the conversion in the presence of r_e is zero for $D = 0.5$.

(c) Sketch the conversion ratio in part (b) and determine its maximum value.

8.14 Inverse of the Watkins–Johnson converter. Identify the PWM switch in the converter shown in Fig. 8.58 and determine its voltage conversion ratio.

8.15 Frequency response of a quadratic converter. An OrCAD/Pspice simulation of the quadratic converter in Fig. 8.64 is shown below in which the control-to-output transfer function is determined by expanding the large-signal model of the PWM switch (VMLSCCM) at a duty cycle of $D = 0.35$. This is accomplished by setting the control voltage at a dc value of 1.35 V with a small-signal ac voltage superimposed on it.

Using the numerical values of the components in Fig. 8.116a, verify the control-to-output transfer function derived in Eq. (8.185) against Fig. 8.116b.

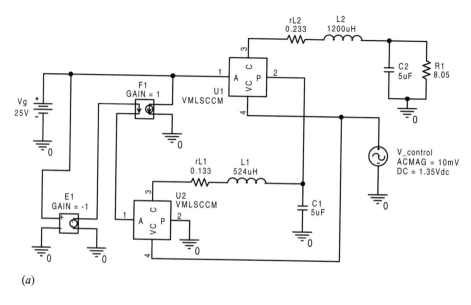

(a)

Figure 8.116

Notes: (a) Pspice simply ignores the actual magnitude of the ac voltage, shown here as 10 mV, and numerically expands the model of the PWM switch at the dc operating point and determines the numerical transfer function of the linearized circuit. (b) In the specification of the parameters of VMLSCCM, the height of the ramp is set to 1 V and the valley voltage at 1 V. Hence, a dc value of 1.35 V in the control voltage results in a duty cycle of 0.35.

8.16 The high-frequency resistance r_e in the model of the PWM switch. (a) Show that $r_e = r_{C_c}$ for the converter in Fig. 8.59. (b) Show that $r_e = (r_C \| R)(1 + n)^2/n$ for the Watkins–Johnson converter in Fig. 8.56. (c) Determine r_e for the converter in Fig. 8.58.

8.17 Storage-time modulation in the Cuk converter.[4,7] The effect of storage-time modulation on the line-to-output transfer function of the Cuk converter can be

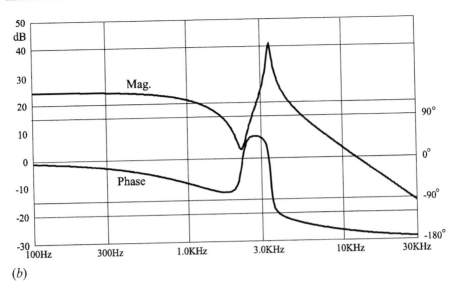

(b)

Figure 8.116 (*cont.*)

easily studied by using the model of the PWM switch in Fig. 8.74 as shown in Fig. 8.117. In particular, we are interested in determining the numerator of the line-to-output transfer function. Using the technique explained in (Section 2.4), show that:

$$\frac{v_o(s)}{v_{in}(s)} = \frac{\left(1 + \dfrac{s}{\omega_{z1}}\right)\left(1 - \dfrac{s}{\omega_{z2}}\right)}{D(s)} \tag{8.280}$$

where:

$$\left.\begin{aligned}
\omega_{z1} &= \frac{1}{r_{C_f} C_f} \\[2em]
\omega_{z2} &= \frac{DD'}{C_c(DD'r_{C_c} + Dr_t + D'r_d + r_{me})}
\end{aligned}\right\} \tag{8.281a, b}$$

in which the modulation resistance is given by:

$$r_{me} = \frac{V_D}{I_{me}} \approx \frac{V_o}{DI_{me}} \tag{8.282}$$

It can be seen that ω_{z2} can be either in the left-half plane or the right-half plane depending on the relative magnitudes of the parasitic resistances and the modulation resistance which can have either a positive or a negative value depending upon the type of base drive used.

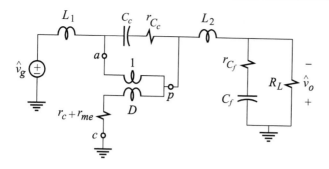

Figure 8.117

REFERENCES

1. S. Cuk and R. D. Middlebrook, "A New Optimum Topology Switching dc-to-dc Converter", *Proceedings of the 1977 IEEE Power Electronics Specialist Conference*, PESC 77 Record,
   pp. 160–179.
2. S. Cuk, "Switching dc-to-dc Converter with Zero Input or Output Current", *Proceedings of the 1978 IEEE Industry and Applications Society Annual Meeting*, October 1–5, Record, pp. 1131–1146.
3. S. Cuk, "A New Zero-Ripple dc-to-dc Converter and Integrated Magnetics", *Proceedings of the 1980 IEEE Power Electronics specialist Conference*, PESC 80 Record, pp. 12–32.
4. V. Vorpérian, "Simplified analysis of PWM Converters Using the Model of the PWM Switch: Parts I and II", *IEEE Transactions on Aerospace Electronic Systems*, Vol. 26, No. 3, 1990,
   pp. 490–505.
5. B. Israelson, J. Martin, C. Reeve and V. Scown, "A 2.5-kV High Reliability TWT Power Supply: Design Techniques for High Efficiency and Low Ripple", *Proceedings of the 1977 IEEE Power Electronics Specialist Conference*, PESC 77 Record, pp. 109–130.
6. D. Maksimovic and S. Cuk, "General Unified Properties and Synthesis or PWM Converters", *Proceedings of the 1989 IEEE Power Electronics Specialist Conference*, PESC 89 Record,
   pp. 515–525.
7. W. Polivka, P. Chetty and R. Middlebrook, "State-Space Average Modelling of Converters with Parasitics and Storage-Time Modulation", *Proceedings of the 1980 IEEE Power Electronics Specialists Conference*, PESC 80 Record.
8. C. Deisch, "Switching Control Method Changes Power Converter into a Current Source", *Proceedings of the 1978 Power Electronics Specialist conference*, PESC 78 Record, pp. 300–306.
9. V. Vorpérian, "Analysis of Current-Controlled PWM Converters Using the Model of the Current-Controlled PWM Switch", *Power Conversion and Intelligent Motion Conference*, 1990, pp. 183–195.

10. R. B. Ridley, "A New Continuous-Time Model for Current-Mode Control", *IEEE Transactions of Power Electronics*, Vol. 6, April 1991, pp. 271–280.

11. V. Vorpérian and S. Cuk, "A Complete DC Analysis of the Series Resonant Converter", *Proceedings of the 1982 Power Electronics Specialist Conference*, PESC 82 Records, pp. 85–100.

12. R. D. Middlebrook, "Input Filter Considerations in the Design and Application of Switching Regulators", *Proceedings of the 1976 IEEE Industry and Applications Society Annual Meeting*, October 11–14, 1976 Record, pp. 366–382.

13. V. Vorpérian, "The Effect of the Magnetizing Inductance on the Small-Signal Dynamics of the Isolated Cuk Converter", *IEEE Transactions on Aerospace and Electronic Systems*, Vol. 32, No. 3, July 1996, pp. 967–983.

Index

anomalous eddy current losses 351–6
antenna, three-winding parallel resonant transformer 344–9
approximate factors of a polynomial 142, 260, 311

band-pass response 49
band-reject response 49
block diagram 17
  common-emitter amplifier 55–6
  feedback system 56
Bode plot 43
  first-order transfer functions 43–8
  second-order transfer function 49–52
boost converter
  basic operation 386–92
  conversion ratio 386–8
  inductor current ripple 386–7
  OrCAD/Pspice sinulation 389–91
  output voltage ripple (heavy load) 388–9
  output voltage ripple (light or no load) 466
  peak current-mode control 446–50
bridge circuit
  input impedance 2–12, 87
  null-impedance calculations 64–8
  phase shifter 101
  RHP zero 83–4
  voltage transfer functions 78–83
bridge circuit with dependent source
  input impedance 4–11
buck converter
  basic operation 370–9
  conversion ratio 371
  discontinuous conduction mode (DCM) 378–9
  inductor current ripple 374–6
  input filter 379–85
  output voltage ripple 372–3, 464
  OrCAD/Pspice simulation 384–5
  single-loop voltage feedback control 433–9
  voltage and current feedback control 453–9
buck-boost converter
  basic operation 392–7
  control-to-output transfer function 404, 408–11
  conversion ratio 393
  inductor current ripple 394–6
  input admittance 404, 408

line-to-output transfer function 404, 407
OrCAD/Pspice simulation 395–7
  output impedance 404, 408
  output voltage ripple (heavy load) 394
  output voltage ripple (light or no load) 466–7
  RHP zero 410–1

Cantor set 354
capacitively tapped parallel resonant circuit 339–44
cascaded resonant circuits 60
cascode amplifier 191–5
cascode MOS amplifier 252–61
  bandwidth 261
CC-PWM (current controlled PWM switch 440, 443–7
  large-signal average model 443
  small-signal average model 444–5
CE–CE amplifier 181–5, 188–91, 240–2, 244
characteristic equation 19
Chebyshev fifth-order low-pass filter 261–4
closed-loop gain, ideal 196, 274
CMLSCCM (current-mode large-signal continuous conduction mode) subcircuit 456–9
Colpitts oscillator 220–5, 248–51
common-emitter amplifier
  collector-to-base feedback resistor 29–30, 57–8
  gain analysis 28–30, 124–30
  Miller effect 53–55
  pole due to emitter bypass capacitor, 28–30, 57–8
complete response of a linear system 34–41
conditional stability 227
conducted emissions 379–80
constant phase angle behavior, and infinite scaling networks 351
conversion ratio
  current 366
  voltage 366
converters *see* boost converter; buck converter; buck-boost converter; Cuk converter; inverse Watkins–Johnson converter; tapped buck converter; Watkins–Johnson converter; quadratic converters, fourth-order converters
coupling coefficient 345
Cuk converter 464–7

low current ripple benefits   398–400
storage time modulation   468–70
with isolation transformer   467
with switches rearranged   415–17
current feedback pair   133–7
current gain transfer function, definition   17
current mixing and input impedance analysis
   196–200
current-mode control   440–6
current-sampling current-mixing feedback
   amplifier   188–91, 240–2
current-sampling voltage-mixing feedback
   amplifier   181–5, 236–9
current sensing and output admittance analysis
   209–13
current source, output impedance   13

dc-to-dc converters   365–470
   basic characteristics   366–9
   battery charger   369–70
   boost converter   386–92, 466
   buck-boost converter   392–7, 466–7
   buck converter   370–85, 464–5, 467
   Cuk converter   397–400, 467–70
   fourth-order converter   415–17
   input resistance   368
   inverse Watkins–Johnson converter   468
   output resistance   462–3
   quadratic converters   417–26
   tapped buck converter   412–15
   Watkins–Johnson converter   414–15
delay networks/equalizers, lattice filters as
   327–35
differential amplifier with adjustable gain
   99–101
discontinuous conduction mode (DCM), buck
   converter   378
driving-point admittance function, definition
   17–19
driving-point analysis, feedback amplifiers   170–5
   basic principle   170–1
   FET amplifier   171–5, 236
   input admittance for voltage mixing   204–9,
      245
   input impedance for current mixing   196–200,
      244
   inside input conductance in shunt-series
      feedback   206–9
   inside output impedance in shunt-series feedback
      202–4
   output admittance for current-sensing   209–13,
      246
   output impedance for voltage sensing   200–4,
      244
driving-point impedance function, definition
   17–19
duty ratio function   371

dynamic load response   439, 459

effective turns ratio   345
efficiency of dc-to-dc converters   366
emitter-follower, output impedance   105
entropy
   high-entropy versus low-entropy expressions
      11–12
   low-entropy expressions   1, 4, 7
equalizer   123–4, 327–35
extra-element theorem (EET)   61–106
   advantages   4
   Bode's contribution to its origins   61
   for dependent sources   88–91
   for impedance elements   74–7
   null double injection   62–4
   see also N-extra element theorem (NEET)
excitation of a transfer function   15
excitation of an impedance function   18–19

fast analytical techniques   1–2
feedback amplifiers see driving point analysis,
   feedback amplifiers; gain analysis, feedback
   amplifiers; loop gain; phase and gain margins;
   stability
feedback control of dc-to-dc converters   432–59
   current feedback control   440–46
   single-loop voltage feedback control   433–9
   two-loop with voltage feedback and peak
      current control   453–9
feedback theorem due to Middlebrook   176
FET amplifier with feedback
   gain analysis   166–70
   input impedance   236
   loop gain   176–9, 236
   output impedance   171–5
filters, passive
   capacitively tapped parallel resonant filter
      339–44
   Chebyshev fifth-order low-pass filter   261–5
   lattice filters   327–35
   parallel resonant   336–9
   RC filters with gain   317–26, 359–61
   three-winding parallel resonant transformer
      344–9, 362
fourth-order converters   415–17
fractal dimension   354
fractal networks   351–7
fractional derivative   357
functional equation, fractal network   353

gain analysis, feedback amplifiers   164–6, 180–95
   CE–CE amplifier with series-shunt feedback
      181–5, 236–9
   closed-loop gain using the EET   164–5
   current-loaded cascode amplifier with
      shunt-shunt feedback   191–5, 242–3

gain analysis, feedback amplifiers (*cont.*)
  differential input followed by a CE amplifier
    stage with shunt-series feedback   157, 185–8,
    239–40
  FET shunt-shunt feedback amplifier   166–70,
    236
  two-stage CE series-series feedback with
    current-sense transformer   188–91, 240–2
gain margin *see* phase and gain margins
general linear element   137

high-frequency and microwave circuits *see* cascode
    MOS amplifier; Chebyshev fifth-order
    low-pass filter; MESFET amplifier; video
    amplifier

ideal closed-loop gain   196, 273
  indeterminacy   157–9, 423
index permutation symbol   138
infinite grid   349–51, 362
infinite ladder   350–1, 363
infinite scaling networks   349–56
  constant phase angle behavior   351
  fractal dimension   354
  fractional derivative   357
  functional equation   353
  power law   356
  Riemann–Liouville fractional integral   357
  unified model for R, L and C 356–7
input filter
  buck-boost   395
  buck converter   379–85, 464
  series damping   380, 466
  shunt or parallel damping   380, 465
input impedance/admittance *see* driving-point
    analysis, feedback amplifiers
inverse gain   89–91, 132, 138
inverse Watkin–Johnson converter   468

lattice fitters   327–35
linear regulator
  series   461–2
  shunt   462
loading of feedback network
  current mixing   197
  current sensing   211
  voltage mixing   206
  voltage sensing   202
loop gain   175–6
  current/voltage loop gain   247–8
  injection at an arbitrary point   213–18, 246
  injection immediately after the amplifying
    element   213
  in buck converter   437, 456
  measurement by injection   236, 250, 437, 456
  low-entropy expressions   1, 4, 7, 11–12

magnitude and phase response   41–3

MESFET amplifier   265–310
  bandwidth   310
microwave and high-frequency circuits *see* cascode
    MOS amplifier; Chebyshev fifth-order
    low-pass filter; MESFET amplifier; video
    amplifier
minimum phase systems   218
MOS amplifier, cascode *see* cascode MOS
    amplifier
mutual inductance   345

N-extra-element-theorem (NEET)   137–9
  2-EET for dependent sources, 130–2
  2-EET for impedance elements   108–10
  determinant form   148
  proof of NEET   147–153
  redundancy relations   139
  reference element function   138
  reference inverse gain function   138
  reference network   108
  reference null inverse gain function   138
  reference state of a port   108–9
  reference transfer function   109–10
negative null impedance   68, 83–4
N-port network   107, 147
notch filter   32–4, 139–47
null double injection   62–4
null-driving point impedance   64
nufl-inverse gain   90, 131, 138
null response in transform circuit   24–5
nulls of a transfer function   24–5
Nyquist plot and stability condition   218–19,
    225–6

open-short theorem   115, 154
operational amplifier, nonideal
  gain in inverting configuration   95–7
  gain in noninverting configuration   97–8
  input impedance in noninverting configuration
    92–5
  output impedance in inverting configuration   91
  pole-zero compensation   58–60
  transimpedance amplifier   228–32
operational amplifier with T-feedback   68–9, 84
opposite state of a port   109
OrCAD/Pspice simulations
  boost converter   389–92
  buck-boost converter   395–7
  buck converter with peak current-mode control
    456–9
  buck converter with voltage feedback   437–9
  cascode amplifier   242–3
  Colpitts oscillator   248–50
  differential input amplifier with voltage sampling
    and voltage mixing   239–40, 244–5
  input filter for buck converter   382–5
  loop gain measurement in a FET amplifier
    237

loop gain measurement in the buck converter
437–8
loop gain measurement in the buck converter
with current mode control 459
quadratic converter 468
three-winding-parallel resonant transformer
362–3
two-stage CE–CE amplifier with current
sampling and current mixing 236–9, 244
two-stage CE–CE amplifier with current
sampling and voltage mixing 240–2, 246
unstable loop-gain measurement 250
video amplifier 311–16
oscillation *see* stability
output impedance/admittance *see* driving-point
analysis, feedback amplifiers

parameter-extraction method 5
passive filters *see* filters, passive
peak current control
model of PWM switch 440–5,
model of the current loop 450–3
subharmonic oscillation 441
permutation symbol, index 138
phase and gain margins 226-32
phase and magnitude response 41–3
phase shifter 83, 101–2
phasor notation 42
poles of a network 17, 19–24
port
opposite state 109
reference state 108
PWM switch
average large-signal and small-signal models
402–3
current-controlled PWM switch 440–4
effect of parasitic elements 426–32
high-frequency resistance across port $a–p$
426–9
invariant terminal characteristics 400–2
on-resistance of the switches 429–30
saturation models of the switches 431–2
storage time modulation 430–1

Q-factor 2–7
high-Q approximation of parallel resonant filter
337–9
tapped parallel resonant filter 343
three-winding parallel resonant filter 347
triple-damped parallel resonant filter 337
quadratic converter 418, 468
control-to-output transfer function 420–6
conversion ratio 420

RC filter with gain *see* filters, passive
redundancy relations 139
reference element function 136
reference inverse gain function 138

reference network 108
reference null inverse gain function 138
reference state of a port 108–9
reference transfer function 109–10
resonant filter *see* filters, passive
Riemman–Liouville fractional integral 357
right-half plane (RHP) zero
in bridge circuits 57, 83–4
in buck-boost converter 410–11
in cascode MOS amplifier 253
in CE amplifier 30, 53–4

scaling equation 354
scaling/fractal networks 351
series-series feedback *see* current-sampling
voltage-mixing feedback
series-shunt feedback *see* current- sampling
current-mixing feedback
shunt-series feedback *see* voltage-sampling
voltage-mixing feedback
shunt-shunt feedback *see* voltage-sampling
current-mixing feedback
stability
conditional 227
feedback sytems 218–26
frequency-domain and time-domain equivalence
35–41
impedance criteria 222–5
Nyquist encirclement stability condition
225–6
stability margin *see* phase and gain, margins
storage time modulation 430, 468–70
subharmonic oscillation 440–6
subtractor with adjustable gain 99

tapped buck converter *see* dc-to-dc converters
three-winding transformer 345–9, 362–3
transadmittance transfer function, definition
17–19
transfer function
contribution of a 1:n transformer to a transfer
function 154–7
determination of poles 19–21
determination of zeros 24–5
first-order 43–8
second-order 48–52
six different types in an electrical circuit,
definition 15–17
transimpedance transfer function, definition
15–17
transimpedance operational amplifier 228–32

unstable systems *see* stability

video amplifier frequency response 311–16
VMLSCCM (voltage mode large-signal continuous
conduction mode) subcircuit for the PWM
switch 437

voltage gain transfer function, definition   17–19
voltage mixing and input admittance *see*
    driving-point analysis, feedback amplifiers
voltage sampling and output impedance *see*
    driving-point analysis, feedback amplifiers
voltage-sampling current-mixing feedback
    amplifier   166–70, 191–5, 236, 242–3
voltage-sampling voltage-mixing feedback

amplifier   157, 185–8, 239–40

wave translation system patent   163
Watkins–Johnson converter   414–15, 467
  inverse converter   468

zeros of a tranfer function   17, 24–5
  determination principle   24–5